T0181193

Studies in Computational Intelligence

Volume 750

Series editor

Janusz Kacprzyk, Polish Academy of Sciences, Warsaw, Poland
e-mail: kacprzyk@ibspan.waw.pl

The series "Studies in Computational Intelligence" (SCI) publishes new developments and advances in the various areas of computational intelligence—quickly and with a high quality. The intent is to cover the theory, applications, and design methods of computational intelligence, as embedded in the fields of engineering, computer science, physics and life sciences, as well as the methodologies behind them. The series contains monographs, lecture notes and edited volumes in computational intelligence spanning the areas of neural networks, connectionist systems, genetic algorithms, evolutionary computation, artificial intelligence, cellular automata, self-organizing systems, soft computing, fuzzy systems, and hybrid intelligent systems. Of particular value to both the contributors and the readership are the short publication timeframe and the world-wide distribution, which enable both wide and rapid dissemination of research output.

More information about this series at http://www.springer.com/series/7092

Olga Kosheleva · Karen Villaverde

How Interval and Fuzzy Techniques Can Improve Teaching

Processing Educational Data: From
Traditional Statistical Techniques
to an Appropriate Combination
of Probabilistic, Interval, and Fuzzy
Approaches

 Springer

Olga Kosheleva
Department of Teacher Education
University of Texas
El Paso, TX
USA

Karen Villaverde
Department of Mathematics
New Mexico State University
Las Cruces, NM
USA

ISSN 1860-949X   ISSN 1860-9503   (electronic)
Studies in Computational Intelligence
ISBN 978-3-662-57256-6   ISBN 978-3-662-55993-2   (eBook)
https://doi.org/10.1007/978-3-662-55993-2

© Springer-Verlag GmbH Germany 2018
Softcover reprint of the hardcover 1st edition 2017
This work is subject to copyright. All rights are reserved by the Publisher, whether the whole or part
of the material is concerned, specifically the rights of translation, reprinting, reuse of illustrations,
recitation, broadcasting, reproduction on microfilms or in any other physical way, and transmission
or information storage and retrieval, electronic adaptation, computer software, or by similar or dissimilar
methodology now known or hereafter developed.
The use of general descriptive names, registered names, trademarks, service marks, etc. in this
publication does not imply, even in the absence of a specific statement, that such names are exempt from
the relevant protective laws and regulations and therefore free for general use.
The publisher, the authors and the editors are safe to assume that the advice and information in this
book are believed to be true and accurate at the date of publication. Neither the publisher nor the
authors or the editors give a warranty, express or implied, with respect to the material contained herein or
for any errors or omissions that may have been made. The publisher remains neutral with regard to
jurisdictional claims in published maps and institutional affiliations.

Printed on acid-free paper

This Springer imprint is published by Springer Nature
The registered company is Springer-Verlag GmbH, DE
The registered company address is: Heidelberger Platz 3, 14197 Berlin, Germany

Dedicated to the memory of Lotfi A. Zadeh

Contents

Chapter 1
Introduction: Need for Interval and Fuzzy Techniques in Math and Science Education

Education is difficult. Most teachers and instructors would agree that while teaching is a very rewarding activity, it is also a very difficult one. It is difficult because teaching is largely an art. There is a lot of advice on teaching, but this advice is usually informal and thus, not easy to follow. Students are different. Whatever worked for one group of students may not work for another group. Students have different preparation level, different motivations, different skills, different attitudes, different relations to other students in the class.

Any idea that can help teachers and students is very welcome.

Where we need help: different aspects of education. There are many aspects of teaching and learning, and in all these aspects, good advices are welcome.

First, before we even start teaching the material, we need to *motivate* the students. This is an extremely important aspect of teaching: if a student does not understand why this material is useful, this student may not be as committed to study. Motivating students is often a difficult task, it is kind of a chicken-and-egg problem:

- for the students to truly understand the use of the class material, they need to have a certain degree of knowledge of this material, but
- to acquire this degree of knowledge, they need motivation.

Once the students are motivated and teaching starts, we need to decide *in what order* we should present the material.

- Some courses first provide the basic ideas of all the topics, and only after all the basics are described, provide the technical details.
- Other courses first deal with one topic, then go to another topic, etc.

Which approach is better? When is each of these approaches better? Within each of these approaches, which topic should be taught first and which next? These are all important questions which can often seriously impact the course results.

Once we decided in what order to present the material, we need to decide *how exactly to teach*, i.e., how to deliver this material. There are many good teaching

© Springer-Verlag GmbH Germany 2018
O. Kosheleva and K. Villaverde, *How Interval and Fuzzy Techniques
Can Improve Teaching*, Studies in Computational Intelligence 750,
https://doi.org/10.1007/978-3-662-55993-2_1

techniques, sometimes, some of these techniques work better, sometimes, other techniques work better, which techniques should we use? when should we use each of these techniques?

Finally, it is important to *assess* (and keep assessing) the results of teaching. We need to assess how well the students learn, how well they master the course material – this is what we do when we grade their homeworks, quizzes, tests, etc.

We also need to assess how well a given teacher did in the class. This is not as straightforward as assessing the level of knowledge and skills developed in the class, because this level depends not only on the teacher, but also on how prepared and how motivated the students were when they started taking this class.

Finally, we also need to assess a given teaching method. This assessment is even more difficult, because when comparing the efficiency of different teaching methods, we need to take into account that these methods were tested by different teachers on different groups of students.

In all these aspects, we need help and advice.

How can we get help and advice? Good news is that many teachers and instructors teach, and many of them teach efficiently. So, a good way to get help and advice is to use this experience.

There are two ways to use this experience. First, we can simply ask good teachers how they teach, and use their recommendations.

Second, we can also take into account that for each class, there is a record of what exactly topic was taught at each class, what method was used for this topic, what were the resulting student grades, etc. All this information can also be used.

- When we deal with a record of a successful class, this information can help us to understand what works.
- Records of less successful classes, however, are also useful: they help us to understand what did not work – and thus, what to avoid in the future.

Need to combine traditional statistical techniques with interval and fuzzy approaches. As we have just mentioned, to understand which teaching techniques work better and when, it is desirable to process the available teaching records. There are many such records, so to process these records, we need to use techniques specifically designed to process large amounts of data. Techniques related to the collection, organization, analysis, interpretation, and presentation of data are known as *statistics*; thus, to be more successful in teaching, we need to use statistical techniques.

One of the main objectives of the statistical analysis is to make conclusions that would be – with high confidence – applicable to other situations as well. There are many statistical techniques that provide such conclusions. In these techniques, we usually:

- formulate a mathematical model of a teaching process,
- check whether this model is consistent with observation and
- if the model is consistent with observations, use this model to predict how similar methods will work in other teaching situations.

This is very similar to how predictions are made in physics and natural science in general. The difference, however, is that:

- In physics, if we know the initial coordinates and velocities of all the bodies, we can predict the future state of the system.
- On the other hand, in education, so many factors beyond the instructor's control affect an individual student's success that it is not realistically possible to predict exactly who will succeed and how exactly, at best, we can predict the *probability* of a student's success.

Because of this, most statistical techniques used in processing educational data are *probabilistic* techniques, techniques based on the probabilistic models.

Need for fuzzy and interval techniques. As we have mentioned, there are two sources of educational recommendations:

- expert statements and
- analysis of the education records.

Let us show that for both sources, it is useful to go beyond the traditional probabilistic techniques.

This argument is the easiest for expert statements. Indeed, experts in general – and education experts are no exception – rarely formulate their statements in precise mathematical terms. A skilled teacher does not say things like "Use 13 min of your class explaining the new material, then call 4 students to the board and let them solve each problem for 17 more minutes, after this 72% of the students will get A on a quiz."Instead, the teacher will say something like "First, use about a third of your class to explain a new material, then call a few students to solve the problems on the board; after that, the vast majority of your students will get a good knowledge of this topic." Such explanations use imprecise ("fuzzy") words from natural language, such as "about a third", "a few", "vast majority", etc. To follow this recommendation, we need to reformulate it in precise terms. The need for such precise reformulation of expert knowledge was well understood in the 1960s, and a special technique – called *fuzzy* technique – was invented by Lotfi A. Zadeh for such a reformulation; see, e.g., [1–3].

The corresponding formalization is based on the fact that in the computer, "true" is usually represented as 1 and "false" as 0. It is therefore reasonable to represent cases when we are not absolutely sure about a statement by using numbers (*degrees of confidence*) from the interval [0, 1]:

- the degree 1 means that we are absolutely sure about the statement;
- the degree 0 means that we are absolutely sure that the given statement is false;
- a degree between 0 and 1 means that we have some confidence that the statement is true, but we realize that it may turn out to be false.

For example, to describe what "a few students" means, we can ask an expert how confident he or she is that calling one student will be OK, that calling two students is OK, etc. An expert can represent his or her degree of confidence in each of the corresponding statements, e.g., by making a mark on a scale. For example, if an expert marks 7 on a scale from 0 to 10, it makes sense to assign the degree of confidence $7/10 = 0.7$.

In short, such fuzzy techniques are useful in processing expert statements. At first glance, usual probabilistic techniques may sound sufficient for processing the second source of information about educational techniques – the teaching records. Indeed, at first glance, we have a bunch of numbers (grades), and traditional statistical methods have been designed specifically to process such numerical databases. Traditional probabilistic methods have indeed been actively used for processing educational data. Moreover, these methods are what vast majority of educational research papers use to process educational data.

However, most of these traditional probabilistic methods do not take into account the subjective character of the educational data. In education, evaluations of the student's knowledge, skills, and abilities are often subjective. Teachers and experts usually make these evaluations by using words from natural language like "good", "excellent". Traditionally, in order to be able to process the evaluation results, these evaluations are first transformed into exact numbers. This transformation, however, ignores the fuzziness of the original estimates.

To get a more adequate picture of the education process and education results, it is therefore desirable to explicitly take this fuzziness into account.

Need to take interval uncertainty into account. A related aspect of uncertainty comes from the fact that even when an instructor keeps a detailed record of each student's achievements and uses this record to generate a detailed numerical grade (e.g., 87.2) which accurately reflects the student's level of knowledge, this detailed grade is usually not transmitted to the instructors who teach the following classes. All these follow-up instructors see is a letter grade.

So, for example, if the previous instructor defines B as any grade between 80 and 90, then, when a student has B in this class, it means that his or her actual numerical grade in that class could be any value from 80 to 89.9.

In other words, the only information that the next instructor gets about the student's performance in the previous class is an *interval* of possible values (in the above example, the interval [80, 90]). When we analyze the relation between the student's success in a future class and the student's achievement in the previous classes, we need to take this interval uncertainty into account.

Fuzzy techniques are already used in education: a brief overview. The idea of using fuzzy techniques to grade a problem first appeared in Zimmermann [4] (p. 246). This idea was used, e.g., in Chang and Sun [5], Chiang and Lin [6], Wu [7].

The main reason for the need to use fuzzy techniques comes from the fact that when instructors assign letter grades to different questions and different aspects of each question, they are often not 100% sure that the appropriate letter grade is, e.g., A.

For such situations, Biswas [8] proposed to allow the instructor to assign different degrees of confidence $d(A) \in [0, 1]$, $d(B) \in [0, 1]$,..., to different letter grades. For example, if an instructor is somewhat inclined to give an A grade but also has reasons to make this grade a B, this instructor may assign a larger degree (e.g., 0.6) to the A grade and a smaller degree (e.g., 0.4) to the B grade. The resulting assignment can be viewed as a fuzzy subset (membership function) of the set $\{A, B, C, D, F\}$ of all letter grades.

Chen and Lee [9] proposed to improve this general idea by allowing instructors to assign degrees not only to the basic letter grades corresponding to "Excellent", "Good", "Satisfactory", etc., but also to intermediate linguistic terms such as "Very Good", "Very Very Good", "Rather Good", etc. This provides a more accurate a more accurate evaluation of the student's knowledge.

In Wang and Chen [10–12] and Almohammadi et al. [13], it is shown that type-2 fuzzy sets provide an even more adequate description of the instructor's evaluation.

This information can be supplemented by the evaluator's degree of confidence in his/her numerical or fuzzy estimate, in the style of Z-numbers; see, e.g., Wang and Chen [14].

Even when an instructor assigns the exact numerical or letter grade, this grade is only an approximate representation of the student's knowledge; in reality, a score of say 8 on a problem worth 10 points does not mean exactly 8, it means *approximately* 8. A better way to understand this score is to use a membership function; see, e.g., Law [15]. In such situations, when it is difficult to directly elicit the corresponding membership functions from the instructor, Cheng, Wei, and Chen [16] proposed to use entropy techniques and rough set techniques.

Once we have evaluated different aspects of the solution – e.g., its correctness, its efficiency, etc. – the next step is to combine all these evaluations into a single grade. Similarly, once we have grades for each problem, we need to combine them into the grade for the test – and then we need to combine the grades for the tests and other assignments into a single grade for the class. Several papers proposed to use fuzzy rules for such a combination. This enables us to take into account:

- the problems' relative difficulty (see, e.g., Weon and Kim [17], Bai and Chen [18, 19], Chen and Li [20], Hameed [21], Hameed and Sorensen [22]),
- importance of each topic (Weon and Kim [17], Chen and Li [20]), etc.

When generating an overall grade for the class, we can use fuzzy techniques to take into account:

- how consistent the student's grades were, how much the student's knowledge improved, etc.; see, e.g., Wilson, Karr, and Freeman [23],
- how efficient was the instructor; see, e.g., Echauz and Vachtsevanos [24]; and
- how strict was the instructor's grading; see, e.g., Bai and Chen [19, 25, 26];

and to combine, e.g., peer evaluations and lecturer's evaluations (Ma and Zhou [27]).

If needed, defuzzification techniques can be used to transform the resulting fuzzy grade into a crisp one.

Fuzzy rules can be used not only to evaluate the studen's performance:

- they can also be used to transform the test result into a description of which concepts turn out to be difficult to learn; see, e.g., Bai and Chen [28];

- fuzzy techniques can be used to automatically construct concept maps – based on the idea that a topic X depends on the topic Y if a low grade in Y implies a low grade in X; see, e.g., Bi and Chen [29].

Several authors also emphasize the importance of teaching fuzzy concepts such as "small", "medium", etc., to help elementary school students get a better understanding of arithmetic (Laski and Siegler [30], Krasa and Shunkwiler [31].

Many authors have used these and other ideas when applying fuzzy techniques to education; see, e.g., Furali [32], Cheng [33], Nolan [34], Rasmani and Shen [35], Wang and Chen [36], Bai and Chen [29], Li and Chen [37], Saleh and Kim [38], Chen and Li [20], Shahbazova [39, 40], and many many other papers.

What we do in this book. There are numerous papers on the use of fuzzy techniques in different aspects of education. We believe that it is time to come up with a unifying view of all these applications. This is what we intend to do in this book. Specifically, in this book, we show that fuzzy techniques can be effectively applied on all the stages of the education process.

Sometimes, the use of these techniques improves the corresponding recommendations, sometimes it simply provides an explanation of some empirical recommendations – with a hope that a further analysis will help to actually improve the recommendations. In all these cases, our results are, at best, preliminary.

Our main objective is to *promote* the use of fuzzy techniques in education, to encourage education researchers to combine these techniques with the traditional probabilistic techniques.

Book structure. We will show that fuzzy and interval techniques help in all aspects of education that we described in the beginning of this chapter, from motivation to details about teaching itself to assessment of the teaching results.

Our main objective is to help teachers:

- in some cases, by providing a straightforward advice,
- in other cases, by providing case studies and/or ideas which will hopefully eventually lead to useful advice.

Because of this practical objective, we will list our ideas and results in the same natural order in which we listed the education aspects. Specifically:

- In Part 1, we explain how interval and fuzzy ideas can help to motivate students.
- In Part 2, we explain how interval and fuzzy ideas can help us decide in what order to present the material.
- In Part 3, we explain how interval and fuzzy ideas can be used to select an appropriate way of teaching the corresponding part of the material.
- Finally, in Part 4, we explain how interval and fuzzy ideas can help with assessment: assessment of students, assessment of teachers, and assessment of teaching techniques.

References

1. G.J. Klir, B. Yuan, *Fuzzy Sets and Fuzzy Logic: Theory and Applications* (Prentice-Hall, Upper Saddle River, 1995)
2. H.T. Nguyen, E.A. Walker, *A First Course in Fuzzy Logic* (CRC Press, Boca Raton, 2006)
3. L.A. Zadeh, Fuzzy sets. Information and Control **8**, 338–353 (1965)
4. H.J. Zimmermann, *Fuzzy Set Theory and Its Applications* (Kluwer, Boston, 1991)
5. D.F. Chang, C.M. Sun, Fuzzy assessment of learning performance of junior school students, in *Proceedings of the First National Symposium on Fuzzy Theory and Applications*, (Hsinchu, Taiwan, Republic of China, 1993), pp. 10–15 (in Chinese)
6. T.T. Chiang, C.M. Lin, Application of fuzzy theory to teaching assessment, in *Proceedings of the Second National Conference on Fuzzy Theory and Applications*, (Taipei, Taiwan, Republic of China, 1994), pp. 92–97
7. M.H. Wu, *A Research on Applying Fuzzy Set Theory and Item Response Theory to Evaluate Learning Performance*, Master's Thesis, Department of Information Management, Chaoyang University of Technology, Wufeng, Taichung County, Taiwan, 2003
8. R. Biswas, An application of fuzzy sets in students' evaluation. Fuzzy Sets Syst. **74**(2), 187–194 (1995)
9. S.M. Chen, C.H. Lee, New methods for students evaluating using fuzzy sets. Fuzzy Sets Syst. **104**(2), 209–218 (1999)
10. H.Y. Wang, S.M. Chen, New methods for evaluating students' answerscripts using fuzzy numbers associated with degrees of confidence, in *Proceedings of the 2006 International Conference on Fuzzy Systems*, (Vancouver, BC, Canada, 2006), pp. 5492–5497
11. H.Y. Wang, S.M. Chen, New methods for evaluating students' answerscripts using vague values, in *Proceedings of the 9th Joint Conference on Information Sciences*, (Kaohsiung, Taiwan, 2006), pp. 1196–1199
12. H.Y. Wang, S.M. Chen, Evaluating students' answerscripts based on the similarity measure between vague sets, in *Proceedings of the 11th Conference on Artificial Intelligence and Its Applications*, (Kaohsiung, Taiwan, 2006), pp. 1539–1545
13. K. Almohammadi et al., A type-2 fuzzy logic recommendation system for adaptive teaching. Soft Comput. **21**, 965–979 (2017)
14. H.Y. Wang, S.M. Chen, Evaluating students' answerscripts using fuzzy numbers associated with degrees of confidence. IEEE Trans. Fuzzy Syst. **16**(2), 403–415 (2008)
15. C.K. Law, Using fuzzy numbers in education grading system. Fuzzy Sets Syst. **83**(3), 311–323 (1996)
16. C.-H. Cheng, L.-Y. Wei, Y.-H. Chen, A new e-learning achievement evaluation model based on rough set and similarity filter. Comput. Intell. **27**(2), 260–279 (2011)
17. S. Weon, J. Kim, Learning achievement evaluation strategy using fuzzy membership function, in *Proceedings of the 31st ASEE/IEEE Frontiers in Education Conference*, vol. 1 (Reno, Nevada, 2001), pp. 19–24
18. S.-M. Bai, S.-M. Chen, Evaluating students' learning achievement using fuzzy membership functions and fuzzy rules. Expert Syst. Appl. **34**, 399–410 (2008)
19. S.-M. Bai, S.-M. Chen, Automatically constructing grade membership functions of fuzzy rules for students' evaluation. Expert Syst. Appl. **35**(3), 1408–1414 (2008)
20. S.-M. Chen, T.-K. Li, A new method to evaluate students learning achievement by automatically generating the importance degrees of attributes of questions, in *Proceedings of the Ninth International Conference on Machine Learning and Cybernetics*, (Qingdao, 11–14 July, 2010), pp. 2495–2499
21. I.A. Hameed, *New Applications and Developments of Fuzzy Systems*, Department of Industrial Engineering, Korea University, 2010
22. I.A. Hameed, C.G. Sorensen, Fuzzy systems in education: a more reliable system for student evaluation, in *Fuzzy Systems*, ed. by A.T. Azar (Intech, Rijeka, 2010), pp. 216–231

23. E. Wilson, C.L. Karr, L.M. Freeman, Flexible, adaptive, automatic fuzzy-based grade assigning system, in *Proceedings of the 1998 Annual North American Fuzzy Information Processing Society Conference NAFIPS'98*, 1998, pp. 334–338
24. J.R. Echauz, G.J. Vachtsevanos, Fuzzy grading system. IEEE Trans. Educ. **38**(2), 158–165 (1995)
25. S.M. Bai, S.M. Chen, Automatically constructing grade membership functions for students' evaluation for fuzzy grading systems, in *Proceedings of the 2006 World Automation Congress*, (Budapest, Hungary, 2006)
26. S.M. Bai, S. M. Chen, A new method for students' learning achievement using fuzzy membership functions, in *Proceedings of the 11th National Conference on Artificial Intelligence*, (Kaohsiung, Taiwan, Republic of China, 2006)
27. J. Ma, D. Zhou, Fuzzy set approach to the assessment of student-centered learning. IEEE Trans. Educ. **43**(2), 237–241 (2000)
28. S.-M. Bai, S.-M. Chen, A new method for learning barriers diagnosis based on fuzzy rules, in *Proceedings of the Seventh International Conference on Machine Learning and Cybernetics*, (Kunming, 12–15 July, 2008), pp. 3090–3095
29. S.-M. Bai, S.-M. Chen, Automatically constructing concept maps based on fuzzy rules for adaptive learning systems. Expert Syst. Appl. **35**(1), 41–49 (2008)
30. E.V. Laski, R.S. Siegler, Is 27 a big number? Correlational and causal connections among numerical categorization, number line estimation, and numerical magnitude comparison. Child Dev. **76**, 1723–1743 (2007)
31. N. Krasa, S. Sunkwiler, *Number Sense and Number Nonsense: Understanding the Challenges of Learning Math* (Paul H. Brookes Publ, Baltimor, 2009)
32. C. Fourali, Using fuzzy logic in educational measurement: the case of portfolio assessment, **11**(3),129–148 (1997)
33. C.H. Cheng, K.L. Yang, Using fuzzy sets in education grading system. J. Chin. Fuzzy Syst. Assoc. **4**(2), 81–89 (1998)
34. J.R. Nolan, An expert fuzzy classification system for supporting the grading of student writing samples. Expert Syst. Appl. **15**(1), 59–68 (1998)
35. K.A. Rasmani, Q. Shen, Data-driven fuzzy rule generation and its application for student academic performance evaluation. Appl. Intell. **25**, 305–319 (2006)
36. H.Y. Wang, S.M. Chen, New methods for evaluating the answerscripts of students using fuzzy sets, in *Proceedings of the 19th International Conference on Industrial, Engineering, and Other Applications of Applied Intelligent Systems*, (Annecy, France, 2006), pp. 442–451
37. T.-K. Li, S.-M. Chen, A new method for students' learning achievement evaluation by automatically generating the weights of attributes with fuzzy reasoning capability, in *Proceedings of the Eighth International Conference on Machine Learning and Cybernetics*, (Baoding, 12–15 July, 2009), pp. 2834–2839
38. I. Saleh, S.-I. Kim, A fuzzy system for evaluating students' learning achievement. Expert Syst. Appl. **36**(3), 6236–6243 (2009)
39. S.N. Shahbazova, Development of the knowledge-based learning system for distance education. Int. J. Intell. Syst. **27**, 343–354 (2012)
40. S.N. Shahbazova, Creation the model of educational and methodical support based on fuzzy logic, in *Recent Developments and New Directions in Soft Computing*, ed. by L.A. Zadeh, A.M. Abbasov, R.R. Yager, S.N. Shahbazova, M.Z. Reformat (Springer, Berlin, 2014)

Part I
How to Motivate Students

Chapter 2
How to Motivate Students: An Overview of Part I

In this book, we describe how interval and fuzzy techniques can help in education. Before we even start teaching the material, we need to *motivate* the students. As we have mentioned, this is an extremely important aspect of teaching: if a student does not understand why this material is useful, this student may not be as committed to study. Motivating students is often a difficult task, it is kind of a chicken-and-egg problem:

- for the students to truly understand the use of the class material, they need to have a certain degree of knowledge of this material, but
- to acquire this degree of knowledge, they need motivation.

In this part of the book, we will describe how interval and fuzzy ideas can help to motivate students.

From the society prospective, the most important reason for education is that education helps in the long run. We need education to stay competitive, to get good jobs, to be productive members of the society. In other words, education is needed because it is useful. In general, students understand this need, that is why they come to a university to study. However:

- while the students usually understand the usefulness of the education in general,
- in many cases, they do not understand the usefulness of each specific topic
- and therefore, are not motivated enough to study some of the related topics.

We instructors know that these topics are useful – this usefulness is why these topics are taught in the first place – but conveying this usefulness is often not easy.

As a result, classes are often taught in an "authoritarian" way (see, e.g., Freire et al. [1–6]): in many cases, when the new material starts, the students do not receive a convincing explanation of why this material is important; instead, the teacher relies on his or her authority as a professional.

This lack of motivation is part of the reason why the existing pedagogical techniques are often not as successful as predicted: no matter how interested and active the students are, they do not reach their full potential simply because they do not

© Springer-Verlag GmbH Germany 2018

O. Kosheleva and K. Villaverde, *How Interval and Fuzzy Techniques Can Improve Teaching*, Studies in Computational Intelligence 750, https://doi.org/10.1007/978-3-662-55993-2_2

fully understand the need to study this particular material. In view of this observation, one of the important keys to success is for a teacher to make sure that the students understand this need – before explaining the material itself.

This need for understanding motivations is especially important for mathematics and related disciplines. In many school disciplines like English, the need to study the language, the need to learn to communicate clearly, may not be always well understood by the students, but it is usually well understood by their parents and the population as a whole. In mathematics, the situation with understandability is much more problematic:

- not only *the students* do not understand the need for mathematics education;
- in general, most people do not understand the significance and importance of mathematics.

Most people see mathematics as a collection of useless rules and formulas that they need to memorize to get the correct answer to the textbook problems, rules and formulas that, in their opinion, have no use in real life. Since students do not understand the need for mathematics education, they feel forced to study it and, as a result, many students develop strong negative attitude towards mathematics ("I hate math").

In Chaps. 3–5, we will describe how uncertainty-related techniques can help to make students better understand the usefulness of the corresponding material.

There are two important issues related to uncertainty.

First, the students need to *understand* the very presence of uncertainty – and, thus, *the need* to take *uncertainty* into consideration. After all, the ultimate objective of learning is to enable students to make good decisions in their future professional life. In most real-life situations, we make decisions based on imprecise, natural-language descriptions and commonsense ideas. For example, suppose that we have two candidates for a faculty position, the first is much better in research but still needs work on his teaching, and the second candidate is slightly more experienced (and thus, slightly better) in teaching, but his record of publications and grants is much worse. In this situation, most probably, the majority will vote for the first candidate. This is a typical way we all make decisions, this is a typical way students make decisions. In many cases, such a commonsense approach leads to successful decisions. So, students often do not understand the need for complex mathematics-based techniques in decision making. To understand this need, we need to emphasize the deficiencies of the common sense approach, we need to explain, to the students, how seemingly reasonable commonsense ideas can inadvertently lead to counter-intuitive erroneous decisions. Good news is that there is often no need to artificially invent such examples – in the history of mathematics and science, it is exactly these types of examples that led researchers to invent techniques which would let us avoid the corresponding paradoxes. This uncertainty-related issue is described in Chap. 3.

Once uncertainty is properly understood and described, once we understand that it *is* important to take uncertainty into consideration, it is necessary to decide *how* to take uncertainty into account. If we properly take uncertainty into account, then the corresponding problem becomes more realistic – and thus, students have more motivation to study it – and at the same, the problem becomes more challenging, so it can serve as a case study for learning to solve complex problems. This issue of uncertainty is described in Chaps. 4 and 5. In these chapters, on the examples from geometry and calculus, we show how taking uncertainty into account leads to realistic challenging problems, problems on whose example students can understand the need for (and the usefulness of) complex techniques that they are studying.

All this helps students understand the usefulness of the material. However, even when a student understands – on the rational level – the usefulness of a certain material, this understanding does not necessarily translate into making students eager and excited about studying this material. A university degree is a challenging multi-year hard work, and it is difficult to keep the same level of excitement for several years. It is therefore important not just to once explain the usefulness on the logical level, but also to keep the flames of excitement going during the teaching itself. In other words, we not only need to explain that the material is *useful*, we also need to make this material *interesting*. This is what we will analyze in Chaps. 6–11.

How can we increase a person's interest in some topic, a person's excitement in studying this topic? Each student already has some topics about which he or she is excited, so a natural way to increase the students' interest in the topic of study is to relate this topic to the topics about which the student is already excited. In other words, if the two topics are properly related, a student's excitement about his or her original topic of excitement will make the topic of study excited as well. The more we relate the topic of study to other topics, the more students will thus get excited. It is great if we can creatively relate the topic of study to something students deeply care about – there are known successes of using, e.g., music to study math or popular movies to study physics. These ideas are great, but it is difficult to come up with a novel good connection. Good news is that most of what we are teaching has already been taught in the past, so we can always borrow ideas, techniques, and relations from this past teaching experience. In Chaps. 6–11, on the example of teaching math, we show how such ideas can be borrowed:

- from ancient traditions (Chaps. 6–9) and
- from modern practices, both usual (Chap. 10) and unusual (Chap. 11).

Finally, in addition to explaining why a *specific* material is useful and interesting, we need to make sure that the students are excited about studying *in general*, that their levels of interest and commitment remain high. If a math instructor convinces the engineering students that they need to learn math, this should not lead to them getting less interested in studying engineering disciplines, ideally they should be excited about all the topics that they study. In Chap. 12, we therefore analyze how to increase this *general* level of interest. It turns out that properly taking uncertainty into account can help with this task as well. In Chap. 13, we show how to best apply financial motivations.

References

1. P. Freire, *Education for Critical Consciousness* (Continuum, New York, 1974)
2. P. Freire, *Letters to Cristina: Reflections on My Life and Work* (Routledge, New York, 1996)
3. P. Freire, *Pedagogy of Hope* (Continuum, New York, 1994)
4. P. Freire, *Pedagogy of the Heart* (Continuum, New York, 2000)
5. P. Freire, *Pedagogy of the Oppressed* (Continuum, New York, 2003)
6. P. Freire, A. Faundez, *Learning to Question: a Pedagogy of Liberation* (Continuum, New York, 1989)

Chapter 3
Need to Understand the Presence of Uncertainty: Emphasizing Paradoxes as a (Seemingly Paradoxical) Way to Enhance the Learning of (Strict) Mathematics

As we have mentioned, there are two important issues related to uncertainty. First, the students need to *understand* the very presence of uncertainty – and, thus, *the need* to take *uncertainty* into consideration. The ultimate objective of learning is to enable students to make good decisions in their future professional life. In most real-life situations, we make decisions based on imprecise, natural-language descriptions and common sense ideas. As a result, students often do not understand the need for complex mathematics-based techniques in decision making. To understand this need, we need to emphasize the deficiencies of the commonsense approach, we need to explain, to the students, how seemingly reasonable commonsense ideas can inadvertently lead to counter-intuitive erroneous decisions. It is therefore important to emphasize the fact that commonsense treatment of the topics leads to contradictions (paradoxes) and therefore, a more formal treatment is needed. Our experience shows that explaining the corresponding paradoxes – like the irrationality of square root of two or paradoxes of set theory – indeed enhances learning.

The results from this chapter first appeared in [1, 2].

Two aspects of mathematics. Students often still lack understanding of why studying mathematics is important. In our opinion, one of the reasons is that when we talk about mathematics, we often somewhat confuse two different aspects of mathematics, aspects differing by degree of certainty.

Mathematics as viewed by scientists and engineers: a certain level of uncertainty is OK. From the viewpoint of scientists and engineers who apply mathematics, crudely speaking, is a science that helps them perform practically useful computations. From this viewpoint, a practicing engineer and a practicing scientist do not distinguish between well-justified proven results and practically helpful heuristic methods that work in many cases – both are part of what they perceive as mathematics.

For example, from the engineering viewpoint, methods of solving non-linear algebraic and differential equations that are based on their linearization are a legitimate and useful mathematical tool – even though a knowledgeable engineer knows that

© Springer-Verlag GmbH Germany 2018
O. Kosheleva and K. Villaverde, *How Interval and Fuzzy Techniques Can Improve Teaching*, Studies in Computational Intelligence 750,
https://doi.org/10.1007/978-3-662-55993-2_3

the linearity assumptions have limitations and that there are practical situations when these assumptions do not work well in practice. Linearized methods work well in the description of normal functioning of mechanical and civil engineering objects, but they are often inadequate in describing different failure or fracture modes.

From the viewpoint of a professional mathematician, however, the main difference between mathematics and all other disciplines is not in practical computations. According to the mathematician's viewpoint, the main idea behind mathematics is the idea that all its statements must be proven and, once proven, they are absolutely correct.

Mathematics as viewed by mathematicians: everything should be absolutely precise. A physical formula may be well justified by experimental data, but this justification is not a proof. In physics, it often happens that further experiments reveal that the original formula is only approximately true, is only true under certain conditions, and therefore, needs to be modified. This happened, for example, with Newton's mechanics. For several centuries, formulas of Newton's mechanics adequately described physical processes ranging from the motion of planets in a Solar system to the working of engines. However, later, it turned out that, strictly speaking, formulas of Newton's mechanics are only approximately true:

- on the other side of the range, to accurately describe the motion of celestial bodies, we need to take into account effects of Special and General Relativity;
- on the other side of the range, to accurately describe the motion of molecules and atoms (and of elementary particles that form atoms), we need to take into account effects of quantum physics.

From this viewpoint, a physical theory developed in the 19th century usually turned out to be only approximately correct, it will be adjusted (or even completely replaced) by a more accurate more adequate theory.

In contrast, in mathematics, once a theorem is proven, it remains correct.

For example, Pythagoras theorem started as an empirical fact, i.e., in effect, as a physical law. Later on, Pythagoras proved that this formula can be derived from the basic axioms (what we now call Euclidean geometry).

- From the physical viewpoint, the Pythagoras formula is still a useful physical law, actively used in computing the distances between points. However, it is now well understood that it is only approximately true. Indeed, according to General Relativity, the actual space-time is curved; hence, in general, the spatial distance between the points with coordinates (x, y, z) and (x', y', z') is, in general, different from the Pythagoras formula $d = \sqrt{(x - x')^2 + (y - y')^2 + (z - z')^2}$ – it is described by more general formulas of Riemannian geometry.
- From the mathematical viewpoint, however, Pythagoras theorem remains a correct theorem in Euclidean geometry.

The difference between useful empirical facts and mathematical theorems was well understood already in the ancient times. For example, to compute the area A of a circular-shaped region, mathematicians from the ancient Egypt, Babylon, and

Greece used approximate expression for π, such as $\pi \approx 3\frac{1}{7}$. They understood, however, that these expressions were only approximately true, and that the actual value of π is different. On the other hand, the formula $A = \pi \cdot r^2$ that related the desired area to the radius r of the circular region is a proven theorem.

In other words, they used an approximate formula $A \approx 3\frac{1}{7} \cdot r^2$, but they understood that the coefficient $3\frac{1}{7}$ is only approximately true, while the coefficient 2 in r^2 cannot be changed. In contrast, in physics, when it turned out that the Newton's formula $F = G \cdot \frac{m}{r^2}$, for a gravitational force caused by a mass m at a distance r turned out to be only approximately true, physicists proposed changing both the gravitational constant G and the parameter 2: e.g., before General Relativity, theories with $F = G \cdot \frac{m}{r^{2+\varepsilon}}$ were the mainstream explanation of the difference between Mercury's predicted and observed motions.

Back to teaching mathematics: what is often lacking is understanding of the need for mathematical rigor. The above analysis explains what is missing in the current students' understanding of why mathematics is useful. In short, students may (somewhat grudgingly) accept that formulas are useful, but not proofs.

Attempts to teach mathematical rigor: examples. In the 1960s, a new set of teaching practices was introduced in the teaching of mathematics. These practices – known as New Math – emphasized mathematical structure through abstract concepts like set theory and number bases other than 10. The original idea of the promoters was that it would drastically speed up the knowledge of mathematics in schools and thus, help reduce the gap between the US and other countries. The original experiments indeed showed this to be a success; however, overall, the level of mathematical knowledge did not increase as drastically as expected, and the perception – that the US schoolchildren are lagging behind in math – is still a serious concern.

This phenomenon is not restricted to the United States. It may be somewhat of an irony that the practices of the New Math were originally introduced as a response to the challenges provided by the perceived supremacy of the Soviet mathematical education system. At the same time, a similar perception of the supremacy of the Western mathematical education system has led renowned Russian mathematicians such as A. N. Kolmogorov to introduce similar ideas of more abstract mathematics into the Soviet mathematical education – starting from the level of elementary mathematics; see, e.g., [3, 4].

Originally, when enthusiastic teachers, including Kolmogorov himself, started using this approach in their teaching, they indeed improved the resulting students' knowledge and skills. However, the overall result of this reform in teaching mathematics turned out to be not as spectacular as Kolmogorov and other predicted (based on their initial success). An (overall positive but) reasonably critical analysis of this situation is given, e.g., in [5].

How to convince the students that rigor is important. What we need to enhance the students' understanding of mathematics is to help them understand that not only engineering formulas are useful, rigorous mathematical proofs have their value. How can we do that?

Good news is that we do not have to invent new reasons: after all, there are reasons why rigorous mathematics was designed in the first place, and why it remains an important (and well-supported) part of science and, more generally, of our quest for knowledge. What we need to do is convincingly convey these reasons to students of mathematics.

In some sense, in mathematics education – like in biology – ontogenesis repeats philogenesis. In biology, an embryo starts as a single cell, then becomes a multi-cell organism, then gets more and more complex – in effect, copying the major steps of biological evolution that started with single-cell organisms, moved to simple multi-cell ones, etc.

Similarly, a young student of mathematics starts with commonsense notions – the same notions that formed the basic mathematics knowledge before mathematics became a science. As the students learn new material, they often repeat the same mistakes and misunderstandings that the students before them made – and, tracing back, the same mistakes that led to the corresponding development of mathematics in the first place.

So, our question is: in the history of mathematics, why was there a need for rigor?

Paradoxes: the main motivation behind rigor in mathematics. History of mathematics (see, e.g., [6]) shows that the main reason why mathematicians started making their methods more rigorous is that heuristic methods sometimes lead to seeming contradictions (*paradoxes*).

Let us remind of some well-known paradoxes (see, e.g., [6]) that led to the development of mathematical rigor.

Heap paradox. Formal approach to mathematics was first developed by the ancient Greeks, and their paradoxes explain why they felt this need for rigor. Probably the first known paradox is the *heap* paradox:

- one little rock does not form a heap;
- if we have a pile of rocks which is not a heap, and add one more rock, then we still do not get a heap;
- by induction, we can thus conclude that no matter how many rocks we add, we will never get a heap;
- on the other hand, everyone knows that heaps are possible.

This paradox can be reformulated in terms of a crowd (one person does not form a crowd, etc.). This paradox, well known to the Greeks, explained the need to restrict ourselves to well-defined notions – in contrast to vaguely defined ("fuzzy") notions such as *heap* or *crowd*.

Heap paradox and the 20th century successes of "fuzzy logic". It is interesting to mention that this same paradox was revived in the 20th century by computer scientists who were interested in making computers understand our commonsense words and terms. To describe such "fuzzy" words in precise terms – understandable to a computer – L. A. Zadeh invented a special technique called *fuzzy logic*. In fuzzy logic, in contrast to the traditional two-valued logic, notions like "is a heap" or "is a crowd" are not necessarily either true or false. In many practical situations, experts may conclude that to some degree, we have a crowd. When one more person enters, this degree of "crowdedness" increases – until this degree attains its largest possible value 1, meaning that all reasonable persons will conclude that this particular collection of people is a crowd.

This solution of a paradox enabled researchers to successfully use informal (seemingly inconsistent) expert knowledge in the design of efficient automated systems for control, data processing, decision making, etc.; see, e.g. [7].

Fuzzy logic was the first example of *soft computing*, attempts to formalize parts of commonsense reasoning that were "left behind" when mathematics became more rigorous. Fuzzy logic and other soft computing techniques are used in many areas of science and engineering; see, e.g., [7]. In particular, some fuzzy logic techniques have been proposed for use in processing pedagogical data; see, e.g., [8, 9]. The relation between fuzzy logic and paradoxes is analyzed, e.g., in [7, 10].

Diagonal of a unit square: a forgotten paradox. Another known paradox is the fact that the diagonal of a square is not commeasurable with its side, i.e., for example, that the diagonal $\sqrt{2}$ of the unit square is not a rational number. Right now, university students view it as a (somewhat boring) theorem, but when this fact was discovered by the Pythagoreans, it was perceived as a paradox – to such an extent that this result was suppressed for quite some time.

This paradox comes from the fact that in practice, all the numbers come from measurements, be it the length of a line or the mass of a body. Measurements are never 100% accurate; as a result, we can only get an approximate value of the measured quantity. Since measurements are not absolutely accurate, it makes sense to present the result of the measurement only with the accuracy corresponding to the accuracy of the measurement. For example, if we measure the distance with an accuracy of ± 1 km, then we can say that the distance from the University of Texas at El Paso to the El Paso International Airport is approximately 16 km – but, based on this measurement, it does not make too much sense to claim that this distance is approximately 16.75 km. As a result of this procedure, all the numbers that we see in practical applications are rational, i.e., numbers of the type $\frac{p}{q}$, where p and q are real numbers. For us, who are accustomed to decimal numbers, these values are usually decimal fractions, i.e., values of the type $\frac{p}{10^q}$; however, in the ancient world, people used other bases as well, so we can safely consider general rational numbers.

Since all numbers coming from practice are rational, it makes sense, in practice-oriented mathematics, to only consider rational numbers. For several centuries, this idea worked well: if we start with a rational length, we can divide it into 2, 3, and more equal pieces, and still get a rational number. Many other reasonable operations also led to rational numbers. Sometimes, mathematicians did not know how exactly a given number can be represented as a fraction $\frac{p}{q}$, but they assumed that this representation is always possible – and made arguments based on this seemingly natural assumption.

Imagine their surprise when it turned out that the quantity as simple-to-define as a diagonal of a unit square is not a rational number! This discovery has lead to the need for a revolutionary reworking of many previous mathematical results and descriptions.

The amazing thing about this paradox – just like about many other paradoxes – is that it is simple to describe (and thus, not that difficult to teach to the students). Indeed, let us assume that $\sqrt{2} = \frac{p}{q}$ for some integers p and q. We can always divide both p and q by their greatest common divisor, so we can safely assume that p and q have no common factors.

By squaring both sides and multiplying both sides by q^2, we conclude that $p^2 = 2q^2$, hence, we conclude that p^2 is even. Thus, p cannot be odd – because then p^2 would be odd as well; so, p is even; in other words, $p = 2k$ for some k. Therefore, $p^2 = 4k^2$ and hence, $q^2 = 2k^2$ is also even – from which we can similarly deduce that q is even as well. Thus, p and q are both even, i.e., they have a common factor 2: a contradiction to our assumption that p and q have no common factors.

Achilles and the Tortoise. A paradox that probably everyone knows is the Achilles and the Tortoise paradox. According to this paradox, fast Achilles cannot catch up with a slow Tortoise no matter how fast he runs: before Achilles reaches the distance separating him from the Tortoise, he needs to cover the first half of this distance – and during that time the Tortoise will move even further away.

Similarly to the $\sqrt{2}$ "paradox", to many people, this argument does not sound like a paradox at all. We know how to solve practical problems of who catched up with whom, when, and where, so this "paradox" may sound artificial. However, in the ancient world, when solutions to all these motion-related problems were not known, this paradox was a serious problem. In effect, it was not resolved until the appearance of calculus in the 17th century enabled us to solve motion-related problems.

Paradoxes related to calculus. Calculus resolved some paradoxes, but – as often happens in the history of science, it led to the discovery of several new paradoxes. An example of such a paradox was related to summation of infinite series. Before calculus, there were very few attempts to build a converging sequence or to add up an infinite series. Calculus provided a general framework within which such a summation became routine. Most functions like $\exp(x)$, $\sin(x)$, $\cos(x)$, where shown to be representable as Taylor series – and this provided a very efficient way of computing these functions, so efficient, that even nowadays, Taylor series are the main techniques used by computers to compute the values of these functions.

For many series, calculus-based ideas led to successful formulas for their sum. However, for some series, the same ideas led to paradoxes. Probably the most well known of these paradoxes was discovered by Euler who tried to compute the sum s of the infinite series

$$1 + (-1) + 1 + (-1) + 1 + \ldots$$

On the one hand, we can combine elements into pairs, and end up with

$$s = (1 + (-1)) + (1 + (-1)) + \ldots = 0 + 0 + \ldots = 0.$$

On the other hand, we can keep the first element in this infinite sum intact and combine elements starting with the second one. Then, we get

$$s = 1 + ((-1) + 1) + ((-1) + 1) + \ldots = 1 + 0 + 0 + \ldots = 1 \neq 0.$$

This paradox and similar paradoxes have led to the need to revisit the foundations of calculus, to introduce rigor. This "revolution of rigor", largely associated with the names of Cauchy and Weierstrass, introduced the modern "epsilon–delta" definitions into calculus. For example, according to Cauchy, an infinite series $a_1 + a_2 + \ldots$ has a sum s if and only if for every $\varepsilon > 0$, there exists an N such that for all $n \geq N$, we have $|(a_1 + \ldots + a_n) - s| \leq \varepsilon$. Euler's paradox is then resolved because according to this definition, the series $1 + (-1) + 1 + (-1) + \ldots$ does not have a sum at all.

The "epsilon–delta" definitions are not easy – but they are needed because without rigorous definitions, we can get paradoxes. Students taking calculus, however, often learn these definitions without understanding why this complexity is needed – and thus, sometimes have trouble understanding the related notions.

Paradoxes of infinity. Infinity has caused many other paradoxes – e.g., the known paradox of Galileo. According to Galileo, there are exactly as many natural numbers as there are squares of natural numbers – because natural numbers can be placed in 1-1 correspondence with their squares: $0 \leftrightarrow 0$, $1 \leftrightarrow 1$, $2 \leftrightarrow 4$, $3 \leftrightarrow 9,\ldots$

On the other hand, since most natural numbers are not squares, the number of squares is much smaller than the number of natural numbers.

This paradox helped mathematicians understand that they have to be careful when dealing with infinite sets – and it was not resolved until Georg Cantor invented set theory in 1879. In Cantor's set theory, Galileo's paradox becomes a theorem – an infinite set can have exactly as many elements as its own proper subset.

Russell's paradox and the 20th century revolution of rigor. Set theory sounded like a convenient and consistent new foundation for mathematics – until the famous philosopher Bertran Russell showed that, in its original form, set theory also has paradoxes. Namely, in set theory, some sets are elements of themselves: e.g., a set U of all possible sets is itself a set and thus, its own element: $U \in U$. Other sets are not elements of themselves: e.g., the set N of all natural numbers is not itself a natural number and thus, it is not its own element: $N \notin N$. It is natural to consider the set $V \stackrel{\text{def}}{=} \{x : x \notin x\}$ of all the sets which are not their own elements. Russell asked a natural question: is this set V its own element or not?

There are only two options: $V \in V$ and $V \notin V$ and in both cases, as Russell has shown, we get a contradiction.

Indeed, if $V \in V$, this, by definition of V as a set of all sets that are not their own elements, means that $V \notin V$. This contradicts to the assumption that $V \in V$.

If $V \notin V$, this means that V is an example of a set that we combine into V, so $V \in V$ – also a contradiction.

This paradox was made even clearer by its "barber" reformulation: a barber attached to a military regiment is ordered to regularly shave those (and only those) who do not shave themselves. The paradox appears when we try to find out whether he is required to shave himself or not. If he does not shave himself, then, according to the order, he has to shave himself. If he shaves himself, then the same order prohibits him from shaving himself – again a contradiction.

The main answer to Russell's paradox was similar to the answer to Galileo's paradox: when we deal with infinite sets, we should not rely on common sense, we should only use formal methods. Several axiomatic approaches to set theory have been developed, and in all of them, Russell's paradox is resolved – usually, by showing that the existence of Russell's set V cannot be deduced from the axioms – moreover, this paradox can be viewed as a theorem proving that such a set cannot exist.

In terms of sets, this non-existence may sound not very intuitive, but in terms of the barber reformulation, this solution makes perfect sense: the above instructions to the barber are clearly inconsistent, so there cannot exist a barber who always follows these instructions.

Paradoxes beyond pure mathematics. Similar paradoxes occur not only in pure mathematics, they also occur in applications of mathematics to physics and other disciplines. The most well known example is quantum mechanics – and its mathematical counterpart, Hilbert spaces and operators in Hilbert spaces. Many counter-intuitive properties of quantum mechanics come, crudely speaking, from the fact that the same elementary particle (e.g., photons, electrons) can exhibit seemingly inconsistent properties. For example, the same particle can sometimes behave as a point particle, with no spatial distribution; on the other hand, sometimes, it can behave as a wave, a spatially distributed process.

It is interesting that in the last two decades, it turned out that these seemingly paradoxical properties of quantum objects can be used to drastically speed up computations and provide additional communication security; see, e.g., [11].

Our recommendation: teaching paradoxes. In short, what we propose is to teach the students paradoxes. The same paradoxes that motivated mathematicians to introduce rigor in the first place can help in teaching students why rigor is necessary.

This proposal is realistic and helpful. We have explained paradoxes to students in Russia and in the US – as part of the after-school program. Our anecdotal evidence – and the experience of several of our Russian colleagues who taught similar paradoxes to students – shows that the paradoxes, with their unpredicted result, do raise the students' interest in the related mathematical concepts.

Teaching paradoxes in mathematics may seem counter-intuitive. For many people who like science and mathematics, one of the main attractions is that science and mathematics are logical, precise, consistent, and clear – in contrast to, e.g., humanities and art, where seemingly inconsistent ideas can (and do) co-exist. From this viewpoint, teaching paradoxes in mathematics may seem a counter-intuitive way of enhancing students interest. Indeed, from this viewpoint, the revelation that mathematics has paradoxes can destroy exactly the motivations that we try to cultivate.

Paradoxes as one of the main sources of evolution in mathematical thought. The viewpoint of science and mathematics as (mostly) paradox-free is usually combined with an understanding of the progress in science and mathematics as a reasonably smooth gradual process. This understanding was shattered by the work of Kuhn [12]. Kuhn's examples were mainly from physics, but history of mathematics confirms that similar revolutionary changes occurred in mathematics as well; see, e.g., [6, 13].

The important role of contradictions and paradoxes in the development of mathematics was actively researched by Russian philosophers of mathematics like A. D. Alexandrov [13–15] (see also [16]).

Paradoxes and Hegel's dialectics. The viewpoint that contradictions and paradoxes are one of the main sources of progress and evolution is not limited to mathematics and science. This viewpoint is one of the main points of Hegel's Dialetics, and this viewpoint was further developed by numerous researchers who transformed Hegel's poetic vision into philosophy of science; see, e.g., [13] and references therein.

Fuzzy techniques as a way of formalizing paradoxes. In some cases, students have no prior idea about some objects or techniques. In this case, all we need to do is to explain these ideas to a student. In such simple cases in which students can be viewed as "tabula rasa", there is probably no need for paradoxes – because paradoxes may only confuse the students.

The need for paradoxes comes from the fact that students are rarely "tabula rasa". Whether it comes to statements about the physical world or mathematical objects, students usually have their own preconceptions. For example, before studying physics, students erroneously believe that a moving body will by itself come to a halt, and we need to constantly apply a force to keep it moving. One of the objectives of teaching is to help students learn the correct statement – law of inertia. In such cases, a paradox approach may be useful.

Why fuzzy techniques: general idea. A paradox means that in some sense, both a statement S and its negation $\neg S$ are true. In other words, a paradox is when a composite statement $S \,\&\, \neg S$ holds. The possibility for such statements to be to some degree true is one of the well-known features of fuzzy logic. It is therefore reasonable to use fuzzy logic to describe such paradoxical statements.

Why fuzzy techniques: details. Our objective is to move a student from a state in which he or she (erroneously) believes in a statement S into a state in which he or she believes in the correct statement $\neg S$. Both original state S and the new state $\neg S$ are often very clear and precise, there is nothing fuzzy about them. The need to consider fuzziness comes from the fact that the students' opinions often evolve continually. So, when a student changes his or her opinion from S to $\neg S$, the student's degree of belief in S changes continuously from 1 (absolute belief in S) to 0 (absolute belief in $\neg S$).

Let us describe this transition in more detail. We start with a student absolutely believing in S; his or her degree of belief in S is exactly 1. Then, we instill some elements of doubt in S in the student's mind. At this stage, the student's degree of belief in S decreases – although at the beginning, it is still close to the "absolute" value 1.

If we stop at this point, the student will go back to believing in S: indeed, if a student needs to choose between S and $\neg S$, then, since the student's degree of belief in S in larger than his or her degree of belief in $\neg S$, the student will choose S. The critical transition happens when the student's degrees of belief in S and $\neg S$ are exactly equal – i.e., when the degree of belief in S is exactly 0.5. After this stage, the degree of belief in S becomes smaller than the degree of belief in $\neg S$. Thus, the student may not have yet fully mastered this subject – i.e., his degree of belief in $\neg S$ in still smaller than 1 – but if given a choice between S and $\neg S$, the student will choose the alternative with the larger degree of belief, i.e., the correct statement $\neg S$.

So, to transit from the original incorrect belief S into the desired correct belief $\neg S$, it is crucial to lead the student to the point when his or her degrees of belief in S and in $\neg S$ are exactly equal – i.e., when his original clear idea will be replaced by a paradox.

How fuzzy techniques can be useful. Fuzzy techniques enable us to gauge the person's degree of belief in different statements and thus, to better understand how close we are to the desired paradoxical situation – situation which will bring us on the path to correct understanding.

Additional idea: mathematics education as a way to oppose oppression. In Chap. 2, we have described the relation between Freire's ideas of opposing oppression and the problems related to teaching mathematics. So far, we have used this relation to enhance the teaching of mathematics.

It is interesting that this same relation was actually used the other way around: namely, in his books, papers, talks, and in his practice, A.D. Alexandrov expressed an idea that teaching mathematics, in turn, liberates the oppressed mind [16, 17]. An example of Alexandrov's (somewhat paradoxical) views are presented at the end of the chapter.

Indeed, teaching mathematics, teaching the need to prove results, emphasizes the need to justify statements not only in mathematics but in everyday life in general – as opposed to trusting uncritically. Not surprisingly, among the brave people who, in the former Soviet Union, led the non-violent struggle for human rights, for democracy, against oppression, mathematicians and mathematically oriented physicists and computer scientists were especially well represented. It suffices to say that the most well known dissident was a physicist Andrei Sakharov, a researcher who actively used complex mathematics in his studies.

A.D. Alexandrov on science and morality. A.D. Alexandrov started a description of his ideas on science and morality by mentioning that Vladimir Soloviev, a prominent Russian philosopher, defines faith (in a 1910s article in the classical Russian Brokhaus Encyclopedia) as something that cannot be acquired or changed by observing facts or by listening to arguments.

According to A. D. Alexandrov, an often repeated statement that faith is necessary for people to behave morally is a shame. Morality should itself be the basis of a human behavior; if a good behavior is not based on the inherent moral feelings, only on faith, then under different circumstances, the same faith can – and often does – justify amoral behavior.

In ancient India, people noticed that everything falls, but the Earth does not fall. Their sages explained that the Earth rests on the backs of four huge elephants. The elephants do not fall because they are standing on the shell of an even larger turtle, the turtle is floating in an ocean, etc. We can continue forever and never come to a final answer. And later on, scientists discovered that the Earth does not rest on anything, it is there all by itself. The same is with morality.

In human society, we often have to work together. Two religious believers who hold different sets of beliefs can never convince each other, because their faiths do not come from facts or from logic and cannot be changed by facts or by logic. So, the only way these two believers can work together is when one forces the other by physical force. Thus, faith leads to violence.

At first, faith was not inconsistent with the search for scientific truth. A pure faith, without prejudice, with readiness to believe in anything – such a faith is, in some sense, equivalent to the scientific search for truth.

Science is different from religious faith. With scientists, you can talk objectively and calmly about everything; why? because when a scientist encounters facts that contradict his previous beliefs, he does not stubbornly stick to his beliefs, he can change them.

According to modern science, all the events in the world are – directly or indirectly – connected to each other. As a result, each individual is – to a smaller or larger extent – connected to everything that is happening in the world. So, the actions of every individual can influence (and does influence) – to a smaller or larger extent – all the events in the world. And if each of us can influence each of these events, this means that each of us is responsible for each of these events.

References

1. O. Kosheleva, Potential application of fuzzy techniques to math education: emphasizing paradoxes as a (seemingly paradoxical) way to enhance the learning of (strict) mathematics, in *Proceedings of the 27th Annual Conference of North American Fuzzy Information Processing Society NAFIPS'2008*, New York, 19–22 May 2008
2. O. Kosheleva, Towards combining Freirean ideas and Russian experience in mathematics education, in *Teaching for Global Community*, ed. by C.A. Rossatto, A. Luykx, H.S. Garcia (Information Age Publishing, Charlotte, 2011), pp. 207–218
3. A.N. Kolmogorov, A.S. Semenovich, R.S. Cherkasov, *Geometry, Textbook for 6th through 8th Grades* (Prosveschenie Publ, Moscow, 1982). (in Russian)
4. A.V. Pogorelov, *Geometry, Textbook for 6th Through 10th Grades* (Prosveschenie Publ, Moscow, 1983). (in Russian)
5. A.D. Alexandrov, *Foundations of Geometry* (Nauka Publ, Moscow, 1987). (in Russian)
6. C.B. Boyer, U.C. Merzbach, *A History of Mathematics* (Wiley, New York, 1991)

7. G.J. Klir, B. Yuan, *Fuzzy Sets and Fuzzy Logic: Theory and Applications* (Prentice-Hall, Upper Saddle River, 1995)
8. R. Aló, O.M. Kosheleva, Optimization techniques under uncertain criteria, and their possible use in computerized education, in *Proceedings of the 25th International Conference of the North American Fuzzy Information Processing Society NAFIPS'2006* Montreal, Quebec, Canada, IEEE Press, Piscataway, New Jersey 3–6 June 2006
9. O.M. Kosheleva, M. Ceberio, Processing educational data: from traditional statistical techniques to an appropriate combination of probabilistic, interval, and fuzzy approaches", *Proceedings of the International Conference on Fuzzy Systems, Neural Networks, and Genetic Algorithms FNG'05*, Tijuana, Mexico, 13–14 October 2005. pp. 39–48
10. H.T. Nguyen, O.M. Kosheleva, V. Kreinovich, Is the success of fuzzy logic really paradoxical? or: towards the actual logic behind expert systems. Int. J. Intell. Syst. **11**, 295–326 (1996)
11. M.A. Nielsen, I.L. Chuang, *Quantum Information and Quantum Computation* (Cambridge University Press, Cambridge, 2000)
12. T.S. Kuhn, *The Structure of Scientific Revolutions* (University of Chicago Press, Chicago, 1962)
13. A.D. Alexandrov, A.N. Kolmogorov, M.A. Lavrentiev, *Mathematics: Its Content, Methods and Meaning* (Dover Publ, New York, 1999)
14. A.D. Alexandrov, A.L. Verner, V.I. Ryzhik, *Geometry, Textbook for 9th and 10th Grades* (Prosveschenie Publ, Moscow, 1984). (in Russian)
15. A.D. Alexandrow, Mathematik und Dialektik, in *Mathematiker Über die Mathematik, Herausgegeben von Michael Otte unter Mitwirkung von H. N. Jahnke*, ed. by Th Mies, G. Schubring (Springer, Berlin, 1974), pp. 47–63
16. O. Kosheleva, He taught us that we are responsible for everything, in *A.D. Alexandrov*, ed. by G.M. Idlis, O.A. Ladyzhenskaya (2002), pp. 125–126. in Russian
17. A.D. Alexandrov, *Scientific Search and Religious Faith* (Nauka Publ, Moscow, 1974). (in Russian)

Chapter 4
Uncertainty-Related Example Explaining Why Geometry is Useful: Geometry of a Plane Rescue

Once we understand that it *is* important to take uncertainty into consideration, it is necessary to decide *how* to take uncertainty into account. If we properly take uncertainty into account, then the corresponding problem becomes more realistic – and thus, students have more motivation to study it – and at the same, the problem becomes more challenging, so it can serve as a case study for learning to solve complex problems. In this chapter, on the example from geometry, we show how taking uncertainty into account leads to realistic challenging problems, problems on whose example students can understand the need for (and the usefulness of) complex techniques that they are studying. In the next chapter, we describe a similar example from calculus.

The results from this chapter first appeared in [1]. The authors are very thankful to Alan Clements from the University of Teesside, England, for attracting their attention to this interesting use of geometry.

Geometry helped to rescue an airplane: a true story. Let us start with describing the actual story of a plane rescue.

On December 21, 1978, a Cessna plane got lost over the Pacific Ocean when its navigation instruments malfunctioned. A Cessna plane is not equipped for a water landing, so if the plane was not located and guided to an airport, Jay Parkins, the pilot, could die. Luckily, a large passenger plane happened to fly in the area, navigated by a New Zealand pilot Gordon Vette, and this plane heard the Cessna's radio signal. Captain Vette used many ideas to locate the plane.

The final location breakthrough came from the fact that the Cessna plane was equipped with a radio transmitter which could only be heard within direct visibility. Thus, Captain Vette could only hear this radio when he was within a certain (geometrically easy to compute) distance r ($r \approx 200$ miles) from the plane. So, he asked the Cessna pilot to circle in place, and flew his plane in a straight line until he lost the radio signal. The point where he lost the signal was exactly r miles from the (unknown) Cessna location.

© Springer-Verlag GmbH Germany 2018
O. Kosheleva and K. Villaverde, *How Interval and Fuzzy Techniques Can Improve Teaching*, Studies in Computational Intelligence 750,
https://doi.org/10.1007/978-3-662-55993-2_4

After that, he turned back, and flew in a somewhat different direction until he lost the radio signal again; this way, he found a second point on the circle. Based on these two points on the circle, Captain Vette found the center of this circle – i.e., the location of the missing plane.

Based on this location, Air Traffic Control gave the pilot directions to the nearby airport (directions had to be given in relation to the Sun, since the navigation instruments did not work). The plane was saved, the pilot landed alive.

This dramatic rescue story is described in detail in [2]; it was even made into a successful TV movie [3].

Why this story is interesting. This story can be used as a good pedagogical example: that seemingly abstract geometry can actually help in very unusual and drastic situations. What makes this problem difficult to solve – in comparison with many textbook geometric problems – is the huge uncertainty: we did not know the actual location of the plane, we did not know the distance to this plane (we could only check whether this distance is smaller than the threshold r), etc.

Towards a related geometric problem. With this situation, comes an interesting geometric problem. The crucial aspect of this plane rescue was time: the plane had to be located before it ran out of fuel. So, the passenger plane had to fly at its maximal speed. With this speed, time is proportional to the distance. So, we must choose the shortest of all the trajectories which would allow us to reach the circle.

The most important thing is to reach one point on a circle. Once we have found it, we can easily find nearby points – e.g., by flying in a small circle around that first point. So, the critical question is finding the shortest path which still guarantees that we find a point on a circle.

Captain Vette chose to fly in a straight line until he lost the radio signal. Was this the best possible decision, or could some other trajectory be better?

Intuitively, going in a straight line makes sense because if we are somewhere inside a circular disk, and we follow a straight line in any direction, then eventually, we will reach the circle – the borderline of the disk. However, it is not intuitively clear whether this is indeed the optimal strategy.

Specifically, for each line, we could go in both directions. If we go in one direction and reach a circle after we flew a distance $\leq r$, then this strategy sounds reasonable. However, if we have flown a distance larger than r and we have not yet reached the circle, this means that we were flying in a wrong direction. So, maybe at this point, a reasonable strategy is to change course?

Let us formulate this problem in precise terms. Let us denote the starting point by O. We know that this point O is inside the disk of given radius r; we do not know where is the center of this disk.

Definition 4.1

- By a *trajectory*, we mean a planar curve of finite length, i.e., a continuous mapping γ from some interval $[0, T]$ into a plane such that $\gamma(0) = O$ and the overall length of this curve is finite.

- Let $r > 0$ be a real number. We say that the trajectory γ is guaranteed to reach any circle of radius r if for every disk of radius r which contains the point O, the curve γ has an intersection with its border (i.e., with the corresponding circle).

Comment. For example, we can restrict ourselves to piece-wise smooth curves. For such curves γ, the length $\ell(\gamma)$ can be described as $\ell(\gamma) = \int \|\dot\gamma\| \, dt$, where $\dot\gamma \overset{\text{def}}{=} \dfrac{d\gamma}{dt}$, and $\|(a_1, a_2)\| = \sqrt{a_1^2 + a_2^2}$ denotes the length of a vector.

Proposition 4.1 *For every real number $r > 0$, the following statements hold:*

- *A straight line segment of length $2r$ is guaranteed to reach any circle of radius r.*
- *Every trajectory γ which is guaranteed to reach any circle of radius r has a length $\ell(\gamma) \geq 2r$.*
- *If γ is a trajectory of length $2r$ which is guaranteed to reach any circle of radius r, then γ is a straight line segment.*

Comments. In other words, the only shortest (= fastest) rescue trajectory is a straight line segment.

Proof 1°. Clearly, the straight line segment of length $2r$ is guaranteed to reach any circle of radius r.

Indeed, inside the circle, the largest distance between the two points is the diameter $2r$. So, once we are inside the disk and we go the distance $2r$, we are outside the disk.

2°. Before we continue with the proof, let us make some remarks about the representation of the curves.

In our definition, we defined a curve as an arbitrary continuous mapping from real numbers to the plane. If we re-scale this curve, i.e., use a function $r(s(t))$, where $s(t)$ is a monotonic function from real numbers to real numbers, then we get the exact same geometric curve. It is convenient to avoid this multiple representation of the same geometric curve by using, e.g., the total length of the path between the point $\gamma(0)$ and $\gamma(t)$ as the new parameter. With this choice of a parameter, the length of a curve from the point $\gamma(0)$ to the point $\gamma(t)$ is equal to t.

In the following text, we will assume that the curve γ is parameterized by length.

3°. Let γ be a curve that is guaranteed to reach a circle of radius r. Let us prove that the length $L = \ell(\gamma)$ of this curve γ is greater than or equal to $2r$.

We will prove this by reduction to a contradiction. Indeed, assume that $L < 2r$. Let us take a circle of radius r with a center in the point $c = \gamma(L/2)$. For every point $\gamma(t)$, the length of the curve between this point and the center is equal to $|t - (L/2)|$. For $0 \leq t < L/2$, this value is equal to $(L/2) - t$ and is, thus, $\leq L/2$. For $L/2 \leq t \leq L$, this value is equal to $t - (L/2)$; since $t \leq L$, this length is $\leq L - (L/2) = L/2$. In both cases, the length is $\leq L/2$.

Since the distance is the shortest possible length of a curve connecting two points, we thus conclude that the distance $\|\gamma(t) - c\|$ between any point $\gamma(t)$ on this curve and the center c does not exceed the length along the curve and hence, does not exceed $L/2$. Since $L < 2r$, we have $(L/2) < r$, so every point on the curve is at a distance $< r$ from the center c. Hence, none of these points is on the circle, contrary to our assumption that the trajectory is guaranteed to reach any circle of radius r.

This contradiction proves that every trajectory which is guaranteed to reach any circle of radius r has length $\geq 2r$.

4°. To complete the proof, let us prove that if γ is a curve of length $\ell(\gamma) = 2r$ which is guaranteed to reach any circle of radius r, then γ is a straight line segment.

4.1°. Let us first prove that at least one of the half-curves $\gamma([0, r])$ and $\gamma([r, 2r])$ is a straight line segment.

We will also prove this by reduction to a contradiction. Let us assume that both half-curves are not straight line segments, and let us prove that under this assumption, the curve γ does not reach a circle of radius r whose center is the midpoint $c = \gamma(r)$.

Indeed, the straight line segment is the only shortest curve connecting two points. Thus, since $\gamma([0, r])$ is not a straight line segment, the distance $\|c - \gamma(0)\|$ between $c = \gamma(r)$ and the point $\gamma(0)$ is smaller than the length r of this half-curve.

For all values $t \in (0, r]$, the distance $\|c - \gamma(t)\|$ between $\gamma(t)$ and $c = \gamma(r)$ is smaller than or equal that the length $r - t$ of the half-curve between these points: $\|c - \gamma(t)\| \leq r - t$. Since $t > 0$, we have $r - t < r$ and hence, $\|c - \gamma(t)\| < r$.

So, for all $t \in [0, r]$, we have $\|c - \gamma(t)\| < r$, i.e., none of the points on this half-curve reaches the circle. Similarly, none of the points on the second half-curve $\gamma([r, 2r])$ reaches the circle. This contradicts to our assumption that γ is guaranteed to reach any circle of radius r. This contradiction proves the statement.

4.2°. We have just proved that at least one of the half-curves $\gamma([0, r])$ and $\gamma([r, 2r])$ is a straight line segment. Without losing generality, let us assume that the first half-curve $\gamma([0, r])$ is a straight line segment (the proof for the case when the second half-curve is a straight line segment is similar).

4.3°. Let us now prove, by reduction to a contradiction, that the second half-curve $\gamma([r, 2r])$ is also a straight line segment.

Indeed, if it is not a straight line segment, then, as we have mentioned in Part 4.1° of this proof, the largest distance d_0 between the midpoint $\gamma(r)$ and points $\gamma(t)$, $r \leq t \leq 2r$, on this half-curve is smaller than r: $d_0 < r$.

Let us now denote the half-difference $(r - d_0)/2$ by δ, and consider a circle of radius r with the center c at the point $\gamma(r - \delta)$. The first half-curve is a straight line segment, so all the points on the first half-curve are located at a distance $\leq r - \delta < r$ from the point c. Thus, none of these points reaches the circle.

For every point $\gamma(t)$ on the second half-curve, we have $\|\gamma(t) - \gamma(r)\| \leq d_0$, hence $\|\gamma(t) - c\| \leq \|\gamma(t) - \gamma(r)\| + \|\gamma(r) - c\| \leq d_0 + \delta$. Since $\delta = (r - d_0)/2 < (r - d_0)$, we have $d_0 + \delta < d_0 + (r - \delta_0) = r$, so $\|\gamma(t) - c\| < r$. Thus, none of the points on this second half-curve reaches the circle either. This contradicts to our assumption that the curve γ is guaranteed to reach any circle of radius r. The contradiction proves that the second half-curve is also a straight line segment.

4.4°. To complete the proof, we must show that the straight line segments $\gamma([0, r])$ and $\gamma([r, 2r])$ continue each other, i.e., that they form a single straight line segment.

Indeed, suppose that they form an angle which is different from 180°. Then, the arc connecting $\gamma(0)$ and $\gamma(2r)$ is not the diameter, hence its half-length ℓ is $< r$. Let us take as the center c of the circle the midpoint $(\gamma(0) + \gamma(2r))/2$ of this arc. For the points $\gamma(0)$ and $\gamma(2r)$, we have $\|c - \gamma(0)\| = \|c - \gamma(2r)\| = \ell < r$. For the midpoint $\gamma(r)$, the distance $\|\gamma(r) - c\|$ is one of the sides of the right triangle $\gamma(0)c\gamma(r)$ of which the side $\gamma(0)\gamma(r)$ is a hypothenuse of length r; hence

$$\|\gamma(r) - c\| < r.$$

By convexity of distance, we thus conclude that for all the points on the curve γ, the distance from c is $<r$ – which contradicts our assumption that the curve γ is guaranteed to reach any circle of radius r. The contradiction proves that the two half-curves form a single straight line segment. The proposition is proven.

Discussion. Our description of the geometric problem was somewhat idealized. The idea of using the second point on a circle in a close vicinity of the first one works if we assume that we can exactly locate the point where we lose the radio signal, i.e., that we can exactly locate the point on an (unknown) circle.

In practice, however, there is always some measurement inaccuracy. If we can only locate a point of intersection with accuracy ε, and we want to find the center of the plane with accuracy δ, then we get an even more complex problem.

References

1. O. Kosheleva, Flight Cessna 771 revisited: geometry of a plane rescue. Geombinatorics **17**(2), 78–84 (2007)
2. S. Stewart, *Emergency! Crisis on the Flight Deck* (Crowood Press, Ramsbury, 2003)
3. R. Young (director), G. Rubino, R. Benedetti (scenario), *Mercy Mission: The Rescue of Flight 771*, Anasazi Productions, TV movie, USA–Australia (1993) (in Australia and New Zealand, it was shown under the title *The Flight from Hell*)

Chapter 5
Uncertainty-Related Example Explaining Why Calculus Is Useful: Example of the Mean Value Theorem

Students learn better when they understand why they need to learn the specific material, why this material is useful in their own discipline. In this chapter, we show how to explain usefulness of simple calculus results to engineering and science students – on the example of such a seemingly theoretical result as the Mean Value Theorem. It turns out that this result is very useful in problems with interval uncertainty.

The results from this chapter first appeared in [1].

Explaining usefulness is important. As we have mentioned in the beginning of this part of the book, many students come to college to become engineers, scientists, etc. When a student wants to be a civil engineer, this desire motivates the student to study civil engineering courses; when a student wants to be a physicist, this desire motivates the student to study physics courses.

However, to get a degree in Civil Engineering, it is not enough to only study Civil Engineering courses: a student also needs to study several courses in mathematics and physics. There is a reason why these courses are needed: the material studied in these courses provides a background needed for the Civil Engineering material. The problem is that when the students take the corresponding classes – such as calculus – they often do not have a clear understanding of why this material is useful for their profession. This lack of understanding inhibits their learning, and makes it harder to teach calculus to engineers and scientists.

To make this teaching easier – and to make students more willing to study – it is therefore desirable to explain to them (as clearly as possible) why the corresponding calculus material is important for their chosen profession.

The more details, the better. As we have just mentioned, it is desirable that the students understand why calculus is important for their profession. Once the students get this understanding, they become more eager to learn. However, often, this eagerness only applies to those parts of the material that the students perceive as

© Springer-Verlag GmbH Germany 2018
O. Kosheleva and K. Villaverde, *How Interval and Fuzzy Techniques Can Improve Teaching*, Studies in Computational Intelligence 750,
https://doi.org/10.1007/978-3-662-55993-2_5

useful, while about other parts of the material – for which the students do not have a good understanding of how useful these parts are – the students remain unmotivated.

It is therefore desirable to provide not only the explanation of why the course as a whole is useful, it is also desirable to provide students with an explanation of why each piece of the material taught in this course is useful.

What we do in this chapter: general idea. In this chapter, we describe the usefulness of a seemingly purely theoretical result: the Mean Value Theorem (its exact formulation is given below).

We will illustrate its usefulness on the example of practical problems related to so-called *interval computations* – techniques for processing uncertainty in data processing in frequent situations when we do not know the probabilities of different approximation errors, we only know the upper bounds on the approximation error. In the description of the usefulness of interval computations, we will follow [2–4].

What we do in this chapter: plan. The usual proof of the Mean Value Theorem is based on the Rolle's Theorem, which is, in turn, based on the basic result that when a differentiable function attains a local minimum or a local maximum, then its derivative at this minimum or maximum point is 0. Because of this dependence, we will start with this basic result; then, we turn to the Rolle's Theorem and after that, to the Mean Value Theorem.

Why it is important to study maxima and minima: main reason. In practice, we are interested in making the best possible decisions. For example, when we invest money, we would like to get the largest possible return on the investment. When we design a bridge, we would like to find a design that – within the given specifications on its stability, durability, etc. – will cost the smallest amount of money. When we design a computer chip, we often want to make sure that its computing rate (as estimated, e.g., by the number of basic arithmetic operations per second) is the largest possible. When a GPS system makes a route for the emergency vehicle to reach the site of an accident, we want to find the path that would take the shortest possible time.

In principle, we can make many different decisions. Let $x = (x_1, \ldots, x_n)$ be a list of all the parameters that are needed to describe a decision. For example, in the financial investment case, x_i may be the proportion of the original investment amount that will be invested in the ith financial instrument. We assume that, once we know the decision x, we can determine the outcome of this decision; let us denote the resulting outcome by $f(x) = f(x_1, \ldots, x_n)$. In these terms, our goal is to find the values of the parameters $x_1, \ldots, x_n$ that lead to the largest possible (or, if appropriate, smallest possible) value of this function $f(x_1, \ldots, x_n)$.

Why it is important to study maxima and minima: need to take uncertainty into account. The above main reason why optimization is useful is well known. However, there is another – less well known – reason why studying minima and maxima is very important for engineering and scientific applications. This reason is related to the fact that in the above motivations, we assumed that we can set the exact values of the corresponding parameters and, when these values are set, we can determine the exact outcome of the corresponding decision.

In practice, there is uncertainty. While it is possible to allocate a fixed amount of money into a certain type of investment, it is not that easy to set an exact size of a bridge width: the actual size may be somewhat different, we need to also set up a tolerance describing how accurately we need to maintain the original sizes: shown we maintain them with accuracy $\pm$ 10 cm? with accuracy $\pm$ 1 cm? Similarly, the outcome may depend not only on the parameters describing our decision, but also on the parameters describing the outside world – and these parameters are also frequently only known with uncertainty.

For example, suppose that we want to design the trajectory of a spaceship that is intended to fly an automatic mission to the Moon. In principle, we know the equations that describe the motion of the spaceship. So, once we know the original orientation and velocity of the spaceship and we know the values of the parameters (e.g., the atmospheric density at different heights), we can predict where the spaceship will be and thus, check whether it will hit the Moon. In reality, it is difficult to maintain the exact values of the initial orientation and velocity, and we only know the atmospheric parameters with some uncertainty.

Interval uncertainty: a practically important situation. Uncertainty means that instead of the exact values of the corresponding quantities x_i, we only know the approximate values $\widetilde{x}_i$. In engineering and science courses, when students deal with uncertainty, it is usually assumed that we know the probability of different possible values of the corresponding approximation errors $\Delta x_i \overset{\text{def}}{=} \widetilde{x}_i - x_i$. In practice, however, we often only know the upper bounds on these errors, i.e., the values Δ_i for which $|\Delta x_i| \leq \Delta_i$.

For example, if we require that the bridge width is 10 m, with tolerance $\pm$ 10 cm (= 0.1 m), this means that the actual width of the corresponding bridge part can be anywhere between $10 - 0.1 = 9.9$ m and $10 + 0.1 = 10.1$ m. In general, the actual (unknown) value of the corresponding quantity x_i can take any value from the interval $\mathbf{x_i} \overset{\text{def}}{=} [\widetilde{x}_i - \Delta_i, \widetilde{x}_i + \Delta_i]$.

Taking into account interval uncertainty necessitates computing minima and maxima. Under interval uncertainty, we want to make sure that for all possible values $x_i \in [\widetilde{x}_i - \Delta_i, \widetilde{x}_i + \Delta_i]$, the corresponding value $f(x_1, \ldots, x_n)$ is within the given bounds $\underline{f}$ and $\overline{f}$ (e.g., that the spaceship hits the Moon and not flies past it). In other words, we need to check that for all possible values $x_i \in [\widetilde{x}_i - \Delta_i, \widetilde{x}_i + \Delta_i]$, we have

$$\underline{f} \leq f(x_1, \ldots, x_n) \leq \overline{f}.$$

To check that *all* the values $f(x_1, \ldots, x_n)$ are smaller than or equal to $\overline{f}$, it is sufficient to check that the *largest* of these values does not exceed $\overline{f}$, i.e., that

$$\max_{x_i \in \mathbf{x_i}} f(x_1, \ldots, x_n) \leq \overline{f}.$$

Similarly, to check that *all* the values $f(x_1, \ldots, x_n)$ are larger than or equal to $\underline{f}$, it is sufficient to check that the *smallest* of these values does not exceed $\overline{f}$, i.e., that

$$\min_{x_i \in \mathbf{x}_i} f(x_1, \ldots, x_n) \geq \underline{f}.$$

In other words, to check the desired condition, it is sufficient to find the minimum and the maximum of the function $f(x_1, \ldots, x_n)$ on the box $\mathbf{x}_1 \times \ldots \times \mathbf{x}_n$, and then check whether the range

$$\mathbf{y} = f(\mathbf{x}_1, \ldots, \mathbf{x}_n) \stackrel{\text{def}}{=} \{f(x_1, \ldots, x_n) : x_1 \in \mathbf{x}_1, \ldots, x_n \in \mathbf{x}_n\} =$$

$$\left[\min_{x_i \in \mathbf{x}_i} f(x_1, \ldots, x_n), \max_{x_i \in \mathbf{x}_i} f(x_1, \ldots, x_n) \right]$$

is indeed contained in the desired interval $\left[\underline{f}, \overline{f} \right]$.

Comment. From this viewpoint, if we cannot compute the exact range $\mathbf{y}$, it is desirable to at least compute the *enclosure* for this range, i.e., an interval $\mathbf{Y} \supseteq \mathbf{y}$. Indeed, if this enclosing (wider) interval $\mathbf{Y}$ is contained in $\left[\underline{f}, \overline{f} \right]$, then we can be sure that the actual range $\mathbf{y}$ is also contained there.

How to find minima and maxima Now that we understand why finding minima and maxima is important, let us recall how calculus helps in this task. It is well known that for functions of one variable, this can be reduced to a problem of finding zeroes of the derivative. Let us recall the corresponding result and its proof – since this result is used in proving the Mean Value Theorem, the main topic of this chapter.

Definition 5.1 By the derivative $f'(x)$ of a function $f(x)$ at a point x we mean the limit

$$\lim_{\Delta x \to 0} \frac{f(x + \Delta x) - f(x)}{\Delta x}.$$

Definition 5.2 A function $f(x)$ is called differentiable at a point x if has a derivative at this point.

Definition 5.3 We say that a function $f(x)$ attains a *local maximum* at the point x if there exists a real number $\varepsilon > 0$ such that $f(x) \geq f(x')$ for all $x' \in (x - \varepsilon, x + \varepsilon)$.

Definition 5.4 We say that a function $f(x)$ attains a local minimum at the point x if there exists a real number $\varepsilon > 0$ such that $f(x) \leq f(x')$ for all $x' \in (x - \varepsilon, x + \varepsilon)$.

Basic Result (case of one variable). *If a function $f(x)$ has a local minimum or a local maximum at a point x, and it is differentiable at this point x, then $f'(x) = 0$.*

Proof Let us proof this result for the case when the function $f(x)$ attains a local maximum; the local minimum case can be handled similarly.

For the local maximum case, by definition of the local maximum, for sufficiently small Δx, we have $f(x) \geq f(x + \Delta x)$, and hence, $f(x + \Delta x) - f(x) \leq 0$.

For $\Delta x > 0$, the division by Δx does not change the sign of the inequality. So, the ratio $\dfrac{f(x + \Delta x) - f(x)}{\Delta x}$ is also non-positive. Thus, when $\Delta x \to 0$, the limit $f'(x)$ is non-positive: $f'(x) \leq 0$.

For $\Delta x < 0$, the division by Δx changes the sign of the inequality. So, the ratio $\dfrac{f(x + \Delta x) - f(x)}{\Delta x}$ is non-negative. Thus, when $\Delta x \to 0$, the limit $f'(x)$ is non-negative: $f'(x) \geq 0$.

From $f'(x) \leq 0$ and $f'(x) \geq 0$, we can now conclude that $f'(x) = 0$. The result is proven.

How this result can be used in optimization: case of a single variable. If we have no restrictions on the value x and we are interested in finding the optimum of a function $f(x)$, then it is sufficient to find the value x for which $f'(x) = 0$. This fact often simplifies the optimization problem.

This result can also be used to find the optimum of a function $f(x)$ on a given interval $[\underline{x}, \overline{x}]$. Indeed, the desired optimum is attained either at one of the endpoints $\underline{x}$ or $\overline{x}$ – or at an interior point. If the optimum is attained at an interior point, this means that it is either a local minimum or a local maximum of the function $f(x)$; thus, we must have $f'(x) = 0$.

So, to find the minimum (or maximum) of a given differentiable function $f(x)$ on a given interval $[\underline{x}, \overline{x}]$, it is sufficient to compute the values of the function at the two endpoints and at all the points x at which $f'(x) = 0$. The largest of these values is the maximum, the smallest of these values is the minimum.

Example Let us use this idea to find the minimum and the maximum of a function $f(x) = x^2$ on the interval $[-2, 1]$. For the endpoints $x = 1$ and $x = -2$, we get values $f(1) = 1^2 = 1$ and $f(-2) = (-2)^2 = 4$.

The equation $f'(x) = 2x = 0$ has only one solution: $x = 0$. At this point $x = 0$, we have $f(0) = 0^2 = 0$. Now, we have computed three values of $f(x)$: 0, 1, and 4. The smallest of these three values is $\min(0, 1, 4) = 0$, the largest of these three values is $\max(0, 1, 4) = 4$. Thus, we can conclude that the smallest value of the given function $f(x) = x^2$ on the interval $[-1, 2]$ is equal to 0, while its largest value is equal to 4.

How this result can be used in optimization: case of several variables. If a function $f(x_1, \ldots, x_n)$ attains local maximum at some point, this means that changing all the values slightly can only decrease the value of this function (or keep the same). In particular, this is true if we only change the value of one variable. This change corresponds to the notion of a partial derivative:

Definition 5.5 By the ith partial derivative $\dfrac{\partial f}{\partial x_i}$ of a function $f(x_1, \ldots, x_n)$ at a point $x = (x_1, \ldots, x_n)$ we mean the limit

$$\lim_{\Delta x_i \to 0} \frac{f(x_1, \ldots, x_{i-1}, x_i + \Delta x_i, x_{i+1}, \ldots, x_n) - f(x_1, \ldots, x_{i-1}, x_i, x_{i+1}, \ldots, x_n)}{\Delta x_i}.$$

Basic Result (case of several variables). *If a function* $f(x_1, \ldots, x_n)$ *has a local minimum or a local maximum at a point* x, *and it is differentiable at this point* x, *then all its partial derivatives at this point are equal to 0:* $\dfrac{\partial f}{\partial x_i} = 0$ *for all* i.

Because of this, if we are interested in finding the maximum or minimum of a function $f(x_1, \ldots, x_n)$ on a given box $[\underline{x}_1, \overline{x}_1] \times \ldots \times [\underline{x}_n, \overline{x}_n]$, then at the point x where this optimum is attained, for each i, one of the following three conditions is satisfied:

- $x_i = \underline{x}_i$;
- $x_i = \overline{x}_i$;
- $\dfrac{\partial f}{\partial x_i} = 0$.

Thus, to find the desired optimum, it is sufficient to try all possible combinations of such conditions. For each variable, we need to check at least two conditions, so overall, we need at least to check 2^n combinations $(x_1, \ldots, x_n)$:

- for $n = 1$, we have two possible values $\underline{x}_1$ and $\overline{x}_1$;
- when we add the second variable, to each of these combinations we add $\underline{x}_2$ or $\overline{x}_2$, so we end up with $2 \times 2 = 2^2 = 4$ possible combinations $(\underline{x}_1, \underline{x}_2)$, $(\underline{x}_1, \overline{x}_2)$, $(\overline{x}_1, \underline{x}_2)$, and $(\overline{x}_1, \overline{x}_2)$;
- when we add the third variable, to each of these combinations we add $\underline{x}_3$ or $\overline{x}_3$, so we end up with $2^2 \times 2 = 2^3 = 8$ possible combinations, etc.

Case of several variables: challenge. For small n, it is possible to check all 2^n combinations. However, in many practical problems, the number n of parameters affecting the result may be in the hundreds. In this case, we need to check 2^{100} combinations. To estimate this number, we can use the known approximate equality $2^{10} = 1024 \approx 10^3$ (this is why Kilobyte is called this way although it has not 1000 bytes but 1024 ones). Due to this approximate equality, we have $2^{100} = \left(2^{10}\right)^{10} \approx \left(10^3\right)^{10} = 10^{30}$. With 10^{12} operations per second, we need $10^{30}/10^{12} = 10^{18}$ seconds. At $\approx 3 \cdot 10^7$ seconds per year, this means that we need $10^{18}/(3 \cdot 10^7) \approx 3 \cdot 10^{10}$ years, 30 billion years, which is longer that the lifetime of the Universe.

This shows that it is not realistically possible to use the above calculus approach to find the exact range. Moreover, the problem of computing the exact range is NP-hard, meaning that most probably it is not possible to reduce this exponentially growing computation time; for exact definitions and proof, see, e.g., [5].

Computing the enclosure: main idea. As we have mentioned earlier, when we cannot easily compute the range exactly, it is reasonable to compute an enclosure for this range. One way to produce such an enclosure is called *straightforward interval computations*.

This method is based on the fact that for the case when $f(x_1, x_2)$ is an arithmetic operation, we can use monotonicity to compute the exact range

$$f(\mathbf{x}_1, \mathbf{x}_2) = \{f(x_1, x_2) : x_1 \in \mathbf{x}_1 \text{ and } x_2 \in \mathbf{x}_2\}.$$

For example, addition $f(x_1, x_2) = x_1 + x_2$ is an increasing function of both variables, and thus:

- its largest value is attained when both x_1 and x_2 attain their largest values $x_1 = \overline{x}_1$ and $x_2 = \overline{x}_2$, and
- its smallest value is attained when both x_1 and x_2 attain their smallest values $x_1 = \underline{x}_1$ and $x_2 = \underline{x}_2$.

Thus, the range has the form

$$[\underline{x}_1, \overline{x}_1] + [\underline{x}_2, \overline{x}_2] = [\underline{x}_1 + \underline{x}_2, \overline{x}_1 + \overline{x}_2].$$

For subtraction $f(x_1, x_2) = x_1 - x_2$, the situation is similar, with the only difference that the difference increases with x_1 and decreases with x_2. Thus, e.g., to find the largest possible value of $x_1 - x_2$, we need to take the largest possible value of x_1 and the smallest possible value of x_2. As a result, we get the following range:

$$[\underline{x}_1, \overline{x}_1] - [\underline{x}_2, \overline{x}_2] = [\underline{x}_1 - \overline{x}_2, \overline{x}_1 - \underline{x}_2].$$

For multiplication $f(x_1, x_2) = x_1 \cdot x_2$, the situation is more complex, since the product is an increasing function of x_1 when $x_2 \geq 0$ and a decreasing function of x_1 when $x_2 \leq 0$. However, in both cases, the dependence on x_1 is linear, and a linear function always attains its maximum and minimum at the endpoints. Thus, to find the maximum and the minimum of the product on the box $[\underline{x}_1, \overline{x}_1] \times [\underline{x}_1, \overline{x}_2]$, it is sufficient to consider all $2 \cdot 2 = 4$ combinations of endpoints. Thus, we get

$$[\underline{x}_1, \overline{x}_1] \cdot [\underline{x}_2, \overline{x}_2] = [\min(\underline{x}_1 \cdot \underline{x}_2, \underline{x}_1 \cdot \overline{x}_2, \overline{x}_1 \cdot \underline{x}_2, \overline{x}_1 \cdot \overline{x}_2), \max(\underline{x}_1 \cdot \underline{x}_2, \underline{x}_1 \cdot \overline{x}_2, \overline{x}_1 \cdot \underline{x}_2, \overline{x}_1 \cdot \overline{x}_2)].$$

Division x_1/x_2 is similarly monotonic in both variables, so unless $0 \in [\underline{x}_2, \overline{x}_2]$ (in which case the ratio can take infinite value) the range can be obtained similarly, as

$$[\underline{x}_1, \overline{x}_1]/[\underline{x}_2, \overline{x}_2] = [\min(\underline{x}_1/\underline{x}_2, \underline{x}_1/\overline{x}_2, \overline{x}_1/\underline{x}_2, \overline{x}_1/\overline{x}_2), \max(\underline{x}_1/\underline{x}_2, \underline{x}_1/\overline{x}_2, \overline{x}_1/\underline{x}_2, \overline{x}_1/\overline{x}_2)].$$

For a general function $f(x_1, \ldots, x_n)$, this method uses the fact that in a computer, each computation consists of a sequence of arithmetic operations. So, to compute the enclosure, we repeat the computations forming the program f step-by-step, replacing each operation with real numbers by the corresponding operation of interval arithmetic (given above). It can be proven, by simple induction, that, as a result, we get an enclosure $\mathbf{Y} \supseteq \mathbf{y}$ for the desired range.

In some cases, this enclosure is exact. In more complex cases (see example below), the enclosure has excess width.

Example. Let us illustrate the above idea on the example of estimating the range of the function $f(x_1) = x_1 - x_1^2$ on the interval $x_1 \in [0, 0.8]$.

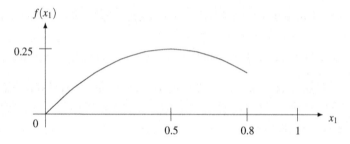

Fig. 5.1 Range of the function $f(x_1) = x_1 - x_1^2$ on the interval $[0, 0.8]$

We start with parsing the expression for the function, i.e., describing how a computer will compute this expression; it will implement the following sequence of elementary operations:

$$r_1 := x_1 \cdot x_1;$$

$$r_2 := x_1 - r_1.$$

According to straightforward interval computations, we perform the same operations, but with *intervals* instead of *numbers*:

$$\mathbf{r}_1 := [0, 0.8] \cdot [0, 0.8] =$$

$$[\min(0 \cdot 0, 0 \cdot 0.8, 0.8 \cdot 0, 0.8 \cdot 0.8), \max(0 \cdot 0, 0 \cdot 0.8, 0.8 \cdot 0, 0.8 \cdot 0.8)] =$$

$$[\min(0, 0, 0, 0.64), \max(0, 0, 0, 0.64)] = [0, 0.64];$$

$$\mathbf{r}_2 := [0, 0.8] - [0, 0.64] = [0 - 0.64, 0.8 - 0] = [-0.64, 0.8].$$

For this function, the actual range is $f(\mathbf{x}_1) = [0, 0.25]$; see Fig. 5.1.

How can we improve this estimate? Interestingly, for that, we can use the Mean Value Theorem. Let us describe this theorem, and then let us describe how it can be used to improve the above estimates.

Mean Value Theorem: reminder. The formulation of the Mean Value Theorem is as follows:

Mean Value Theorem. *Let $f(x)$ be a continuous function on the interval $[a, b]$ which is differentiable everywhere on the open interval (a, b). Then, there exists a value $\eta \in (a, b)$ for which*

$$\frac{f(b) - f(a)}{b - a} = f'(\eta).$$

Comment. Since the ratio $\dfrac{f(b) - f(a)}{b - a}$ does not change if we swap a and b, the same result holds if we consider $a > b$. In other words, we can reformulate the Mean Value Theorem as follows:

Mean Value Theorem. *Let a and b be real numbers, and let a function $f(x)$ be continuous for all values x between a and b and differentiable for all values x which are strictly between a and b. Then, there exists a value η which is strictly between a and b and for which*

$$\frac{f(b) - f(a)}{b - a} = f'(\eta).$$

The Mean Value Theorem can also be equivalently reformulated as

$$f(b) = f(a) + f'(\eta) \cdot (b - a).$$

It is worth mentioning that this formula is similar to the Taylor series formula

$$f(b) = f(a) + f'(a) \cdot (b - a) + \frac{1}{2} \cdot f''(a) \cdot (b - a)^2 + \ldots;$$

the main difference is that we use the derivative not at the point a but at some point $\eta \in (a, b)$, which helps us to make the Mean Value Theorem formula exact with the first two terms only.

From the geometric viewpoint, the ratio $\dfrac{f(b) - f(a)}{b - a}$ is the slope of the secant, i.e., of a straight line connecting the points $(a, f(a))$ and $(b, f(b))$ from the function's graph, and the value $f'(\eta)$ is the slope of the tangent

$$y = f(\eta) + f'(\eta) \cdot (x - \eta)$$

to the graph. In these terms, the Mean Value Theorem claims that there exists a point η at which the tangent has the same slope as the secant, i.e., at which the tangent is parallel to the secant.

The proof of the Mean Value Theorem is usually based on the following Rolle's Theorem:

Rolle's Theorem. *For every continuous function $f(x)$ on the interval $[a, b]$ which is differentiable everywhere on the open interval (a, b) and for which $f(a) = f(b)$, there exists a point $\eta \in (a, b)$ for which $f'(\eta) = 0$.*

Similarly to the Mean Value Theorem, Rolle's Theorem can be reformulated in the following form:

Rolle's Theorem. *Let a and b be real numbers, and let a function f for which $f(a) = f(b)$ be continuous for all values x between a and b and differentiable for all values x which are strictly between a and b. Then, there exists a point η which is strictly between a and b and for which $f'(\eta) = 0$.*

In the following text, we will use the following multi-dimensional generalization of the Mean Value Theorem:

Mean Value Theorem. (multi-D version) *Let $a_1, \ldots, a_n$ and $b_1, \ldots, b_n$ be real numbers, and let a function $f(x_1, \ldots, x_n)$ be continuous for all values x_i which are between a_i and b_i and differentiable for all values x_i which are strictly between a_i and b_i. Then, there exists values η_i which are strictly between a_i and b_i and for which*

$$f(b_1, \ldots, b_n) = f(a_1, \ldots, a_n) + \sum_{i=1}^{n} \frac{\partial f}{\partial x_i}(\eta_1, \ldots, \eta_n) \cdot (b_i - a_i).$$

Proof of the Rolle's Theorem. To prove this theorem, we consider two possible cases: (1) when $f(x)$ is a constant, and (2) when $f(x)$ is not a constant.

1°. If the function $f(x)$ is a constant, i.e., if $f(x) = f(x')$ for all $x, x' \in [a, b]$, then, by definition, the derivative is always 0, so we can take any $\eta \in (a, b)$ as the desired point.

2°. If the function $f(x)$ is not a constant, this means that there exists a value $x_0 \in [a, b]$ for which $f(x_0) \neq f(a)$ – because otherwise, the function $f(x)$ would be a constant. Since $f(x_0) \neq f(a)$, we have either $f(x_0) < f(a)$ or $f(x_0) > f(a)$. Let us analyze these two cases one by one.

2.1°. If $f(x_0) > f(a)$, this means that the maximum of $f(x)$ on the interval $[a, b]$ is larger than $f(a) = f(b)$ and thus, that this maximum is attained at a point $\eta \in (a, b)$. According to the basic result, at this maximum point, we have $f'(\eta) = 0$.

2.2°. If $f(x_0) < f(a)$, this means that the minimum of $f(x)$ on the interval $[a, b]$ is smaller than $f(a) = f(b)$ and thus, that this minimum is attained as a point $\eta \in (a, b)$. According to the basic result, at this minimum point, we have $f'(\eta) = 0$.

In all possible cases, we have found a value $\eta \in (a, b)$ for which $f'(\eta) = 0$. The statement is proven.

Proof of the 1-D Mean Value Theorem. To prove this result, let us consider an auxiliary function

$$g(x) \stackrel{\text{def}}{=} f(x) - \frac{f(b) - f(a)}{b - a} \cdot (x - a).$$

One can easily check that this function is differentiable, with

$$g'(x) = f'(x) - \frac{f(b) - f(a)}{b - a}.$$

At the endpoints a and b of the interval $[a, b]$, the new function $g(x)$ takes the values

$$g(a) = f(a) - \frac{f(b) - f(a)}{b - a} \cdot (a - a) = f(a)$$

and

$$g(b) = f(b) - \frac{f(b) - f(a)}{b - a} \cdot (b - a) = f(b) - (f(b) - f(a)) = f(a).$$

So, we have $g(a) = g(b)$. Hence, due to the Rolle's Theorem, there exists a point $\eta \in (a, b)$ at which $g'(\eta) = 0$. Substituting the above expression for $g'(x)$ into this equality, we conclude that

$$f'(\eta) - \frac{f(b) - f(a)}{b - a} = 0,$$

i.e., that

$$f'(\eta) = \frac{f(b) - f(a)}{b - a}.$$

The statement is proven.

Proof of multi-D Mean Value Theorem. To prove this result, let us consider an auxiliary function

$$g(\lambda) = f(a_1 + \lambda \cdot (b_1 - a_1), \ldots, a_n + \lambda \cdot (b_n - a_n)).$$

For each $\lambda \in [0, 1]$, the corresponding values $a_i + \lambda \cdot (b_i - a_i)$ are between a_i and b_i. For $\lambda = 0$, we get $g(0) = f(a_1, \ldots, a_n)$, and for $\lambda = 1$, we get

$$g(1) = f(a_1 + (b_1 - a_1), \ldots, a_n + (b_n - a_n)) = f(b_1, \ldots, b_n).$$

The derivative of this function is equal to

$$g'(\lambda) = \sum_{i=1}^{n} \frac{\partial f}{\partial x_i} (a_1 + \lambda \cdot (b_1 - a_1), \ldots, a_n + \lambda \cdot (b_n - a_n)) \cdot (b_i - a_i).$$

Due to the 1-D version of the Mean Value Theorem, we conclude that there exists a value $\lambda \in (0, 1)$ for which

$$g(1) = g(0) + g'(\lambda) \cdot (1 - 0) = g(0) + g'(\lambda).$$

Substituting the above expressions for $g(0)$, $g(1)$, and $g'(x)$ into this formula, we conclude that

$$f(b_1, \ldots, b_n) = f(a_1, \ldots, a_n) +$$

$$\sum_{i=1}^{n} \frac{\partial f}{\partial x_i} (a_1 + \lambda \cdot (b_1 - a_1), \ldots, a_n + \lambda \cdot (b_n - a_n)) \cdot (b_i - a_i),$$

i.e., the desired formula for $\eta_i = a_i + \eta \cdot (b_i - a_i)$. The statement is proven.

Usefulness of the Mean Value Theorem At first glance, it may seem that the Mean Value Theorem is a purely theoretical statement: indeed, all we conclude is that there exists some values η_i, without any idea on how to find these values η_i. However, as we will see, this result is actually very useful in interval computations.

The use of the Mean Value Theorem in interval computations. We start by representing each interval $\mathbf{x}_i = [\underline{x}_i, \overline{x}_i]$ in the form $[\widetilde{x}_i - \Delta_i, \widetilde{x}_i + \Delta_i]$, where $\widetilde{x}_i = (\underline{x}_i + \overline{x}_i)/2$ is the midpoint of the interval $\mathbf{x}_i$ and $\Delta_i = (\overline{x}_i - \underline{x}_i)/2$ is the half-width of this interval.

The Mean Value Theorem implies that

$$f(x_1, \ldots, x_n) = f(\widetilde{x}_1, \ldots, \widetilde{x}_n) + \sum_{i=1}^{n} \frac{\partial f}{\partial x_i}(\eta_1, \ldots, \eta_n) \cdot (x_i - \widetilde{x}_i),$$

where each η_i is some value from the interval $\mathbf{x}_i$.

Since $\eta_i \in \mathbf{x}_i$, the value of the ith derivative belongs to the interval range of this derivative on these intervals. We also know that $x_i - \widetilde{x}_i \in [-\Delta_i, \Delta_i]$. Thus, we can conclude that

$$f(\mathbf{x}_1, \ldots, \mathbf{x}_n) \subseteq f(\widetilde{x}_1, \ldots, \widetilde{x}_n) + \sum_{i=1}^{n} \frac{\partial f}{\partial x_i}(\mathbf{x}_1, \ldots, \mathbf{x}_n) \cdot [-\Delta_i, \Delta_i].$$

To compute the ranges of the partial derivatives, we can use, e.g., straightforward interval computations.

Example. Let us illustrate this method on the above example of estimating the range of the function $f(x_1) = x_1 - x_1^2$ over the interval $[0, 0.8]$. For this interval, the midpoint is $\widetilde{x}_1 = 0.4$; at this midpoint, $f(\widetilde{x}_1) = 0.24$. The half-width is $\Delta_1 = 0.4$. The only partial derivative here is $\dfrac{\partial f}{\partial x_1} = 1 - 2x_1$, its range on $[0, 0.8]$ is equal to

$$1 - 2 \cdot [0, 0.8] = [1, 1] - [2, 2] \cdot [0, 0.8] =$$

$$[1, 1] - [\min(2 \cdot 0, 2 \cdot 0.8, 2 \cdot 0, 2 \cdot 0.8), \max(2 \cdot 0, 2 \cdot 0.8, 2 \cdot 0, 2 \cdot 0.8)] =$$

$$[1, 1] - [\min(0, 1.6, 0.1.6), \max(0.1.6, 0, 1.6)] =$$

$$[1, 1] - [0, 1.6] = [1 - 1.6, 1 - 0] = [-0.6, 1].$$

Thus, we get the following enclosure $\mathbf{Y} \supseteq \mathbf{y}$ for the desired range $\mathbf{y}$ of the function $f(x_1)$ on the interval $[0, 08]$:

$$\mathbf{Y} = 0.24 + [-0.6, 1] \cdot [-0.4, 0.4] = 0.24 + [-0.4, 0.4] = [-0.16, 0.64].$$

This enclosure is narrower than the straightforward enclosure $[-0.64, 0.8]$, but it still contains excess width.

How can we get better estimates? In this approach, we, in effect, ignored quadratic and higher order terms, i.e., terms of the type $\dfrac{\partial^2 f}{\partial x_i \partial x_j} \cdot \Delta x_i \cdot \Delta x_j$. When the estimate is not accurate enough, it means that this ignored term is too large. There are two ways to reduce the size of the ignored term:

- we can try to decrease this quadratic term, or
- we can try to explicitly include higher order terms in the Taylor expansion formula, so that the remainder term will be proportional to say Δx_i^3 and thus, be much smaller.

Let us describe these two ideas in detail.

First idea: bisection. Let us first describe the situation in which we try to minimize the second-order remainder term. In the above expression for this term, we cannot change the second derivative. The only thing we can decrease is the difference $\Delta x_i = x_i - \widetilde{x}_i$ between the actual value and the midpoint. This value is bounded by the half-width Δ_i of the box. So, to decrease this value, we can subdivide the original box into several narrower subboxes. Usually, we divide into two subboxes, so this subdivision is called *bisection*.

The range over the whole box is equal to the union of the ranges over all the subboxes. The widths of each subbox are smaller, so we get smaller Δx_i and hopefully, more accurate estimates for ranges over each of this subbox. Then, we take the union of the ranges over subboxes.

Example. Let us illustrate this idea on the above $x_1 - x_1^2$ example. In this example, we divide the original interval $[0, 0.8]$ into two subintervals $[0, 0.4]$ and $[0.4, 0.8]$. For both intervals, $\Delta_1 = 0.2$.

In the first subinterval, the midpoint is $\widetilde{x}_1 = 0.2$, so $f(\widetilde{x}_1) = 0.2 - 0.04 = 0.16$. The range of the derivative is equal to

$$1 - 2 \cdot [0, 0.4] = 1 - [0, 0.8] = [0.2, 1],$$

hence we get an enclosure

$$0.16 + [0.2, 1] \cdot [-0.2, 0.2] = [-0.04, 0.36].$$

For the second interval, $\widetilde{x}_1 = 0.6$, $f(0.6) = 0.24$, the range of the derivative is

$$1 - 2 \cdot [0.4, 0.8] = [-0.6, 0.2],$$

hence we get an enclosure

$$0.24 + [-0.6, 0.2] \cdot [-0.2, 0.2] = [0.12, 0.36].$$

The union of these two enclosures is the interval $[-0.04, 0.36]$. This enclosure is much more accurate than before.

Further bisection leads to even more accurate estimates – the smaller the subintervals, the more accurate the enclosure.

Bisection: general comment. The more subboxes we consider, the smaller Δx_i and thus, the more accurate the corresponding enclosures. However, once we have more boxes, we need to spend more time processing these boxes. Thus, we have a trade-off between computation time and accuracy: the more computation time we allow, the more accurate estimates we will be able to compute.

Additional idea: monotonicity checking. If the function $f(x_1, \ldots, x_n)$ is monotonic over the original box $\mathbf{x}_1 \times \ldots \times \mathbf{x}_n$, then we can easily compute its exact range. Since we used the Mean Value Theorem for the original box, this probably means that on that box, the function is not monotonic: for example, with respect to x_1, it may be increasing at some points in this box, and decreasing at other points.

However, as we divide the original box into smaller subboxes, it is quite possible that at least some of these subboxes will be outside the areas where the derivatives are 0 and thus, the function $f(x_1, \ldots, x_n)$ will be monotonic. So, after we subdivide the box into subboxes, we should first check monotonicity on each of these subboxes – and if the function is monotonic, we can easily compute its range.

In calculus terms, if a function is increasing with respect to x_i, then its partial derivative $d_i \stackrel{\text{def}}{=} \dfrac{\partial f}{\partial x_i}$ is a limit of non-negative numbers and is, thus, non-negative everywhere on this subbox (we used a similar argument in the proof of the Basic Result). Vice versa, if the derivative $f'(x)$ is everywhere non-negative $f'(x) \geq 0$, then we can use the 1-D Mean Value Theorem $f(b) = f(a) + f'(\eta) \cdot (b - a)$ to prove that if $a < b$, then $f(a) \leq f(b)$, i.e., that the function is increasing.

Similarly, a differentiable function is decreasing with respect to x_i if and only if the corresponding partial derivative is everywhere non-positive.

Thus, to check monotonicity, we should find the range $\left[\underline{d}_i, \overline{d}_i\right]$ of this derivative (we need to do it anyway to use the Mean Value Theorem to estimate the expression):

- if $\underline{d}_i \geq 0$, this means that the derivative is everywhere non-negative and thus, the function f is increasing in x_i;
- if $\overline{d}_i \leq 0$, this means that the derivative is everywhere non-positive and thus, the function f is decreasing in x_i.

If $\underline{d}_i < 0 < \overline{d}_i$, then we have to use the Mean Value Theorem.

If the function is monotonic (e.g., increasing) only with respect to some of the variables x_i, then

- to compute $\overline{y}$, it is sufficient to consider only the value $x_i = \overline{x}_i$, and
- to compute $\underline{y}$, it is sufficient to consider only the value $x_i = \underline{x}_i$.

For such subboxes, we reduce the original problem to two problems with fewer variables, problems which are thus easier to solve.

Example. For the example $f(x_1) = x_1 - x_1^2$, the partial derivative is equal to $1 - 2 \cdot x_1$.

On the first subbox $[0, 0.4]$, the range of this derivative is $1 - 2 \cdot [0, 0.4] = [0.2, 1]$. Thus, the derivative is always non-negative, the function is increasing on this subbox, and its range on this subbox is equal to $[f(0), f(0.4)] = [0, 0.16]$.

On the second subbox $[0.4, 0.8]$, the range of the derivative is $1 - 2 \cdot [0.4, 0.8] = [-0.6, 0.2]$. Here, we do not have guaranteed monotonicity, so we can use the Mean Value Theorem to get the enclosure $[0.12, 0.36]$ for the range.

The union of these two enclosures is the interval $[0, 0.36]$, which is slightly more accurate than before. Further bisection leads to even narrower enclosures.

Towards general Taylor techniques. As we have mentioned, another way to get narrower enclosures is to use so-called *Taylor techniques*, i.e., to explicitly consider second-order and higher-order terms in the Taylor expansion; see, e.g., [6, 7], and references therein. Let us illustrate the main ideas of Taylor analysis on the case when we allow second order terms.

Formulas involving quadratic and higher-order terms can also be obtained based on the Mean Value Theorem. Specifically, we need a generalization of the Mean Value Theorem originally proved by Cauchy, one of the 19 century founders of modern mathematical analysis.

Cauchy Mean Value Theorem. *Let $F(x)$ and $G(x)$ be continuous on the interval $[a, b]$ and differentiable everywhere on the open interval (a, b), and let $G'(x) \neq 0$ for all $x \in (a, b)$. Then, there exists a value $\eta \in (a, b)$ for which*

$$\frac{F(b) - F(a)}{G(b) - G(a)} = \frac{F'(\eta)}{G'(\eta)}.$$

Similarly to the original Mean Value Theorem, this result can be equivalently reformulated as follows:

Cauchy Mean Value Theorem. *Let a and b be real numbers, and let functions $F(x)$ and $G(x)$ be continuous for all values x between a and b and differentiable for all values x strictly between a_i and b_i, and let $G'(x) \neq 0$ for all such x. Then, there exists a value η which is strictly between a and b and for which*

$$\frac{F(b) - F(a)}{G(b) - G(a)} = \frac{F'(\eta)}{G'(\eta)}.$$

Proof Similarly to the proof of the original Mean Value Theorem, the proof of this result follows when we apply the Rolle's Theorem to the auxiliary function $g(x) = f(x) - h \cdot (g(x) - g(a))$, where the parameter h is chosen in such a way that $g(b) = g(a)$.

Taylor's Theorem. (1-D quadratic case) *Let $f(x)$ be continuous on the interval $[a, b]$ and twice differentiable everywhere on the open interval (a, b). Then, there exists a point $\eta \in (a, b)$ for which*

$$f(b) = f(a) + f'(a) \cdot (b-a) + \frac{1}{2} \cdot f''(\eta) \cdot (b-a)^2.$$

Proof To prove this result, we apply Cauchy Mean Value Theorem to two auxiliary functions: $F(x) = f(x) + f'(x) \cdot (b-x)$ and $G(x) = (x-b)^2$. Here,

$$F(b) = f(b) + f'(b) \cdot (b-x) = f(b),$$

$$F(a) = f(a) + f'(a) \cdot (b-a),$$

so

$$F(b) - F(a) = f(b) - f(a) - f'(a) \cdot (b-a).$$

Similarly,

$$G(b) - G(a) = (b-b)^2 - (b-a)^2 = -(b-a)^2.$$

For the derivatives, we get

$$F'(x) = f'(x) - f'(x) + f''(x) \cdot (b-x) = f''(x) \cdot (b-x)$$

and

$$G'(x) = 2 \cdot (x-b),$$

so

$$\frac{F'(x)}{G'(x)} = \frac{f''(x) \cdot (b-x)}{2 \cdot (x-b)} = -\frac{1}{2} \cdot f''(x).$$

Thus, the Cauchy Mean Value Theorem implies there exists a value η for which

$$\frac{f(b) - f(a) - f'(a) \cdot (b-a)}{-(b-a)^2} = -\frac{1}{2} \cdot f''(\eta).$$

Multiplying both sides by the denominator of the left-hand side, we get the desired equality. The statement is proven.

By applying this result to the same auxiliary function

$$g(\lambda) = f(a_1 + \lambda \cdot (b_1 - a_1), \ldots, a_n + \lambda \cdot (b_n - a_n))$$

that we used to prove the multi-D version of the Mean Value Theorem, we conclude that

$$f(x_1, \ldots, x_n) = f(\widetilde{x}_1, \ldots, \widetilde{x}_n) + \sum_{i=1}^{n} \frac{\partial f}{\partial x_i}(\widetilde{x}_1, \ldots, \widetilde{x}_n) \cdot (x_i - \widetilde{x}_i) +$$

$$\frac{1}{2} \cdot \sum_{i=1}^{n} \sum_{j=1}^{m} \frac{\partial^2 f}{\partial x_i \partial x_j} (\eta_1, \ldots, \eta_n) \cdot (x_i - \widetilde{x}_i) \cdot (x_j - \widetilde{x}_j).$$

Thus, we get the enclosure

$$f(x_1, \ldots, x_n) \subseteq \mathbf{Y} \stackrel{\text{def}}{=} f(\widetilde{x}_1, \ldots, \widetilde{x}_n) + \sum_{i=1}^{n} \frac{\partial f}{\partial x_i} (\widetilde{x}_1, \ldots, \widetilde{x}_n) \cdot [-\Delta_i, \Delta_i] +$$

$$\frac{1}{2} \cdot \sum_{i=1}^{n} \sum_{j=1}^{m} \frac{\partial^2 f}{\partial x_i \partial x_j} (\mathbf{x}_1, \ldots, \mathbf{x}_n) \cdot [-\Delta_i, \Delta_i] \cdot [-\Delta_j, \Delta_j].$$

Example. Let us illustrate this idea on the above example of $f(x_1) = x_1 - x_1^2$. Here, $\Delta_1 = 0.4$, $\widetilde{x}_1 = 0.4$, so $f(\widetilde{x}_1) = 0.24$ and $\frac{\partial f}{\partial x_1}(\widetilde{x}_1) = 1 - 2 \cdot 0.4 = 0.2$. The second derivative is equal to -2, so the Taylor estimate takes the form

$$\mathbf{Y} = 0.24 + 0.2 \cdot [-0.4, 0.4] - [-0.4, 0.4]^2.$$

Strictly speaking, if we interpret Δx_1^2 as $\Delta x_1 \cdot \Delta x_1$ and use the formulas of interval multiplication, we get the interval

$$[-0.4, 0.4]^2 = [-0.4, 0.4] \cdot [-0.4, 0.4] = [-0.16, 0.16]$$

and thus, the enclosure

$$\mathbf{Y} = 0.24 + [-0.08, 0.08] - [-0.16, 0.16] = [0.16, 0.32] - [-0.16, 0.16] = [0, 0.48]$$

for the desired range. However, we can view x^2 as a special function, for which the range over $[-0.4, 0.4]$ is known to be $[0, 0.16]$. In this case, the above enclosure takes the form

$$\mathbf{Y} = 0.24 + [-0.08, 0.08] - [0, 0.16] = [0.16, 0.32] - [0, 0.16] = [0, 0.32]$$

which is much closer to the actual range $[0, 0.25]$.

Taylor methods: general comment. The more terms we consider in the Taylor expansion, the smaller the remainder term and thus, the more accurate the corresponding enclosures. However, once we have more terms, we need to spend more time computing these terms. Thus, for Taylor methods, we also have a trade-off between computation time and accuracy: the more computation time we allow, the more accurate estimates we will be able to compute.

References

1. O. Kosheleva, How to make sure that the grading scheme encourages students to learn all the material: fuzzy-motivated solution and its justification. Int. J. Intell. Technol. Appl. Stat. (IJITAS) **10**(2), 7–19 (2017)
2. L. Jaulin, M. Kieffer, O. Didrit, E. Walter, *Applied Interval Analysis, with Examples in Parameter and State Estimation*, Robust Control and Robotics (Springer, London, 2001)
3. V. Kreinovich, Interval computations and interval-related statistical techniques: tools for estimating uncertainty of the results of data processing and indirect measurements, in *Data Modeling for Metrology and Testing in Measurement Science*, ed. by F. Pavese, A.B. Forbes (Birkhauser-Springer, Boston, 2009), pp. 117–145
4. R.E. Moore, R.B. Kearfott, M.J. Cloud, *Introduction to Interval Analysis* (SIAM Press, Philadelphia, Pennsylviania, 2009)
5. V. Kreinovich, A. Lakeyev, J. Rohn, P. Kahl, *Computational Complexity and Feasibility of Data Processing and Interval Computations* (Kluwer, Dordrecht, 1998)
6. M. Berz, G. Hoffstätter, Computation and application of taylor polynomials with interval remainder bounds. Reliab. Comput. **4**(1), 83–97 (1998)
7. A. Neumaier, Taylor forms. Reliab. Comput. **9**, 43–79 (2002)

Chapter 6
How to Enhance Student Motivations by Borrowing from Ancient Tradition: Egyptian Fractions

How can we increase a person's interest in some topic, a person's excitement in studying this topic? Each student already has some topics about which he or she is excited, so a natural way to increase the students' interest in the topic of study is to relate this topic to the topics about which the student is already excited. In other words, if the two topics are properly related, a student's excitement about his or her original topic of excitement will make the topic of study exciting as well. The more we relate the topic of study to other topics, the more students will thus get excited.

It is great if we can creatively relate the topic of study to something students deeply care about – there are known successes of using, e.g., music to study math or popular movies to study physics. These ideas are great, but it is difficult to come up with a novel good connection. Good news is that most of what we are teaching has already been taught in the past, so we can always borrow ideas, techniques, and relations from this past teaching experience. In this and following chapters, on the example of teaching math, we provide examples of such relations. Specifically, in Chaps. 6–9, we show how such ideas can be borrowed from ancient traditions, and in Chaps. 10 and 11, we show how these ideas can be borrowed from modern practices.

In this chapter, we go back to ancient Egypt. It is well known that the ancient Egyptians represented each fraction as a sum of unit fractions – i.e., fractions with unit numerators; this is how they, e.g., divided loaves of bread. What is not clear is why they used this representation. In this chapter, we describe a possible explanation: crudely speaking, that the main idea behind the Egyptian fractions provides an optimal way of dividing the loaves. We also analyze the related properties of fractions.

The results from this chapter first appeared in [1].

What are Egyptian fractions. It is known that people of ancient Egypt represented fractions as sums of unit fractions – i.e., fractions of the type $\frac{1}{n}$. This representation

© Springer-Verlag GmbH Germany 2018
O. Kosheleva and K. Villaverde, *How Interval and Fuzzy Techniques Can Improve Teaching*, Studies in Computational Intelligence 750,
https://doi.org/10.1007/978-3-662-55993-2_6

is described, in detail, in the Rhind (Ahmes) Papyrus, the most extensive Egyptian mathematical papyrus; see, e.g., [2]. According to the papyrus, this was a method recommended, e.g., for dividing loaves of bread between several people.

For example, the number $\frac{5}{6}$ can be represented as

$$\frac{5}{6} = \frac{1}{2} + \frac{1}{3}.$$

In other words,

$$5 = 6 \cdot \left(\frac{1}{2} + \frac{1}{3}\right) = 6 \cdot \frac{1}{2} + 6 \cdot \frac{1}{3}.$$

So, according to the method described in the Rhind Papyrus, if we want to divide 5 loaves between 6 people, we must:

- divide $6 \cdot \frac{1}{2} = 3$ loaves into two equal parts each, and
- divide $6 \cdot \frac{1}{3} = 2$ loaves into three equal parts each.

As a result, we get six half-loaves and six third-loaves. Each of the six people receives one half and one third:

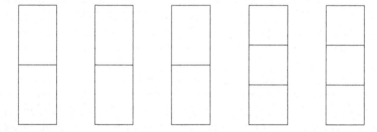

Why Egyptian fractions? A question. Most algorithms with Egyptian fractions are so complicated that it is puzzling why they were used in the first place. For example, according to [3], R. Graham (who wrote his Ph.D. dissertation on unit fractions) asked André Weil why, and A. Weil answered "They took a wrong turn".

Why Egyptian fractions: A possible answer. Let us assume that we can only divide loaves into equal pieces. One cut divides a loaf of bread into 2 equal pieces. In general, to divide a loaf into q equal pieces, we need $q - 1$ cuts.

If we want to divide 5 loaves between 6 people, to give each of them $\frac{5}{6}$ of a loaf, then a natural way to do it is to divide each of 5 loaves into 6 equal pieces. To divide each loaf, we need $6 - 1 = 5$ cuts, so to divide all 5 loaves, we need $5 \cdot 5 = 25$ cuts.

On the other hand, in the Egyptian fraction approach, we need to divide 3 loaves in half (1 cut each) and 2 loaves into three equal pieces (2 cuts each), to the total of $3 + 2 \cdot 2 = 7 \ll 25$ cuts.

General question: what is the smallest number of cuts? Suppose that we want to divide a large number of loaves in such a way that every person gets $\frac{p}{q}$-th of a loaf. In other words, for some large number N, we have N people, and we want to distribute

$$N \cdot \frac{p}{q}$$

loaves between these people.

The straightforward way would be to divide each loaf into q equal parts. A more general approach is to divide some loaves into q_1 equal parts, some loaves into q_2 equal parts, etc., and some loaves into q_k equal parts. Then, we give each person some pieces from each of these divisions:

- some number of parts $\frac{1}{q_1}$; we will denote this number by p_1;
- some number of parts $\frac{1}{q_2}$; we will denote this number by p_2;
- etc.

In other words, we represent the desired ratio $\frac{p}{q}$ as a sum

$$\frac{p}{q} = \frac{p_1}{q_1} + \ldots + \frac{p_k}{q_k} \tag{6.1}$$

for natural numbers p_i and q_i.

Each representation of this type corresponds to a possible way of cutting loaves of bread. To find out how we can minimize the number of cuts, let us find out how many cuts per loaf correspond to a representation (6.1). For N people, we need

$$N \cdot \frac{p}{q}$$

loaves, out of which:

- $N \cdot \frac{p_1}{q_1}$ loaves are divided into q_1 equal pieces,
- $N \cdot \frac{p_2}{q_2}$ loaves are divided into q_2 equal pieces,
- etc.

To divide a loaf into q_i pieces, we need $q_i - 1$ cuts, so the overall number of cuts is equal to

$$N \cdot \frac{p_1}{q_1} \cdot (q_1 - 1) + \ldots + N \cdot \frac{p_k}{q_k} \cdot (q_k - 1) =$$

$$N \cdot (p_1 + \ldots + p_k) - N \cdot \left(\frac{p_1}{q_1} + \ldots + \frac{p_k}{q_k} \right).$$

Due to (6.1), we conclude that the for N persons, the overall number of cuts is equal to

$$N \cdot (p_1 + \ldots + p_k) - N \cdot \frac{p}{q}.$$

Thus, the average number of cuts per person is equal to

$$p_1 + \ldots + p_k - \frac{p}{q}.$$

So, the average number of cuts is the smallest if and only if the sum $p_1 + \ldots + p_k$ of the numerators in the representation (6.1) attains the smallest possible value.

For every positive rational number

$$r = \frac{p}{q},$$

let us denote, by $\|r\|$, the smallest possible sum $p_1 + \ldots + p_k$ among all representations of type (6.1).

In these terms, the smallest possible number of cuts per person is equal to $\|r\| - r$. What are the properties of this function $\|r\|$?

1. For every rational number, we have

$$\|r\| \geq r.$$

2. For every integer n, we have

$$\|n\| = n.$$

3. For every rational number r and for every integer n, we have

$$\left\| \frac{r}{n} \right\| \leq \|r\|.$$

4. For every two rational numbers r and r', we have

$$\|r + r'\| \leq \|r\| + \|r'\|$$

and

$$\|r \cdot r'\| \leq \|r\| \cdot \|r'\|.$$

Proof Since $\|r\| - r$ is the average number of cuts, i.e., a non-negative number, we have $\|r\| \geq r$. The first property is proven.

For integers n, we do not need any cuts, so $\|n\| - n = 0$ and $\|n\| = n$. Thus, the second property is proven as well.

Let us prove the third property. Let

$$r = \frac{p_1}{q_1} + \ldots + \frac{p_k}{q_k}$$

be a representation corresponding to $\|r\|$, i.e., representations for which

$$\|r\| = p_1 + \ldots + p_k.$$

Then,

$$\frac{r}{n} = \frac{p_1}{n \cdot q_1} + \ldots + \frac{p_k}{n \cdot q_k}.$$

For this representation of $\dfrac{r}{n}$, the sum of the numerators is the same, i.e., it is equal to $\|r\|$. Thus, the smallest possible sum $\left\|\dfrac{r}{n}\right\|$ of the numerators in the representation of $\dfrac{r}{n}$ cannot exceed $\|r\|$. The third property is proven.

Let us prove the fourth property. Let

$$r = \frac{p_1}{q_1} + \ldots + \frac{p_k}{q_k}$$

and

$$r' = \frac{p'_1}{q'_1} + \ldots + \frac{p'_{k'}}{q'_{k'}}$$

be representations corresponding to $\|r\|$ and $\|r'\|$, i.e., representations for which

$$\|r\| = p_1 + \ldots + p_k$$

and

$$\|r'\| = p'_1 + \ldots + p'_{k'}.$$

Then, for the sum of these representations, we get

$$r + r' = \frac{p_1}{q_1} + \ldots + \frac{p_k}{q_k} + \frac{p'_1}{q'_1} + \ldots + \frac{p'_{k'}}{q'_{k'}}$$

with

$$p_1 + \ldots + p_k + p'_1 + \ldots + p'_{k'} = \|r\| + \|r'\|.$$

Thus, the smallest possible sum $\|r + r'\|$ of the numerators in the representation of $r + r'$ cannot exceed $\|r\| + \|r'\|$.

Similarly, for the product

$$r \cdot r' = \left(\frac{p_1}{q_1} + \ldots + \frac{p_k}{q_k}\right) \cdot \left(\frac{p'_1}{q'_1} + \ldots + \frac{p'_{k'}}{q'_{k'}}\right) =$$

$$\sum_{i,j} \frac{p_i \cdot p'_j}{q_i \cdot q'_j},$$

the sum of the numerators is equal to

$$\sum_{i,j} (p_i \cdot p'_j) = \sum_i p_i \cdot \sum_j p'_j = \|r\| \cdot \|r'\|,$$

so $\|r \cdot r'\| \leq \|r\| \cdot \|r'\|$.

The fourth property is proven, so the proposition is proven.

Comment. We can thus say that $\| \cdot \|$ is an integer-valued additive and multiplicative norm on the set of all positive rational numbers.

Computing the smallest number of cuts: formulation of the problem. How can we actually compute the smallest number of cuts, i.e., the norm $\| \cdot \|$? This problem is not trivial, because, from the Egyptian papyri, it is known that even for small q, we often need large numbers q_k.

We will show that an algorithm for computing $\|r\|$ is possible.

Reduction to an auxiliary algorithm. In order to compute $\|r\|$, we will first design a sequence of auxiliary algorithms.

Specifically, for every integer n, we will design an algorithm A_n that checks whether $\|r\| \leq n$. By definition of $\|r\|$, this means that this algorithm A_n checks whether there exists a representation of a given fraction r as a sum of

$$p_1 + \ldots + p_k \leq n$$

unit fractions $\frac{1}{q_i}$.

For every fraction r, the value $\|r\|$ is a positive integer. Thus, once we have designed the auxiliary algorithms A_n, we can compute $\|r\|$ as follows:

- First, we use the algorithm A_1 to check whether $\|r\| \leq 1$.

  - If the algorithm A_1 concludes that $\|r\| \leq 1$, then, since $\|r\|$ is a positive integer, we have $\|r\| = 1$. In this case, we return the value $\|r\| = 1$ and stop the computations.
  - If the algorithm A_1 concludes that $\|r\| > 1$, then we continue.

- Next, we apply the algorithm A_2 to check whether $\|r\| \leq 2$.

- If the algorithm A_2 concludes that $\|r\| \leq 2$, then, since we already know that $\|r\| > 1$, we have $\|r\| = 2$. In this case, we return the value $\|r\| = 2$ and stop the computations.
- If the algorithm A_2 concludes that $\|r\| > 2$, then we continue.

- After that, we apply the algorithm A_3 to check whether $\|r\| \leq 3$.

  - If the algorithm A_3 concludes that $\|r\| \leq 3$, then, since we already know that $\|r\| > 2$, we have $\|r\| = 3$. In this case, we return the value $\|r\| = 3$ and stop the computations.
  - If the algorithm A_3 concludes that $\|r\| > 3$, then we continue.

- etc.

For every fraction r, by applying the auxiliary algorithms A_1, A_2, etc., we eventually reach the smallest integer n for which $\|r\| \leq n$. This integer n is the desired norm $\|r\|$.

To complete the construction of the desired algorithm, we therefore need to construct the auxiliary algorithms A_n.

Designing the auxiliary algorithms A_n. We will build the algorithms A_n by induction over n.

For $n = 1$, the algorithm A_1 follows from the fact that the only way to get

$$p_1 + \ldots + p_k = \|r\| = 1$$

is to have $p_1 = 1$ and $k = 1$. So, only fractions of the type $\dfrac{1}{n}$ have $\|r\| = 1$. Thus, to check whether $\|r\| = 1$, it is sufficient to check whether r is a unit fraction.

Let us now suppose that we have already designed an algorithm A_n. Let us use this algorithm to design a new algorithm A_{n+1} for checking whether

$$\|r\| \leq n + 1.$$

To construct this new algorithm, we will use the fact that $\|r\| \leq n + 1$ means that the given fraction $r = \frac{p}{q}$ can be represented as the sum of $\leq n + 1$ unit fractions

$$\frac{p}{q} = \frac{1}{q_1} + \ldots + \frac{1}{q_M}$$

for some $M \leq n + 1$.

Without losing generality, we can assume that $q_1 \leq q_2 \leq \ldots \leq q_M$. Thus,

$$\frac{1}{q_i} \leq \frac{1}{q_1}$$

for all i and hence,

$$\frac{1}{q_1} \leq \frac{p}{q} \leq \frac{M}{q_1}.$$

Since $M \leq n + 1$, we conclude that

$$\frac{1}{q_1} \leq \frac{p}{q} \leq \frac{n+1}{q_1}.$$

Thus,

$$\frac{q}{p} \leq q_1 \leq (n+1) \cdot \frac{q}{p}.$$

There are only finitely many integers q_1 in the interval

$$\left[\frac{q}{p}, (n+1) \cdot \frac{q}{p} \right].$$

So, to check whether $\|r\| \leq n + 1$, it is sufficient to try all the integers

$$q_1 \in \left[\frac{q}{p}, (n+1) \cdot \frac{q}{p} \right],$$

and for each of them, check whether the corresponding difference

$$d(r, q_1) \stackrel{\text{def}}{=} \frac{p}{q} - \frac{1}{q_1}$$

can be represented as a sum of $\leq n$ unit fractions, i.e., whether

$$\left\| \frac{p}{q} - \frac{1}{q_1} \right\| \leq n.$$

For each q_1, this auxiliary checking can be done by the algorithm A_n. Thus, the algorithm A_{n+1} is designed.

Comment. As we will see from the following examples, these algorithms do not only compute the value $\|r\|$ for a given fraction r, they also produce the corresponding representation of the fraction r as the sum of $\|r\|$ unit fractions.

Computing the smallest number of cuts: example. Let us illustrate this algorithm on the example of the fraction

$$r = \frac{p}{q} = \frac{4}{5}.$$

In accordance with our general algorithm, we first apply the auxiliary algorithm A_1. For the given fraction, q does not divide p, so $\frac{4}{5}$ is not a unit fraction and thus

$$\left\| \frac{4}{5} \right\| > 1.$$

Let us now check whether $\left\|\dfrac{4}{5}\right\| \leq 2$. According to the above algorithm A_2, the integer value q_1 must satisfy the inequality

$$\frac{5}{4} \leq q_1 \leq 2 \cdot \frac{5}{4},$$

i.e., $1.25 \leq q_1 \leq 2.5$. There is only one integer in the corresponding interval, namely, the integer $q_1 = 2$.

We must now use the algorithm A_1 to check whether the difference

$$d(r, q_1) = \frac{p}{q} - \frac{1}{q_1} = \frac{4}{5} - \frac{1}{2} = \frac{3}{10}$$

has $\|d(r, q_1)\| \leq 1$. This difference $d(r, q_1) = \dfrac{3}{10}$ is not a unit fraction, so $\|d(r, q_1)\| > 1$, hence $\left\|\dfrac{4}{5}\right\| > 2$.

Let us now check whether $\left\|\dfrac{4}{5}\right\| \leq 3$. According to the above algorithm A_3, the integer value q_1 must satisfy the inequality

$$\frac{5}{4} \leq q_1 \leq 3 \cdot \frac{5}{4},$$

i.e., $1.25 \leq q_1 \leq 3.75$. There are two integers in the corresponding interval, $q_1 = 2$ and $q_1 = 3$. We must now apply the algorithm A_2 to the corresponding differences

$$d(r, q_1) = \frac{p}{q} - \frac{1}{q_1}.$$

For $q_1 = 2$, the corresponding difference $d(r, 2)$ is equal to

$$d(r, 2) = \frac{4}{5} - \frac{1}{2} = \frac{3}{10}.$$

Let us apply the algorithm A_2 to this difference $r' \overset{\text{def}}{=} \dfrac{3}{10}$.

According to the algorithm A_2, we must select an integer q_1' for which

$$\frac{10}{3} \leq q_1' \leq 2 \cdot \frac{10}{3},$$

i.e., we must consider $q_1' = 4, q_1' = 5$, and $q_1' = 6$. Already for $q_1' = 4$, the difference

$$d(r', 4) = \frac{3}{10} - \frac{1}{4} = \frac{1}{20}$$

is a unit fraction, so

$$\left\|\frac{3}{10}\right\| = 2,$$

with the corresponding representation

$$\frac{3}{10} = \frac{1}{4} + \frac{1}{20}.$$

Since

$$\frac{3}{10} = \frac{4}{5} - \frac{1}{2},$$

we thus conclude that

$$\left\|\frac{4}{5}\right\| = 3,$$

with

$$\frac{4}{5} = \frac{1}{2} + \frac{1}{4} + \frac{1}{20}.$$

Comment. This is not the only possible representation of the fraction $\frac{4}{5}$ as a sum of three unit fractions: for $q_1' = 5$, we get an alternative representation with the same number of unit fractions:

$$\frac{4}{5} = \frac{1}{2} + \frac{1}{5} + \frac{1}{10}.$$

It is worth mentioning that, as one can check, the remaining value $q_1' = 6$ (and the values q_1' corresponding to $q_1 = 3$) do not lead to a sum of three fractions.

Discussion: Are Egyptian fractions optimal? A natural question is: are actual Egyptian fraction representations – as given in the Rhind Papyrus – optimal? Not always.

For example, the Egyptians did not allow identical unit fractions in their representations and had other unclear preferences. As a result, e.g., instead of

$$\frac{2}{13} = \frac{1}{13} + \frac{1}{13},$$

they used a representation

$$\frac{2}{13} = \frac{1}{8} + \frac{1}{52} + \frac{1}{104}.$$

From the viewpoint of the smallest number of cuts, this representation does not make sense: it replaces a representation corresponding to

$$p_1 + \ldots + p_k = 2$$

with a representation for which

$$p'_1 + \ldots + p'_{k'} = 1 + 1 + 1 = 3 > 2,$$

i.e., with a representation with more cuts.

So, we do not claim that the ancient Egyptian always had it right, what we claim is that their general idea of reducing the sum of the numerators as much as possible seems to be right. From this viewpoint, it will be interesting to further analyze the properties of the norm $\| \cdot \|$.

How we can use these results and ideas in education. In elementary and middle school mathematics, the concept of fractions is one of the most difficult concepts. To master fractions, students have to avoid and overcome numerous misconceptions. Because of this difficulty, researchers in mathematics educations have developed a large number of innovative ways of teaching fractions, to increase the students' interest and motivation. Egyptian fractions, with their non-traditional structure and easy-to-understand techniques, are known to have led to many successful examples of such innovative activities; see, e.g., [4–10].

Since these examples work so well when teaching schoolchildren, we have taught the corresponding activities to future teachers, in the methods courses at the University of Texas at El Paso. Interestingly, not only the future teachers learned to use these examples, but they got very much interested in these examples themselves. Egyptian fraction ideas helped the future teachers better understand the variety of possible ideas behind the seemingly simple school mathematics, in particular, the multi-faceted nature of the concept of fraction.

The interesting fact about the Egyptian approach to fractions is that, as we have seen in this chapter, this approach leads very fast from simple arithmetic to reasonably complex algorithms – when we want not just to represent every fraction in this way, but to get an *optimal* representation, with the smallest possible number of terms. When presented with these complex algorithms, more computer-advanced students started thinking about possible implementations of these algorithms.

The students were especially intrigued by the fact that while the resulting representation of each fraction is optimal, the algorithm itself may be not be optimal – in the sense that no one has proved that the same representations cannot be obtained by a much faster algorithm. Looking for a faster algorithm, investigating general property of optimal representations – these are the tasks that students are eager to do by programming these problems and by running these programs on different examples.

The interest that these problems raised when teaching computer-skilled future teachers make us believe that this topic can also be of interest to students studying computer science. Their interest may be further enhanced by the following relevant observation about the hardware implementation of arithmetic operations in the modern computers.

Inside the computers, addition is usually performed by using, in effect, the same algorithm that we learn at school. The main difference between our way of adding numbers and computer implementation is that we add decimal numbers, while the computers perform addition on numbers represented in the binary code.

Similarly, inside the computers, multiplication is performed in a way which is similar to how we multiply:

- multiplying by different digits and then
- adding the results.

In contrast, in the modern computers, division is performed completely differently from how we normally divide. Namely, the computer calculates $\frac{x}{y}$ as the product $x \cdot \frac{1}{y}$, where

- for several basic values y, the ratios $\frac{1}{y}$ are pre-computed and stored, and
- the ratios $\frac{1}{y}$ for other values y are computed based on the known (stored) values.

We can summarize this implementation by saying that modern computers use unit fractions – exactly the same ideas as the Egyptians pioneered.

This analogy can go one step further. We have mentioned that it is still not known which algorithms for computing the Egyptian fractions representations are the fastest. Even simple computer experiments, easily started by students, have the potential of leading to interesting observations and results. Similarly, it is not known which algorithms for division lead to the fastest computer implementations. There is also room for computer simulations and experimentations. And who knows, maybe experiments with the millennia-old Egyptian fractions can lead to stimulating new ideas that will help make modern computers faster?

References

1. O. Kosheleva, V. Kreinovich, Egyptian fractions revisited. Inf. Edu. **8**(1), 35–48 (2009)
2. C.B. Boyer, U.C. Merzbach, *A History of Mathematics* (Wiley, New York, 1991)
3. P. Hoffman, *The Man Who Loved Only Numbers: The Story of Paul Erdős and the Search for Mathematical Truth* (Hyperion, New York, 1988)
4. D. Eppstein, Egyptian fractions website. http://www.ics.uci.edu/~eppstein/numth/egypt/
5. O. Kosheleva, V. Kreinovich. Egyptian fractions revisited, in *Abstracts of the 2005 Meeting of the Southwestern Section of the Mathematical Association of America (MAA)*, 1–2 April 2005, p. 6
6. R. Aló, O. M. Kosheleva. Optimization techniques under uncertain criteria, and their possible use in computerized education, in *Proceedings of the 25th International Conference of the North American Fuzzy Information Processing Society NAFIPS'2006*, Montreal, Quebec, Canada, IEEE Press, Piscataway, New Jersey, June 3–6 2006
7. O. Kosheleva, I. Lyublinskaya, Can Egyptian papyrus enrich our students' understanding of fractions?, in *Abstracts of the Annual Meeting of the National Council of Teachers of Mathematics NCTM "Mathematics: Representing the Future"*, Atlanta, Georgia, 21–24 March 2007, p. 40

8. O. Kosheleva, I. Lyublinskaya, Using innovative fraction activities as a vehicle for examining conceptual understanding of fraction concepts in pre–service elementary teachers mathematical education, in *Proceedings of the 29th Annual Meeting of the North American Chapter of the International Group for the Psychology of Mathematics Education PME–NA 2007*, ed. by T. Lamberg, L.R. Wiest, Stateline (Lake Tahoe), Nevada, 25–28 October 2007, University of Nevada, Reno, p. 36–38
9. National Council of Teachers of Mathematics NCTM, *Egyptian Fractions*, NCTM Student Math Notes, 2009
10. L. Streefland, *Fractions in Realistic Mathematics Education: A Paradigm Of Developmental Research* (Kluwer Academic Publishers, Dodrecht, The Netherlands, 1991)

Chapter 7
How to Enhance Student Motivations by Borrowing from Ancient Tradition: Mayan and Babylonian Arithmetics

In the previous chapter, we showed how to use ideas from ancient Egypt when teaching math. In this chapter (and in the following chapter), we consider ideas from Mayan and Babylonian mathematics.

Most number systems use a single base – e.g., 10 or 2 – and represent each number as a combination of powers of the base. However, historically, there were two civilizations that used a more complex system to represent numbers. They also used bases: Babylonians used 60 and Mayans used 20, but for each power, instead of a single digit, they used two. For example, number 19 was represented by the Babylonians as $19_B = 1 \cdot 10 + 9$ and by the Mayans as $34_M = 3 \cdot 5 + 4$. In this chapter, we show that such a representation is not just due to historic reasons: for the corresponding large bases, such a representation is actually optimal – in some reasonable sense.

The results from this chapter first appeared in [1]. The authors are thankful to Judith Munter for the useful discussion of Mayan mathematics.

Traditional numerical systems use a single base. Most numerical systems are based on using a single number as a base. For example, in the decimal system, each natural number is represented by a sequence of decimal digits such as 2017, so that:

- the last digit 1 means ones (i.e., multiples of 10^0);
- the next digit means 10s (i.e., multiples of 10^1);
- the next digit means 100s (i.e., multiples of 10^2);
- the next digit means 1000s (i.e., multiples of 10^3),
- etc.

and the whole number means

$$2017_{10} = 2 \cdot 10^3 + 0 \cdot 10^2 + 1 \cdot 10^1 + 7 \cdot 10^0.$$

© Springer-Verlag GmbH Germany 2018
O. Kosheleva and K. Villaverde, *How Interval and Fuzzy Techniques Can Improve Teaching*, Studies in Computational Intelligence 750, https://doi.org/10.1007/978-3-662-55993-2_7

Similarly, in a binary system, with base 2, the number 1011_2 means

$$1 \cdot 2^3 + 0 \cdot 2^2 + 1 \cdot 2^1 + 1 \cdot 2^0 = 11_{10}.$$

Mayan and Babylonian systems were more complex. However, there are two exceptional systems, systems used by two civilizations that actually did a lot of computations, especially related to astronomy. In general, they use the base system: the Mayan used base $b = 20$ and the Babylonian used base $b = 60$ (our division of an hour into 60 min and a minute into 60 s goes back to the Babylonians). However, their systems were more complex than the base systems: namely, to represent each digit from 0 to $b - 1$, they used a representation using a different "sub-base":

- the Mayans used 5; see, e.g., [2–6], while
- the Babylonians used 10; see, e.g., [2, 4, 5, 7–9].

Thus, e.g., the number 19 was represented:

- by the Mayans as $34_M = 3 \cdot 5 + 4$,
- and by the Babylonians as $19_B = 1 \cdot 10 + 9$.

As a result, numbers had a complex representation. For example, to get the Mayan representation of 386_{10}, we:

- first represent it in base 20, as $386 = 19 \cdot 20^1 + 6 \cdot 20^0$, and
- then represent each "digit" of this representation in a base-5 system: $19 = 3 \cdot 5 + 4 = 34_M$ and $6 = 1 \cdot 5 + 1 = 11_M$,

resulting in $386_{10} = 3411_M$.

In other words, we get $386_{10} = 3 \cdot 5 \cdot 20^1 + 4 \cdot 20^1 + 1 \cdot 5 \cdot 20^0 + 1 \cdot 20^0$. In general, we have a representation of the type

$$l_n s_n \dots l_0 s_{0,M} = l_n \cdot 5 \cdot 20^n + s_n \cdot 20^n + \dots + l_0 \cdot 5 \cdot 20^0 + s_0 \cdot 20^0.$$

For the Babylonian system, we similarly have

$$l_n s_n \dots l_0 s_{0,B} = l_n \cdot 10 \cdot 60^n + s_n \cdot 60^n + \dots + l_0 \cdot 10 \cdot 60^0 + s_0 \cdot 60^0.$$

A natural question: why this complexity? At first glance, these complex numerical systems seem to require more computations than simply using the systems with base 20 and 60: we will show that addition and multiplication become more complicated.

Indeed, when we add two digits in the decimal system (or in any other traditional system), the result – what goes into the corresponding digit of the result and whether we have a carry – does not depend on whether we are adding the last digits or the next-to-last digits or the first digits. In contrast, in the Mayan system, the result of adding two digits depends on whether we are adding last digits – in this case, 5 and above means a carry – or the next-to-last digit – in this case, 4 and above means a carry. Rules for multiplications are even more complicated. So, at first glance, these system may seem too complex to be practically useful.

What we do in this chapter. In this chapter, we analyze the computational complexity of the most complex of the two basic arithmetic operations – multiplication – in different notations, and we show that for large bases like 20 and 60, a complex system indeed leads to faster computations than simply using a positional system. Moreover, we show that for 20, the optimal complex system should be based (as it was) on the fact that $20 = 4 \cdot 5$, and for 60, the system should be based on the fact that $60 = 6 \cdot 10$.

The computational complexity will be estimated not based on any modern sophisticated algorithms, but rather based on the usual algorithms for multiplication, algorithms that use multiplication tables.

Why multiplication. Multiplication is much more complex than addition – because, crudely speaking, it consists of several additions. Thus, if we have a computational procedure that has approximately as many multiplications as additions, the largest amount of time is spent on multiplications. Hence, when we select a way to represent numbers, it makes sense to select the way for which multiplication requires the smallest amount of computation. So, to make the desired selection, let us analyze the complexity of multiplication.

Multiplication in the usual representation: analysis. Let us start by a representation in which we fix a base b, and represent each number as a sequence of digits ranging from 0 to $b - 1$. In this representation, multiplication reduces to digit-by-digit multiplication – followed by summation of the results. Crudely speaking, each digit-by-digit multiplication means that we look up the result in the multiplication table, and then place the result into the new table – in which we prepare everything for addition.

A multiplication table consists of $(b - 1)^2$ results of multiplying each of the non-zero digits $1, \ldots, b - 1$ by every other digit (multiplication by 0 always gives 0, so multiplication tables only contain non-zero digits).

The time to access this information is proportional to the largest distance to the farthest bit containing this information. In the brain, neurons are mostly located on the surface (see, e.g., [10]), so all the information is, in effect, contained in a two-dimensional area. To minimize the worst-case time, it is desirable to place the data in the most compact way, where the worst-case distance is the smallest. The most compact way of representing $(b - 1)^2$ values is to store them in a 2-D square array $(b - 1) \times (b - 1)$, then the worst access time is proportional to $b - 1$.

Placing the result into a place ready for addition requires one more computational step, so overall, the number of steps is b.

Multiplication in the more complex representation. Let us now assume that instead of storing the digits from 0 to $b - 1$, we select a number b_s that divides b, i.e., for which $b = b_s \cdot b_l$ for some integer b_l, and then store two "sub-digits": a sub-digit that goes from 0 to $b_s - 1$ and a sub-digit that goes from 0 to $b_l - 1$. For example:

- the Mayans used $b_s = 5$ for $b = 20$; in this case, $b_l = 4$;
- the Babylonians used $b_s = 10$ for $b = 60$; in this case, $b_l = 6$.

Each original digit d is now represented by two "sub-digits" $d = d_l d_s$: a "small" sub-digit d_s (i.e., a sub-digit corresponding to values $0, \ldots, b_s - 1$ describing smaller numbers), and a "large" sub-digit d_l – i.e., a sub-digit corresponding to values $0, \ldots,$ $b_l - 1$ describing larger numbers. Thus, to describe the product of two original digits $d = d_l d_s$ and $d' = d'_l d'_s$, we need to compute four products:

- one product of "large" sub-digits $d_l d'_l$,
- one product of "small" sub-digits $d_s d'_s$; and
- two products of sub-digits of different type $d_l d'_s$ and $d_s d'_l$.

Thus, we need three multiplication tables:

- a table for "large" sub-digits – it will be used once for each original digit;
- a table for "small" sub-digits – it will also be used once for each original digit, and
- a multiplication table describing results of multiplying a "large" sub-digit and a "small" sub-digit: this table will be used twice.

The first table contains the results of multiplying each of $b_l - 1$ "large" sub-digits by every "large" sub-digit. This table contains $(b_l - 1)^2$ results and therefore, access to this table requires time $b_l - 1$. Storing this result requires one more step, so we need time b_l to perform this sub-digit multiplication.

Similarly, the second table contains the results of multiplying each of $b_s - 1$ "small" sub-digits by every "small" sub-digit. This table contains $(b_s - 1)^2$ results and therefore, access to this table requires time $b_s - 1$. Storing this result requires one more step, so we need time b_s to perform this sub-digit multiplication.

Finally, the third table contains $(b_l - 1) \cdot (b_s - 1)$ results of multiplying two different types of sub-digits. As we have mentioned, in general, when we store N results, we need access time $\sqrt{N}$. Thus, for this table, the access time is $\sqrt{(b_l - 1) \cdot (b_s - 1)}$. We need to access this table twice, so we need access time $2 \cdot \sqrt{(b_l - 1) \cdot (b_s - 1)}$. Storing the corresponding two results requires two more steps, so we need time $2 \cdot \sqrt{(b_l - 1) \cdot (b_s - 1)} + 2$ to perform these two sub-digit multiplications.

By combining all these times, we conclude that the overall computation time is equal to

$$b_l + b_s + 2 \cdot \sqrt{(b_l - 1) \cdot (b_s - 1)} + 2.$$

So, we arrive at the following conclusion:

Analysis: conclusion. For each base b, we select a complex method if there exists a value b_s that divides b and for which

$$b_l + b_s + 2 \cdot \sqrt{(b_l - 1) \cdot (b_s - 1)} + 2 < b.$$

If such a value b_s exists, as an optimal value, we select a value b_s for which the sum

$$b_l + b_s + 2 \cdot \sqrt{(b_l - 1) \cdot (b_s - 1)} + 2.$$

is the smallest possible.

Comment. Once we have the original base b represented as a product of two integers, we can have the first one as b_s and the second one as b_l – or vice versa. For example, for $b = 20 = 4 \cdot 5$, we have two options:

- we can have $b_l = 4$ and $b_s = 5$, and
- we can also have $b_l = 5$ and $b_s = 4$.

From the above formulas, we can see that the computational complexity does not depend on which of the two numbers we take as b_s and which as b_l, Thus, in our selection based on the computational complexity, we will only select two factors of b, without specifying which one corresponds to "larger" sub-digits and which one corresponds to "smaller" sub-digits.

Explanation of Mayan and Babylonian Systems. Let us now apply to the above analysis to the cases $b = 20$ and $b = 60$ corresponding to the Mayan and the Babylonian number systems.

Case of $b = 20$: explanation of the Mayan system. For $b = 20$, the computational complexity (= number of computational steps) of the original method is equal to $b = 20$.

There are only two ways to represent 20 as a product of two non-trivial natural numbers (i.e., natural numbers different from 1):

- as $20 = 2 \cdot 10$, and
- as $20 = 4 \cdot 5$.

In the first case, without losing generality, we can assume that $b_s = 2$ and $b_l = 10$. Thus,

$$\sqrt{(b_l - 1) \cdot (b_s - 1)} = \sqrt{(2 - 1) \cdot (10 - 1)} = \sqrt{1 \cdot 9} = \sqrt{9} = 3$$

and so,

$$b_l + b_s + 2 \cdot \sqrt{(b_l - 1) \cdot (b_s - 1)} + 2 = 2 + 10 + 2 \cdot 3 + 2 = 20.$$

So, in this case, we do not gain anything: computations become more complex, but the computation time remains the same.

In the second case, without losing generality, we can assume that $b_s = 4$ and $b_l = 5$. Thus,

$$\sqrt{(b_l - 1) \cdot (b_s - 1)} = \sqrt{(4 - 1) \cdot (5 - 1)} = \sqrt{3 \cdot 2} = 2 \cdot \sqrt{3} \approx 3.46$$

and so,

$$b_l + b_s + 2 \cdot \sqrt{(b_l - 1) \cdot (b_s - 1)} + 2 \approx 4 + 5 + 2 \cdot 3.46 + 2 = 17.92 < 20.$$

So, this case indeed leads to the fastest computations – which is probably the reason why the Mayans selected this case for their computations.

Case of $b = 60$: explanation of Babylonian system. For $b = 60$, the computational complexity (= number of computational steps) of the original method is equal to $b = 60$.

There are several ways to represent 60 as a product of two non-trivial natural numbers (i.e., natural numbers different from 1):

- as $60 = 2 \cdot 30$,
- as $60 = 3 \cdot 20$,
- as $60 = 4 \cdot 15$,
- as $60 = 5 \cdot 12$, and
- as $60 = 6 \cdot 10$.

In the first case, without losing generality, we can assume that $b_s = 2$ and $b_l = 30$. Thus,

$$\sqrt{(b_l - 1) \cdot (b_s - 1)} = \sqrt{(2 - 1) \cdot (30 - 1)} = \sqrt{1 \cdot 29} = \sqrt{29} \approx 5.4,$$

and so,

$$b_l + b_s + 2 \cdot \sqrt{(b_l - 1) \cdot (b_s - 1)} + 2 \approx 2 + 30 + 2 \cdot 5.4 + 2 = 44.8 < 60.$$

So, in this case, some selection is sub-digits leads to faster computations. Let us see which values b_s and $b + l$ lead to the fastest computations.

In the second case, without losing generality, we can assume that $b_s = 3$ and $b_l = 20$. Thus,

$$\sqrt{(b_l - 1) \cdot (b_s - 1)} = \sqrt{(3 - 1) \cdot (20 - 1)} = \sqrt{2 \cdot 19} = \sqrt{38} \approx 6.2,$$

and so,

$$b_l + b_s + 2 \cdot \sqrt{(b_l - 1) \cdot (b_s - 1)} + 2 \approx 3 + 20 + 2 \cdot 6.2 + 2 = 37.4.$$

In the third case, without losing generality, we can assume that $b_s = 4$ and $b_l = 15$. Thus,

$$\sqrt{(b_l - 1) \cdot (b_s - 1)} = \sqrt{(4 - 1) \cdot (15 - 1)} = \sqrt{3 \cdot 14} = \sqrt{42} \approx 6.5,$$

and so,

$$b_l + b_s + 2 \cdot \sqrt{(b_l - 1) \cdot (b_s - 1)} + 2 \approx 4 + 15 + 2 \cdot 6.5 + 2 = 34.$$

In the fourth case, without losing generality, we can assume that $b_s = 5$ and $b_l = 12$. Thus,

$$\sqrt{(b_l - 1) \cdot (b_s - 1)} = \sqrt{(5 - 1) \cdot (12 - 1)} = \sqrt{4 \cdot 11} = \sqrt{44} \approx 6.7,$$

and so,

$$b_l + b_s + 2 \cdot \sqrt{(b_l - 1) \cdot (b_s - 1)} + 2 \approx 5 + 12 + 2 \cdot 6.7 + 2 = 32.4.$$

Finally, in the last case, without losing generality, we can assume that $b_s = 5$ and $b_l = 6$. Thus,

$$\sqrt{(b_l - 1) \cdot (b_s - 1)} = \sqrt{(5 - 1) \cdot (6 - 1)} = \sqrt{4 \cdot 5} = \sqrt{20} \approx 4.5,$$

and so,

$$b_l + b_s + 2 \cdot \sqrt{(b_l - 1) \cdot (b_s - 1)} + 2 \approx 5 + 6 + 2 \cdot 4.5 + 2 = 22.$$

By comparing all the computational complexity results, we conclude that the case $b_s = 5$ and $b_l = 6$ indeed leads to the fastest computations – which is probably the reason why the Babylonians selected this case for their computations.

Case of $b = 10$. To doublecheck that our explanation is reasonable, let us show for the usual decimal system $b = 10$, we do not get any need for sub-digits. Indeed, for $b = 10$, the computational complexity (= number of computational steps) of the original method is equal to $b = 20$.

There are only one way to represent 10 as a product of two non-trivial natural numbers (i.e., natural numbers different from 1): as $10 = 2 \cdot 5$. In this case, without losing generality, we can assume that $b_s = 2$ and $b_l = 5$. Thus,

$$\sqrt{(b_l - 1) \cdot (b_s - 1)} = \sqrt{(2 - 1) \cdot (5 - 1)} = \sqrt{1 \cdot 4} = \sqrt{4} = 2$$

and so,

$$b_l + b_s + 2 \cdot \sqrt{(b_l - 1) \cdot (b_s - 1)} + 2 = 2 + 5 + 2 \cdot 2 + 2 = 11 > 10,$$

i.e., using sub-digits would slow down computations.

So, in the case of $b = 10$, the traditional representation (without sub-digits) is indeed the best one.

References

1. O. Kosheleva, Mayan and Babylonian arithmetics can be explained by the need to minimize computations. Appl. Math. Sci. **6**(15), 697–705 (2012)
2. C.B. Boyer, U.C. Merzbach, *A History of Mathematics* (Wiley, New York, 1991)
3. M.D. Coe, *The Maya* (Thames and Hudson, London, New York, 2011)
4. G. Ifrah, *A Universal History of Numbers: From Prehistory to the Invention of the Computer* (London, 1998)
5. D.E. Knuth, *Seminumerical Algorithms* (Addison Wesley, Reading, Massachusetts, 1981)
6. G. Morley, *Ancient Maya* (Stanford University Press, Stanford, 1946)
7. O. Neugebauer, *The Exact Sciences in Antiquity* (Dover Publ, New York, 1969)
8. B.L. van der Waerden, *Science Awakening* (P. Noorhoff, Groningen, 1954)
9. B.L. van der Waerden, *Geometry and Algebra in Ancient Civilizations* (New York, 1983)
10. P.M. Milner, *Physiological Psychology* (Holt, New York, 1971)

Chapter 8
How to Enhance Student Motivations by Borrowing from Ancient Tradition: Babylonian Method of Computing the Square Root

In the previous chapter, we showed how to use ideas from Mayan and Babylonian arithmetic when teaching math. In this chapter, we consider Babylonian mathematical ideas beyond simple arithmetic, namely, the ideas of computing the square root. When computing a square root, computers still, in effect, use an iterative algorithm developed by the Babylonians millennia ago. This is a very unusual phenomenon, because for most other computations, better algorithms have been invented – even division is performed, in the computer, by an algorithm which is much more efficient that division methods that we have all learned in school. What is the explanation for the success of the Babylonians' method? One explanation is that this is, in effect, Newton's method, based on the best ideas from calculus. This explanations works well from the mathematical viewpoint – it explains why this method is so efficient, but since the Babylonians were very far from calculus, it does not explain why this method was invented in the first place. In this chapter, we provide two possible explanations for this method's origin. We show that this method naturally emerges from fuzzy techniques, and we also show that it can be explained as (in some reasonable sense) the computationally simplest technique.

The results from this chapter first appeared in [1].

How to compute the square root: an iterative formula going back to the ancient Babylonians. How can we compute the square root $x = \sqrt{a}$ of a given number a? Historically the first method for computing the square root was invented by the ancient Babylonians; see, e.g., [2] and references therein. In this "Babylonian" method, we start with an arbitrary positive number x_0, and then apply the following iterative process:

$$x_{n+1} = \frac{1}{2} \cdot \left(x_n + \frac{a}{x_n} \right). \tag{8.1}$$

© Springer-Verlag GmbH Germany 2018
O. Kosheleva and K. Villaverde, *How Interval and Fuzzy Techniques
Can Improve Teaching*, Studies in Computational Intelligence 750,
https://doi.org/10.1007/978-3-662-55993-2_8

To be more precise, Babylonians rarely described their mathematical procedures in algorithmic form: they usually described them by presenting several examples. Only in the Greek mathematics these procedures were reformulated in the general abstract form. This is true for the square root procedure as well: the first person to describe this procedure in general abstract terms was Heron of Alexandria, mostly known as the inventor of the steam engine; see, e.g., [3]. Because of this, the above algorithm is also known as "Heron's method".

Properties of the Babylonian method. If we start with the value x_0 which is already equal to the square root of a, $x_0 = \sqrt{a}$, then, as one can easily check, the next iteration is the exact same value:

$$x_1 = \frac{1}{2} \cdot \left(x_0 + \frac{a}{x_0} \right) = \frac{1}{2} \cdot \left(\sqrt{a} + \frac{a}{\sqrt{a}} \right) =$$

$$\frac{1}{2} \cdot \left(\sqrt{a} + \sqrt{a} \right) = \sqrt{a}. \tag{8.2}$$

It has been proven that this method converges: $x_n \to \sqrt{a}$ for all possible starting values x_0. It is intuitively clear that the closer the original approximation x_0 to the square root, the faster the convergence.

Convergence is not so straightforward to prove, but it is straightforward to prove that if the sequence x_n converges to some value x, then this limit value x is indeed the desired square root. Indeed, by tending to a limit $x_n \to x$ and $x_{n+1} \to x$ in the formula (8.1), we conclude that

$$x = \frac{1}{2} \cdot \left(x + \frac{a}{x} \right). \tag{8.3}$$

Multiplying both sides of this equality by 2, we get

$$2x = x + \frac{a}{x}. \tag{8.4}$$

By subtracting x from both sides, we conclude that $x = \frac{a}{x}$. Multiplying both sides of this equality by x, we get $a = x^2$; this is exactly the defining equation of the square root.

Babylonian method: a numerical example. The Babylonian method for computing the square root is very efficient. To illustrate how efficient it is, let us illustrate it on the example of computing the square root of 2, when $a = 2$.

Let us start with the simplest possible real number $x_0 = 1$. Then, according to the Babylonian algorithm, we compute the next approximation x_1 to $\sqrt{2}$ as

$$x_1 = \frac{1}{2} \cdot \left(x_0 + \frac{a}{x_0} \right) = \frac{1}{2} \cdot \left(1 + \frac{2}{1} \right) = \frac{1}{2} \cdot 3 = 1.5. \tag{8.5}$$

The next approximation is

$$x_2 = \frac{1}{2} \cdot \left(x_1 + \frac{a}{x_1} \right) = \frac{1}{2} \cdot \left(1.5 + \frac{2}{1.5} \right) =$$

$$\frac{1}{2} \cdot (1.5 + 1.3333\ldots) = \frac{1}{2} \cdot 2.8333\ldots = 1.4166\ldots. \tag{8.6}$$

So, after two simple iterations, we already get 3 correct decimal digits of $\sqrt{2}$:

$$\sqrt{2} = 1.41\ldots$$

Babylonian method is very efficient. The above example is typical. In general, the Babylonian method for computing the square root converges very fast.

In fact, it is so efficient that most modern computers use it for computing the square roots. The computers also take advantage of the fact that inside the computer, all the numbers are represented in the binary code. As a result, division by 2 simply means shifting the binary point one digit to the left: just like in the standard decimal code, division by 10 simply means shifting the decimal point one digit to the left, e.g., from 15.2 to 1.52.

The longevity of the Babylonian method is very unusual. The fact that the Babylonian method for computing the square root has been preserved intact and is used in the modern computers is very unusual.

Even for simple arithmetic operations such as division, the traditional numerical procedures that people have used for centuries turned out to be not as efficient as newly designed ones. For example, in most computers, subtraction and operations with negative numbers are not done as we do it, but by using the 2 s complement representation; see, e.g., [4]. Similarly, division is not performed the way we do it, but rather by using a special version of Newton's method, etc.

In contrast, the Babylonian method for computing the square root remains widely used. What is the reason for this longevity? How could Babylonians come up with a method which is so efficient?

Newton's explanation of the efficiency of the Babylonian method. Historically the first natural explanation of the efficiency of the Babylonian method was proposed by Isaac Newton. Newton showed that this method is a particular case of a general method for solving non-linear equations, a method that we now call Newton's method.

Specifically, suppose that we want to solve an equation

$$f(x) = 0, \tag{8.7}$$

and we know an approximate solution x_n. How can we find the next iteration?

We assumed that the known value x_n is close to the desired solution x. So, we can describe this solution as $x = x_n + \Delta x$, where the correction $\Delta x \stackrel{\text{def}}{=} x - x_n$ is relatively small. In terms of Δx, the Eq. (8.7) takes the form

$$f(x_n + \Delta x) = 0. \tag{8.8}$$

Since Δx is small, we can use the derivative $f'(x_n)$. Specifically, the derivative $f'(x)$ is defined as the limit

$$f'(x_n) = \lim_{h \to 0} \frac{f(x_n + h) - f(x_n)}{h}. \tag{8.9}$$

The limit means that the smaller h, the closer is the ratio

$$\frac{f(x_n + h) - f(x_n)}{h} \tag{8.10}$$

to the derivative $f'(x_n)$. Since Δx is small, the ratio

$$\frac{f(x_n + \Delta x) - f(x_n)}{\Delta x} \tag{8.11}$$

is close to the derivative $f'(x_n)$:

$$f'(x_n) \approx \frac{f(x_n + \Delta x) - f(x_n)}{\Delta x}. \tag{8.12}$$

We know that $f(x_n + \Delta x) = f(x) = 0$; thus, (8.8) implies that

$$f'(x_n) \approx -\frac{f(x_n)}{\Delta x}. \tag{8.13}$$

and hence,

$$\Delta x \approx -\frac{f(x_n)}{f'(x_n)}. \tag{8.14}$$

So, as the next approximation to the root, it is reasonable to take the value $x_{n+1} = x_n + \Delta x$, i.e., the value

$$x_{n+1} = x_n - \frac{f(x_n)}{f'(x_n)}. \tag{8.15}$$

Finding the square root $x = \sqrt{a}$ means finding a solution to the equation $x^2 - a = 0$. This equation has the form $f(x) = 0$ for $f(x) = x^2 - a$. Substituting this function $f(x)$ into the general formula (8.15), we get

$$x_{n+1} = x_n - \frac{x_n^2 - a}{2 \cdot x_n}. \tag{8.16}$$

Explicitly dividing each term in the right-hand side expression by x_n, we get

$$x_{n+1} = x_n - \left(\frac{x_n}{2} - \frac{a}{2 \cdot x_n} \right). \tag{8.17}$$

Opening parentheses, we get

$$x_{n+1} = x_n - \frac{x_n}{2} + \frac{a}{2 \cdot x_n}. \tag{8.18}$$

Replacing $x_n - \dfrac{x_n}{2}$ with $\dfrac{x_n}{2}$, and moving the common divisor 2 outside the sum, we get the Babylonian formula

$$x_{n+1} = \frac{1}{2} \cdot \left(x_n + \frac{a}{x_n} \right). \tag{8.19}$$

There should be a more elementary explanation of the Babylonian formula. Newton's explanation explains why the Babylonian method is so efficient – because it is a particular case of the efficient Newton's method for solving nonlinear equations.

However, Newton's explanation does not explain how this method was invented in the first place, since

- the main ideas of Newton's method are heavily based on calculus, while
- the Babylonians were very far away from discovering calculus ideas.

We therefore need a more elementary explanation of the Babylonian formula. Two such explanations are provided in this chapter.

Explanation based on fuzzy techniques: main idea. We are looking for a square root $x = \sqrt{a}$, i.e., for the value for which

$$x = \frac{a}{x}. \tag{8.20}$$

Instead of the exact value x, we only know an approximate value $x_n \approx x$. Since $x_n \approx x$, we conclude that

$$\frac{a}{x_n} \approx \frac{a}{x}. \tag{8.21}$$

Because of (8.20), we can conclude that

$$\frac{a}{x_n} \approx x. \tag{8.22}$$

Thus, the desired square root x must satisfy two requirements:

- the first requirement is that $x \approx x_n$;

- the second requirement is that $x \approx y_n$ where we denoted

$$y_n \stackrel{\text{def}}{=} \frac{a}{x_n}.$$ (8.23)

Thus, we must find x from the following requirement:

$$(x \approx x_n) \,\&\, (x \approx y_n).$$ (8.24)

Fuzzy logic is a method for formalizing statements like this; see, e.g., [5, 6]. In fuzzy logic, to find the value x from the above requirement, we must follow the following steps:

First, we select the membership function $\mu_{\approx}(z)$ for describing the "approximate" relation $\approx$. After this selection,

- the degree of confidence in a statement $x \approx x_n$ is equal to $\mu_{\approx}(x - x_n)$; and
- the degree of confidence in a statement $x \approx y_n$ is equal to $\mu_{\approx}(x - y_n)$.

Next, we select a t-norm $f_{\&}(d, d')$ to describe the effect of the "and" operator on the corresponding degrees. After this selection, for each possible real number x, the degree $\mu(x)$ to which this number satisfies the above requirement can be computed as

$$\mu(x) = f_{\&}(\mu_{\approx}(x - x_n), \mu_{\approx}(x - y_n)).$$ (8.25)

Finally, we need to select a defuzzification procedure that transforms the membership function $\mu(x)$ into a single most appropriate value. For example, as this x, it is natural to take the value x for which the degree $\mu(x)$ is the largest possible.

Explanation based on fuzzy techniques: analysis. Let us consider different possible selections of a membership function and of the t-norm.

As an example, let us take a Gaussian membership function to describe "approximate"

$$\mu_{\approx}(z) = \exp\left(-\frac{z^2}{2\sigma^2}\right)$$ (8.26)

for some $\sigma > 0$, and the product as a t-norm:

$$f_{\&}(d, d') = d \cdot d'.$$ (8.27)

In this case, we have

$$\mu(x) = \mu_{\approx}(x - x_n) \cdot \mu_{\approx}(x - y_n) = \exp\left(-\frac{(x - x_n)^2}{2\sigma^2}\right) \cdot \exp\left(-\frac{(x - y_n)^2}{2\sigma^2}\right) =$$

$$\exp\left(-\frac{(x - x_n)^2 + (x - y_n)^2}{2\sigma^2}\right).$$ (8.28)

Due to monotonicity of the exponential function, this value attains the largest possible
value when the expression

$$(x - x_n)^2 + (x - y_n)^2 \tag{8.29}$$

is the smallest possible. Differentiating this expression with respect to x and equating
the derivative to 0, we conclude that $2(x - x_n) + 2(x - y_n) = 0$, i.e., that

$$x = \frac{x_n + y_n}{2}. \tag{8.30}$$

If for the same Gaussian membership function for "approximate", we choose
$f_\&(d, d') = \min(d, d')$ as the t-norm, we get a different expression for $\mu(x)$:

$$\mu(x) = \min(\mu_\approx(x - x_n), \mu_\approx(x - y_n)) =$$

$$\min\left(\exp\left(-\frac{(x - x_n)^2}{2\sigma^2}\right), \exp\left(-\frac{(x - y_n)^2}{2\sigma^2}\right)\right). \tag{8.31}$$

Due to monotonicity of the exponential function, this value is equal to

$$\mu(x) = \exp\left(-\frac{\max((x - x_n)^2, (x - y_n)^2)}{2\sigma^2}\right), \tag{8.32}$$

and this value is the largest when the expression

$$\max((x - x_n)^2, (x - y_n)^2) \tag{8.33}$$

is the smallest possible. This expression, in its term, can be rewritten as

$$\max(|x - x_n|^2, |x - y_n|^2).$$

Due to the fact that the absolute values are always non-negative and the square
function is monotonic on non-negative values, this expression has the form

$$(\max(|x - x_n|, |x - y_n|))^2,$$

and its minimum is attained when the simpler expression $\max(|x - x_n|, |x - y_n|)$
is the smallest possible. Let us show that this expression also attains its smallest
possible value at the midpoint (8.32).

Indeed, in geometric terms, the minimized expression $\max(|x - x_n|, |x - y_n|)$
is simply the largest of the distances $|x - x_n|$ and $|x - y_n|$ between the desired
point x and the given points x_n and y_n. Due to the triangle inequality, we have
$|x - x_n| + |x - y_n| \geq |x_n - y_n|$. Thus, it is not possible that both distances $|x - x_n|$
and $|x - y_n|$ are smaller than $\dfrac{|x_n - y_n|}{2}$ – because then their sum would be smaller

than $|x_n - y_n|$. So, at least one of these distances has to be larger than or equal to $\dfrac{|x_n - y_n|}{2}$, and therefore, the largest of these distances $\max(|x - x_n|, |x - y_n|)$ is always larger than or equal to $\dfrac{|x_n - y_n|}{2}$. The only way for this largest distance is to be equal to $\dfrac{|x_n - y_n|}{2}$ is when both distance $|x - x_n|$ and $|x - y_n|$ are equal to $\dfrac{|x_n - y_n|}{2}$, i.e., when the desired point x is exactly at a midpoint (8.32) between the two given points x_n and y_n.

One can show that we get the exact same answer (8.32) if we use triangular membership functions, symmetric piece-wise quadratic membership functions, different t-norms, etc.

Explanation based on fuzzy techniques: result. Our analysis shows that for many specific selections of the membership function for "approximate" and of the t-norm, we get the same answer:

$$x = \frac{x_n + y_n}{2}.$$

In other words, for many specific selections, as the next approximation x_{n+1} to the square root x, we take exactly the value x_{n+1} from the Babylonian procedure.

Thus, fuzzy techniques indeed explain the selection of the Babylonian method.

The important role of symmetry. The fact that different choices lead to the same result x can be explained by the symmetry of the problem. Indeed, the problem is symmetric with respect to reflection

$$x \to (x_n + y_n) - x \tag{8.34}$$

that swaps the values x_n and y_n. Thus, if we get the unique solution x, this solution must be invariant with respect to this symmetry – otherwise, the symmetric point would be another solution [7].

This invariance means that $x = (x_n + y_n) - x$ and thus, that $x = \dfrac{x_n + y_n}{2}$.

This application of fuzzy techniques is rather unusual. The fact that fuzzy techniques can be useful is well known [5, 6]. However, usually, fuzzy techniques lead to a good *approximation*: to an ideal control, to an ideal clustering, etc. What is unusual about the Babylonian algorithm is that here, fuzzy techniques lead to *exactly* the correct algorithm.

Explanation based on computational complexity: main claim. In any iterative procedure for computing the square root, once we have the previous approximation x_n, we can use:

- this approximation x_n,
- the value a, and
- (if needed) constants

to compute the next approximation $x_{n+1} = f(x_n, a)$ for an appropriate expression f.

For the iterative method to be successful in computing the square root, the expression $f(x_n, a)$ should satisfy the following natural properties:

- first, if we start with the value x_n which is already the square root $x_n = \sqrt{a}$, then this procedure should not change this value, i.e., we should have $f(\sqrt{a}, a) = \sqrt{a}$;
- second, this iterative method should converge.

In the Babylonian method, the computation of the corresponding expression $f(x_n, a)$ involves three arithmetic operations:

- a division

$$\frac{a}{x_n};\tag{8.35}$$

- an addition

$$x_n + \frac{a}{x_n};\tag{8.36}$$

and
- a multiplication by 0.5 – which, for binary numbers, is simply a shift by one bit:

$$0.5 \cdot \left(x_n + \frac{a}{x_n} \right).\tag{8.37}$$

Our claim is that this is the simplest possible operation.

In other words, our claim is that it is not possible to find an expression $f(x_n, a)$ which would be computable in 0, 1, or 2 arithmetic operations.

Let us prove this claim.

It is not possible to have no arithmetic operations. If we are not allowing any arithmetic operations at all, then as $x_{n+1} = f(x_n, a)$, we should return

- either the value x_n,
- or the value a,
- or some constant c.

In all three cases, we do not get any convergence to the square root:

- in the first case, the values x_n remain the same and never converge to $\sqrt{a}$:

$$x_0 = x_1 = x_2 = \ldots = x_n = \ldots;\tag{8.38}$$

- in the second case, we start with some initial value x_0 and then repeatedly return the values equal to a:

$$x_1 = x_2 = \ldots = x_n = \ldots = a;\tag{8.39}$$

- in the second case, we start with a value x_0 and then repeatedly return the values equal to the constant c:

$$x_1 = x_2 = \ldots = x_n = \ldots = c. \tag{8.40}$$

It is not possible to have one arithmetic operation. The arithmetic operation is either addition, or subtraction, or multiplication, or division. Let us consider these four cases one by one. All operations involve the values x_n and a and a possible constant(s) c and c'.

For addition, depending on what we add, we get $x_n + x_n$, $x_n + a$, $x_n + c$, $a + a$, $a + c$, and $c + c'$. In all these cases, for $x_n = \sqrt{a}$, the result is different from $\sqrt{a}$. So, the expression $f(x_n, a)$ involving only one addition does not satisfy the condition $f(\sqrt{a}, a) = \sqrt{a}$.

For subtraction, depending on what we subtract, we get the expressions $x_n - a$, $a - x_n$, $x_n - c$ with $c \neq 0$, $c - x_n$, $a - c$, $c - a$, and $c - c'$ (we dismiss the trivial expressions of the type $a - a = 0$). In all these cases, for $x_n = \sqrt{a}$, the result is different from $\sqrt{a}$. So, the expression $f(x_n, a)$ involving only one subtraction does not satisfy the condition $f(\sqrt{a}, a) = \sqrt{a}$.

For multiplication, depending on what we multiply, we get $x_n \cdot x_n$, $x_n \cdot a$, $x_n \cdot c$, $a \cdot a$, $a \cdot c$, and $c \cdot c'$. In all these cases, for $x_n = \sqrt{a}$, the result is different from $\sqrt{a}$. So, the expression $f(x_n, a)$ involving only one multiplication does not satisfy the condition $f(\sqrt{a}, a) = \sqrt{a}$.

For division, depending on what we divide, we get the expressions $\dfrac{x_n}{a}$, $\dfrac{a}{x_n}$, $\dfrac{x_n}{c}$ with $c \neq 1$, $\dfrac{c}{x_n}$, $\dfrac{a}{c}$, $\dfrac{c}{a}$, and $\dfrac{c}{c'}$ (we dismiss the trivial expressions of the type $\dfrac{a}{a} = 1$). In all these cases, except for the case $\dfrac{a}{x_n}$, for $x_n = \sqrt{a}$, the result is different from $\sqrt{a}$. So, the expression $f(x_n, a)$ corresponding to all these cases does not satisfy the condition $f(\sqrt{a}, a) = \sqrt{a}$.

In the remaining case

$$f(x_n, a) = \frac{a}{x_n}, \tag{8.41}$$

we do have

$$f(\sqrt{a}, a) = \frac{a}{\sqrt{a}} = \sqrt{a}, \tag{8.42}$$

but we do not satisfy the second condition: of convergence. Indeed, in this case,

$$x_1 = \frac{a}{x_0}, \tag{8.43}$$

then

$$x_2 = \frac{a}{x_1} = \frac{a}{a/x_0} = x_0, \tag{8.44}$$

and then again,

$$x_3 = \frac{a}{x_2} = \frac{a}{x_0} = x_1, \tag{8.45}$$

etc. So, here, we have

$$a_0 = a_2 = \ldots = a_{2n} = \ldots, \tag{8.46}$$

$$a_1 = a_3 = \ldots = a_{2n+1} = \ldots \tag{8.47}$$

and no convergence.

It is not possible to have two arithmetic operations. Similarly, one can prove that it is not possible to have two arithmetic operations. This can be proven by enumerating all possible sequences of arithmetic operations and then checking that for all possible inputs (x_n, a, or c) the resulting expression $f(x_n, a)$:

- either does not satisfy the requirement $f(\sqrt{a}, a) = \sqrt{a}$;
- or does not lead to the convergence to the square root.

For example, if we first add, and then multiply, then we get the expression $(e + e') \cdot e''$. Replacing each of the possible inputs e, e', and e'' with one of the possible values x_n, a, or c, we get all possible expressions $f(x_n, a)$ corresponding to this case.

For example, if we take $e = x_n$ and $e' = e'' = a$, we get the expression $f(x_n, a) = (x_n + a) \cdot a$. This expression clearly does not satisfy the requirement $f(\sqrt{a}, a) = \sqrt{a}$. If we take $e = x_n$, $e' = a$, and $e'' = c$, then we get the expression $f(x_n, a) = (x_n + a) \cdot a$ which also does not satisfy the same requirement.

By considering all possible cases, we can thus prove that no expression with 2 arithmetic operations is indeed possible.

Possible pedagogical use of this proof. In our opinion, this proof provides a good pedagogical example of a simple easy-to-handle arithmetic problem that is actually not a toy problem at all: it is related to an efficient algorithm for computing the square root.

For each combination of operations and inputs, it is relatively easy to come up with an explanation of why this particular combination will not work. With a sufficiently large number of students in a class, and a sufficient time allocated for this exercise, students can actually check all the possibilities – and thus get a sense of achievement.

Indeed, we have 4 possible operations coming first, 4 possible operations coming second, so we have $4^2 = 16$ possible sequences of operations. So, if we have 16 students in the class, each student can handle one of these combinations. If we have 32 students, then we can divide students into pairs so that each pair of students handles one combination of operations.

For each of the these combinations of operations, we have 3 options for each of the inputs, so totally, we have a manageable number of $3^3 = 27$ possible combinations.

The Babylonian formula is the simplest formula among all that require three arithmetic operations. In principle, there are other iterative procedure for computing the square root that require three arithmetic operations on each iteration, e.g.,

$x_{n+1} = \dfrac{a + x_n}{1 + x_n}$. When this procedure converges $x_n \to x$, we get $x = \dfrac{a + x}{1 + x}$ hence $x \cdot (1 + x) = x + x^2 = a + x$ and $x^2 = a$.

Let us show that among all such operations, the Babylonian procedure is the fastest possible.

Indeed, in the above discussion, to estimate how fast each computation is, we simply counted the number of arithmetic operations. This counting makes sense as a good approximation to the actual computation time, but it implicitly assumes that all arithmetic operations require the exact same computation time. In reality, in the computer, different arithmetic operations actually require different computation times. Specifically, in the computer:

- a shift (i.e., multiplication by a power of 2) is the are the simplest (hence fastest) operation;
- addition and subtraction are next simplest (hence next fastest);
- multiplication, in effect, consists of several additions – of the results of multiplying the first number by different digits of the second one; thus, multiplication takes longer than addition or subtraction;
- finally, division, in effect, consists of several multiplications – and thus requires an even longer time than multiplication.

Each iteration of the Babylonian algorithm consists of one shift, one addition, and one division. Since division is the most time-consuming of the arithmetic operations, to prove that the Babylonian algorithm is the fastest, we must first prove that no algorithm without division is possible. Indeed, if we only use addition, subtraction, or multiplication, then the resulting expression is a polynomial. Once we have a polynomial $f(x_n, a)$, the requirement $f(\sqrt{a}, a) = \sqrt{a}$ can only be satisfied for a linear function $f(x_n, a) = x_n$ for which there is no convergence.

Thus, each expression must have at least one division, i.e., at least as many as the Babylonian expression. It can still be faster than the Babylonian formula if we have exactly 1 division and no additions or subtractions, just a shift. By enumerating all possibilities, one can conclude that such an expression is impossible.

Thus, every expression must have at least one division, and at least one addition or subtraction. So, it is not possible to have an expression which is faster than the Babylonian one, but it may be potentially possible to have an expression which is exactly as fast as the Babylonian one, i.e., that consists of:

- one division,
- one addition, and
- one shift.

Again, one can enumerate all possible combinations of these three operations and see that the Babylonian expression is the only one that leads to convergence to the square root.

References

1. O. Kosheleva, Babylonian method of computing the square root: justifications based on fuzzy techniques and on computational complexity, in *Proceedings of the 28th North American Fuzzy Information Processing Society Annual Conference NAFIPS'09*, Cincinnati, Ohio, 14–17 June 2009
2. D. Flannery, *The Square Root of Two* (Springer, Berlin, 2005)
3. T. Heath, *A History of Greek Mathematics* (Clarendon Press, Oxford, 1921)
4. R.E. Bryant, D.R. O'Hallaron, *Computer Systems: A Programmer's Perspective* (Prentice Hall, Upper Saddle River, 2003)
5. G.J. Klir, B. Yuan, *Fuzzy Sets and Fuzzy Logic: Theory and Applications* (Prentice-Hall, Upper Saddle River, 1995)
6. H.T. Nguyen, E.A. Walker, *A First Course in Fuzzy Logic* (CRC Press, Boca Raton, 2006)
7. H.T. Nguyen, V. Kreinovich, *Applications of Continuous Mathematics to Computer Science* (Kluwer, Dordrecht, 1997)

Chapter 9
How to Enhance Student Motivations by Borrowing from Ancient Tradition: Russian Peasant Multiplication Algorithm

In the previous chapter, we showed how to use ideas from ancient Egyptian, Mayan and Babylonian mathematics when teaching math. In this chapter, we consider the use of a more recent computational tradition, namely, of a Russian peasant multiplication algorithm.

It turns out that this algorithm is closely related to the fact that when people make crude estimates, they usually feel reasonably comfortable choosing between alternatives which differ by a half order of magnitude (HOM). This fact was studied by J. Hobbs who provided an explanation for this level of granularity based on the need for the resulting crude estimates to represent both the original data and the result of processing this data. According to this explanation, HOM are optimal – when we limit ourselves to these first crude estimates. In many practical situations, we do not stop with the original estimate, we refine it one or more times by using granules of smaller and smaller size. In this chapter, we show that the need to optimally process such refined estimates leads to the same HOM granularity. Thus, we provide a new explanation for this level of granularity.

The results from this chapter first appeared in [1].

Half-orders of magnitude: empirical fact. People often need to make crude estimates of a quantity, e.g., estimating the size of a crowd or someone's salary. In [2–4], it was observed that when people make these crude estimates, they usually feel reasonably comfortable choosing between alternatives which differ by a half order of magnitude (HOM).

For example, a person can reasonably estimate whether the size of a crowd was closer to 100, or to 300, or to 1000. If we ask for an estimate on a more refined scale, e.g., 300 or 350, people will generally be unable to directly come up with such estimates. On the other hand, if we ask for an estimate on a coarser scale, e.g., 100 or 1000, people may be able to answer, but they will feel their answer is uninformative.

An interesting example of HOM is presented by coinage and currency. Most countries have, in addition to denominations for the powers of ten, one or two coins

© Springer-Verlag GmbH Germany 2018
O. Kosheleva and K. Villaverde, *How Interval and Fuzzy Techniques Can Improve Teaching*, Studies in Computational Intelligence 750, https://doi.org/10.1007/978-3-662-55993-2_9

or bills between every two powers of ten. Thus, in the United States, in addition to coins or bills for $.01, $.10, $1.00, $10.00, and $100.00, there are also coins or bills in common use for $.05, $.25, $5.00, $20,00, and $50.00. These latter provide rough HOM measures for monetary amounts.

Half-orders of magnitude: the existing explanation. In [3, 4], an explanation is given for this level of granularity based on the need for the resulting crude estimates to represent both the original data and the result of processing this data. According to this explanation, HOM are optimal – when we limit ourselves to these first crude estimates.

Towards a new explanation. In many practical situations, we do not stop with the original estimate, we refine it one or more times by using granules of smaller and smaller size. In this chapter, we show that the need to optimally process such refined estimates leads to the same HOM granularity. Thus, we provide a new explanation for this level of granularity.

Estimating versus data processing: main difference. Estimation is a one-time process which provides a crude estimate for the quantity of interest. In many practical situations, this estimate is quite sufficient for decision making.

In other situations, however, the original crude estimate is not sufficient, and we must refine it. Let us describe this refinement in precise terms.

Refined estimates: a description. What does it mean to have a value m as a granularity level? Crudely speaking, this means that we consider granules of the sizes 1, $m, m^2, \ldots, m^k, \ldots$

A rough estimate means that we simply compare the actual value v with the sizes of these granules. The largest granule m^k for which $m^k \leq v$ is then used as a rough estimate of the quantity v: $m^k \leq v < m^{k+1}$. This rough-estimate granule means that we can estimate v from below by using granules of size m^k, but not by using larger granules.

Once we know that the granules of size m^k *can* be used to estimate v, a natural next question is *how many* granules of this size we can fit within v. Of course, we can only have $c_k < m$ granules. (Otherwise, we would be able to fit $m \cdot m^k$ values in v, and we would have $v \geq m^{k+1}$, i.e., we would conclude that the next granule also fits within v – contrary to our choice of m^k as the largest granule that fits within v.) So, in this next approximation, we are looking for the value $c_k < m$ for which $c_k \cdot m^k \leq v < (c_k + 1) \cdot m^k$. The resulting value $c_k \cdot m^k$ – i.e., the size k plus the value c_k – provides a more accurate description of v than simply the size k of the largest granule.

The difference between the actual value v and the estimate $c_k \cdot m^k$ cannot be fitted with granules of size m^k. Thus, to get an even more accurate description of v, we must use granules of next smaller size m^{k-1} to cover this difference. In other words, we must find the largest value c_{k-1} for which $c_{k-1} \cdot m^{k-1}$ is contained in the difference $v - c_k \cdot m^k$, i.e., for which $c_{k-1} \cdot m^{k-1} \leq v - c_k \cdot m^k < (c_{k-1} + 1) \cdot m^{k-1}$. This is equivalent to selecting c_{k-1} for which

$$c_k \cdot m^k + c_{k-1} \cdot m^{k-1} \le v < c_k \cdot m^k + (c_{k-1} + 1) \cdot m^{k-1}.$$

A further refinement of this estimate means that we use granules of even smaller size m^{k-2} to estimate the difference between the actual value v and the estimate-so-far $c_k \cdot m^k + c_{k-1} \cdot m^{k-1}$, etc. One can see that this refined estimation process leads to an m-ary representation of integers:

$$v = c_k \cdot m^k + c_{k-1} \cdot m^{k-1} + \ldots + c_1 \cdot m^1 + c_0.$$

Example. For example, to represent the number $v = 256$ with decimal granules 1, $m = 10$, 100, 1000, etc., we first find the largest granule which fits within 256: the granule 100. This granule is our first (order-of-magnitude) representation of the number 256.

To get a better representation, we can describe how many times this granule fits within 256, i.e., approximate 256 as $2 \cdot 100$.

To get an even more accurate representation, we need to use granules of next smaller size 10 to represent the difference $256 - 200 = 56$ between the original number 256 and its approximate value 200. We can fit this granule 5 times, so we get an approximation $5 \cdot 10$ for the difference and correspondingly, the approximation $2 \cdot 100 + 5 \cdot 10 = 250$ for the original number 256. With this approximation, we still have an un-approximated difference $256 - 250 = 6$.

To get a more accurate approximation, we use the granules of smaller size 1. Within 6, this granule fits 6 times, so we get a representation $2 \cdot 100 + 5 \cdot 10 + 6 \cdot 1$ for the original number.

Conclusion: selecting granularity level means, in effect, selecting a base for number representation. The above general description and example both show that the use of a certain granule size m means, in effect, that we use m-ary system to represent numbers.

Which value m is the best for m-ary number representation? In view of the above observation, the question of which granule size is the best can be reformulated as follows: for which m, the m-ary representation is the best?

Aren't binary numbers the best? They are used in computers. Normally, people use decimal numbers, with $m = 10$, and computers use binary numbers, with $m = 2$. It may seem that the fact that well-designed and well-optimized computational devices such as computers use binary numbers is an indication that (at least empirically) $m = 2$ is the best choice.

However, this is not necessarily true. The computer engineering choice of $m = 2$ is largely motivated by specific electronic hardware technologies, in which it is easier to manufacture an electronic switch with 2 possible states than with 3 or 10. Our objective is to explain *human* behavior, and for human data processing, these hardware considerations do not apply.

Binary numbers have been used in human data processing as well: Russian peasant multiplication algorithm. Binary numbers for electronic computers are a recent (20 century) phenomenon. However, it is worth mentioning that binary numbers were, in effect, used in data processing for several millennia. According to [5], binary-related algorithm for multiplication was used by ancient Egyptian mathematicians as early as 1800 B.C.E. This method is called *Russian peasant multiplication algorithm* because it was first observed in the 19 century by the Western visitors to Russia – where this method was widely used by the common folks (i.e., mainly peasants) [5, 6]. Later, a similar method was found (and decoded) in an ancient Egyptian papyrus.

This algorithm is especially useful if we want to multiply different numbers x by a given number n. This happens, e.g., if a merchant wants to compute the price of different amount of the items that he is selling: in this example, n is the price of a single item, and x is the number of such items.

In this procedure, we first transform the fixed number n into the binary code, i.e., represent n as a sum of powers of two. Interestingly, the transition to binary code was computed in the ancient Egypt in exactly the same way as it is done now: by sequentially dividing a number by 2 and then reading the remainders from bottom up.

Once such a binary representation is found, we can compute the product $n \cdot x$ as follows:

- first, we add x to itself, resulting in $2x$;
- then, we add $2x$ to itself, resulting in $4x = 2^2 \cdot x$;
- after that, we add $2^2 \cdot x$ to itself, then getting $8x = 2^3 \cdot x$, etc.
- once we have the values $2^i \cdot x$, we add those values which correspond to the representation of n as the sum of powers of 2, thus getting $n \cdot x$.

Example. For example, $n = 13$ is represented in binary code as

$$1101_2 = 2^3 + 2^2 + 2^0 = 8 + 4 + 1.$$

For $n = 13$, the conversion to binary is performed as follows:

```
13 / 2 = 6 rem 1
 6 / 2 = 3 rem 0
 3 / 2 = 1 rem 1
 1 / 2 = 0 rem 1
```

Reading remainders from bottom up, we get the binary representation 1101_2.

Now, to compute $13x$, we consequently compute $2x$, $4x$, $8x$, and then add

$$x + 4x + 8x.$$

This method is often faster than using decimal numbers. To compute $13x$, we need 3 additions (namely, doubling) to compute all three powers of two, and then 2 more additions to compute $x + 4x$ and then $13x$ as $(x + 4x) + 8x$. Overall, we need 5 additions.

This number is much smaller than what we would have needed is we decided to reduce multiplication to addition in the standard decimal representation, in which we would need to compute $x, 2x, 3x, ..., 10x$, and then add $3x + 10x$, to the overall of 11 additions.

A similar method is used in cryptosystems. The efficiency of binary-based multiplication prompted the use of a similar technique in cryptosystems. In particular, in the most widely used RSA techniques (see, e.g., [7]), techniques which are used every time we access a secure webpage or make financial transactions online. Cryptosystems make computer communications secure by encoding messages, largely by raising a number x (representing a message) to a given power n (to be more precise, they compute the power x^n modulo some large number N). The efficiency of RSA and similar cryptosystems is based on the fact that it is computationally efficient to compute x^n but (unless we know factors of N) it is very computationally difficult to recover x from the transmitted message $M \stackrel{\text{def}}{=} x^n$. This exponentiation is time-consuming, it forms the dominant part of cryptoalgorithms running time; see, e.g., [8]. So, to make cryptosystems more efficient, it is important to compute x^n fast.

At present, exponentiation is mainly done by using the binary representation of n. Namely, we use multiplication to compute $x^2 = x \cdot x$, $x^4 = x^2 \cdot x^2$, $x^8 = x^4 \cdot x^4$, ..., and then we multiply the powers corresponding to the powers of 2 that are present in the binary expansion of n.

For example, to compute x^{13}, we compute x^2, x^4, x^8, and then multiply $x \cdot x^4 \cdot x^8$. Overall, just like we need 5 additions to multiply a given number by 13, we need 5 multiplications to raise a given number x to the 13-th power.

Binary-based methods are widely used but they are not always optimal. In practice, binary techniques are so much faster than decimal-based ones that it was originally conjectured that they are optimal for all n. Specifically, it was conjectured that if we want to compute a product $n \cdot x$ by using only additions (or, equivalently, compute the power x^n by using only multiplications), then the above binary-based procedure is optimal.

This turned out to be only true for $n \leq 14$. For $n = 15$, the binary procedure requires that we compute $2x, 4x, 8x$, and then compute $x + 2x + 4x + 8x$, to the total of 6 additions. However, we can compute $15x$ in only 5 additions: $2x = x + x$, $3x = x + 2x$, $6x = 3x + 3x$, $9x = 6x + 3x$, and $15x = 6x + 9x$; see, e.g., [5].

Fastest known methods: methods based on m-ary number representations. At present, the fastest known algorithms for multiplication via addition (or, equivalently, for fast multiplication) are based on the use of m-ary number representations for an appropriate m (not necessarily $m = 2$) [5, 8]. Specifically, once we have an m-ary representation

$$n = c_k \cdot m^k + c_{k-1} \cdot m^{k-1} + ... + c_1 \cdot m^1 + c_0,$$

we can compute $n \cdot x$ as follows:

```
Compute 2x = x + x, 3x = 2x + x, ..., (m − 1) · x = ((m − 2) · x) + x.
a ← 0
for i = k to 0 by −1
    a ← m · a
    a ← a + (c_i · x)
return a.
```

Let us briefly explain this algorithm. At first, we take $a = 0$ and $i = k$. For this value i, we first get $a \leftarrow m \cdot 0 = 0$ and then $a \leftarrow 0 + c_k \cdot x$, so after this iteration, we get $a = c_k \cdot x$.

On the next iteration, we take $i = k - 1$. On this iteration, we first multiply the current value of a by m, resulting in $a = c_k \cdot m \cdot x$, and then add $c_{k-1} \cdot x$. So, after this iteration, we get $a = (c_k \cdot m + c_{k-1}) \cdot x$.

Similarly, after the next iteration corresponding to $i = k - 2$, we get $a = (c_k \cdot m^2 + c_{k-1} \cdot m + c_{k-2}) \cdot x$, ..., and after the last iteration corresponding to $i = 0$, we get the desired value $a = (c_k \cdot m^k + c_{k-1} \cdot m^{k-1} + \ldots + c_0) \cdot x = n \cdot x$.

Similarly, we can compute x^n as follows:

```
Compute x² = x · x, x³ = x² · x, ..., x^{m−1} = x^{m−2} · x.
a ← 1
for i = k to 0 by −1
    a ← a^m
    a ← a · x^{c_i}
return a.
```

Let us briefly explain this algorithm. At first, we take $a = 1$ and $i = k$. For this value i, we first get $a \leftarrow 1^m = 1$ and then $a \leftarrow 1 \cdot x^{c_k}$, so after this iteration, we get $a = x^{c_k}$.

On the next iteration, we take $i = k - 1$. On this iteration, we first raise the current value of a to the m-th power, resulting in $a = (x^{c_k})^m = x^{c_k \cdot m}$, and then multiply by $x^{c_{k-1}}$. So, after this iteration, we get $a = x^{c_k \cdot m + c_{k-1}}$.

Similarly, after the next iteration corresponding to $i = k - 2$, we get $a = x^{c_k \cdot m^2 + c_{k-1} \cdot m + c_{k-2}}$, ..., and after the last iteration corresponding to $i - 0$, we get the desired value

$$a = x^{c_k \cdot m^k + c_{k-1} \cdot m^{k-1} + \ldots + c_0} = x^n.$$

These methods are mainly used when $m = 2^p$, because then computing $m \cdot a$ requires only p additions (doublings) and, correspondingly, computing a^m requires only p multiplications (squarings).

Computational complexity (running time) of m-ary methods with $m = 2^p$. For $m = 2$, the above method requires $\lfloor \log_2(n) \rfloor$ doublings and $\leq \lfloor \log_2(n) \rfloor$ additions. So, in the worst case, we need $2\lfloor \log_2(n) \rfloor$ additions.

In practice, if $c_i = 0$, then we do not need to add the corresponding value $2^i \cdot x$. On average, for each digit c_i, all m possible values $0, 1, \ldots, m - 1$ are equally probable. In particular, with the probability $1/m$, we get $c_i = 0$, in which case we do not need to add the corresponding term. For $m = 2$, this probability is 1/2, so on average, we need $\lfloor \log_2(n) \rfloor$ doublings and $(1/2) \cdot \lfloor \log_2(n) \rfloor$ additions, to the overall of $(3/2) \cdot \lfloor \log_2(n) \rfloor$ additions.

For $m = 2^p$, we need $2^p - 2$ additions to compute $2x, 3x, \ldots, (m-1) \cdot x$, $\lfloor \log_2(n) \rfloor$ doublings (to compute a_m), and at most $\lfloor \log_2(n) \rfloor / p$ additions of $c_i \cdot x$. The overall worst-case complexity is thus $2^p - 2 + (1 + 1/p) \cdot \lfloor \log_2(n) \rfloor$ additions.

In the average case, we only need the addition of $c_i \cdot x$ when $c \neq 0$, i.e., with probability $1 - 1/m = 1 - 1/2^p$. Thus, the average-case complexity is equal to

$$2^p - 2 + \left(1 + \frac{1}{p} \cdot \left(1 - \frac{1}{2^p}\right)\right) \cdot \lfloor \log_2(n) \rfloor$$

additions [8].

It is known that we get the asymptotically fastest computations for

$$p = \log_2(\log_2(n)) - 2 \log_2(\log_2(\log_2(n))).$$

When are methods with $m = 2$, $m = 4$, and $m = 8$ actually better? Analysis based on worst-case complexity. In some practical situations, it is important to guarantee that the computation finishes on time. In this case, it is desirable to minimize the worst-case complexity, because this is the complexity which provides the desired guarantee. Let us therefore compare the worst-case complexity t_p corresponding to different values $m = 2^p$.

For $p = 1$, we get $t_1 = 2\lfloor \log_2(n) \rfloor$. For $p = 2$, we get $t_2 = 2 + 1\frac{1}{2} \cdot \lfloor \log_2(n) \rfloor$. For $p = 3$, we get $t_3 = 6 + 1\frac{1}{3} \cdot \lfloor \log_2(n) \rfloor$.

The value $m = 2$ corresponding to $p = 1$ is optimal when $t_1 \leq t_2$, i.e., when $2\lfloor \log_2(n) \rfloor \leq 2 + 1\frac{1}{2} \cdot \lfloor \log_2(n) \rfloor$. This is equivalent to $\frac{1}{2} \cdot \lfloor \log_2(n) \rfloor \leq 2$, i.e., to $\lfloor \log_2(n) \rfloor \leq 4$ and $n < 2^5 = 32$.

The value $m = 4$ corresponding to $p = 2$ is optimal when $t_1 > t_2$ (i.e., when $n \geq 32$) and $t_2 \leq t_3$, i.e., when $2 + 1\frac{1}{2} \cdot \lfloor \log_2(n) \rfloor \leq 6 + 1\frac{1}{3} \cdot \lfloor \log_2(n) \rfloor$. This condition is equivalent to $\frac{1}{6} \cdot \lfloor \log_2(n) \rfloor \leq 4$, i.e., to $\lfloor \log_2(n) \rfloor \leq 24$ and $n < 2^{25} \approx 3 \cdot 10^7$.

Thus, for the values n which do not exceed 30 million (i.e., in practice, in all practical cases when we need estimates), the granularity values of $m = 2$ and $m = 4$ are optimal – and $m = 2$ is only optimal for small values n, when we do not really need any estimation. Crudely speaking, we can say that the worst-case complexity corresponds to $m = 4$.

When are methods with $m = 2$, $m = 4$, and $m = 8$ actually better? Analysis based on average-case complexity. In some practical situations, we need to perform several computations, with several different values x; in some such situations, the individual computation time is not crucial, what is important is that the overall computation time be as small as possible. In such situations, it makes sense to consider the average time complexity $\bar{t}_p$ as an optimality criterion.

For $p = 1$, we get $\bar{t}_1 = \dfrac{3}{2} \cdot \lfloor \log_2(n) \rfloor$. For $p = 2$, we get

$$\frac{1}{2} \cdot \left(1 - \frac{1}{4}\right) = \frac{1}{2} \cdot \frac{3}{4} = \frac{3}{8},$$

so $\bar{t}_2 = 2 + 1\dfrac{3}{8} \cdot \lfloor \log_2(n) \rfloor$. For $p = 3$, we get

$$\frac{1}{3} \cdot \left(1 - \frac{1}{8}\right) = \frac{1}{3} \cdot \frac{7}{8} = \frac{7}{24},$$

so $\bar{t}_3 = 6 + 1\dfrac{7}{24} \cdot \lfloor \log_2(n) \rfloor$.

In this case, the granularity value $m = 2$ corresponding to $p = 1$ is optimal when $\bar{t}_1 \leq \bar{t}_2$, i.e., when $\dfrac{3}{2} \cdot \lfloor \log_2(n) \rfloor \leq 2 + 1\dfrac{3}{8} \cdot \lfloor \log_2(n) \rfloor$. This condition is equivalent to $\left(\dfrac{1}{2} - \dfrac{3}{7}\right) \cdot \lfloor \log_2(n) \rfloor \leq 2$, i.e., to $\dfrac{1}{14} \cdot \lfloor \log_2(n) \rfloor \leq 2$, $\lfloor \log_2(n) \rfloor \leq 28$, and $n < 2^{29} \approx 5 \cdot 10^8$. Thus, for all practical values, the granularity value $m = 2$ is optimal.

Conclusion. J. Hobbs has observed that for human experts, it is natural to express their rough estimates in terms of half-orders of magnitude (HOM), when there are approximately two possible estimates within each order of magnitude (i.e., within each factor of 10). For example, when estimating a size of a crowd, a human naturally distinguishes between "low hundreds", "high hundreds", "low thousands", "high thousands", etc. How can we explain this granule size?

In this chapter, we use the ideas behind the Russian peasant multiplication algorithm to show that for values appropriate for human estimation, from the viewpoint of data processing under refined granularity, the optimal granule size is either $m = 4$ (for the more typical case of individual problems), or $m = 2$ (for mass problems). In both cases, we have a granule size which is similar to half-order of magnitude. So, we get a new theoretical explanation for the HOM phenomenon observed by J. Hobbs.

References

1. J.I. Vargas, O. Kosheleva, Russian peasant multiplication algorithm, RSA cryptosystem, and a new explanation of half-orders of magnitude. J. Uncertain Syst. **1**(3), 178–184 (2007)
2. J.R. Hobbs, Half orders of magnitude, In: L. Obrst, I. Mani (eds.), in *Proceeding of the Workshop on Semantic Approximation, Granularity, and Vagueness, A Workshop of the Seventh Interna-*

tional Conference on Principles of Knowledge Representation and Reasoning KR'2000, Breck-enridge, Colorado, 11 April 2000, pp. 28–38

3. J.R. Hobbs, V. Kreinovich, Optimal choice of granularity in commonsense estimation: why half–orders of magnitude, in *Proceedings of the Joint 9th World Congress of the International Fuzzy Systems Association and 20th International Conference of the North American Fuzzy Information Processing Society IFSA/NAFIPS 2001*, Vancouver, Canada, 25–28 July 2001, pp. 1343–1348

4. J. Hobbs, V. Kreinovich, Optimal choice of granularity in commonsense estimation: why half-orders of magnitude. Int. J. Intell. Syst. **21**(8), 843–855 (2006)

5. D.E. Knuth, *The Art of Computer Programming*, (Seminumerical Algorithms (Addison Wesley, Reading, 1969)

6. L.N.H. Bunt, P.S. Jones, J.D. Bedient, *The Historical Roots of Elementary Mathematics* (Dover, New York, 1988)

7. ThH Cormen, C.E. Leiserson, R.L. Rivest, C. Stein, *Introduction to Algorithms* (MIT Press, Cambridge, 2001)

8. D.M. Gordon, A survey of fast exponentiation methods. J. Algorithms **27**, 129–146 (1998)

Chapter 10
How to Enhance Student Motivations by Borrowing from Modern Practices: Geometric Approach to Error-Less Counting

In the previous chapters, we showed how to use ancient traditions when teaching math. In this chapter (and in the following chapter), we show how ideas borrowed from modern practices can also help.

One of such easy-to-explain modern practices is that to decrease counting errors, it is often reasonable to arrange the counted objects into rectangles and/or parallelepipeds. In this chapter, we describe how to design optimal arrangements of this type.

The results from this chapter first appeared in [1].

A geometric approach to counting: a description. Once, we went to buy 24 cups of easy-to-prepare "instant lunch" soup. When we counted these cups ourselves, we had to count several times to make sure that we did not make a mistake. A salesperson counted them very easily: she grouped them into a nice parallelepiped of length 4, width 3, and height 2, and then multiplied these three numbers to get $4 \times 3 \times 2 = 24$.

Why it is interesting. Of course, everyone knows that the volume of a parallelepiped is the product of the sides, and that, similarly, the area of a rectangle is the product of its sides. The corresponding geometric "area" approach is one of the main methods of teaching multiplication of integers (and fractions); see, e.g., Chaps. 10 and 13 from [2–6], and references therein. What was interesting is that the salesperson used the same geometric arrangement not for multiplication, but for error-less counting.

What we do in this chapter. We explain why this geometric approach reduced the errors, and what is the best way to use this approach if we want to decrease counting errors.

Main idea. Every time we perform an arithmetic operation, there is a probability that we make a mistake.

- When we count a new object – i.e., when we add one to the previous total – we can make a mistake.

© Springer-Verlag GmbH Germany 2018

O. Kosheleva and K. Villaverde, *How Interval and Fuzzy Techniques Can Improve Teaching*, Studies in Computational Intelligence 750, https://doi.org/10.1007/978-3-662-55993-2_10

- When we multiply two numbers, we can make a mistake.
- When we add two numbers, we can make a mistake.

Thus, to minimize the probability of a mistake, we should minimize the number of arithmetic operations.

This idea explains why the geometric approach leads to fewer errors. If we simply count 24 objects one after another, we start with the first one and then perform $24 - 1 = 23$ arithmetic operations – operations of adding 1. Thus, we have 23 chances of making a mistake.

On the other hand, once we have a $4 \times 3 \times 2$ arrangement, we need to count all the sides and then multiply the results. Counting 4 elements in the length requires $4 - 1 = 3$ computational steps, counting 3 elements in the width requires $3 - 1 = 2$ steps, counting 2 elements in the height requires $2 - 1 = 1$ steps, and we also need two multiplications. Thus, totally, we need $3 + 2 + 1 + 2 = 8$ arithmetic operations. Since $8 \ll 23$, we have much fewer chances to make a mistake.

General case: towards a precise formulation of the problem. With $n = 24$ objects, we were lucky, since 24 can be easily represented as a product of three numbers. With a prime number like $n = 23$, this would not have been possible – but we can still arrange $n = 23$ into several parallelepipeds and thus, decrease the total number of computational steps.

When the number of objects is small, instead of (or in addition to) parallelepipeds, it may be useful to combine them into rectangles.

If we know that we need to count n objects, what is the best arrangement that minimizes the number of computational steps?

For each parallelepiped of sizes $a_i \times b_i \times c_i$, counting the volume of each parallelepiped means that

- we first count the length a_i; this requires $a_i - 1$ counting steps;
- then, we count the width b_i; this requires $b_i - 1$ counting steps;
- after that, we count the height c_i; this requires $c_i - 1$ counting steps;
- then, we need two more multiplication steps: to multiply a_i by b_i and to multiply the result by c_i.

Thus, overall, we need $(a_i - 1) + (b_i - 1) + (c_i - 1) + 2 = a_i + b_i + c_i - 1$ steps.

Similarly, for each rectangle of sizes $a_i \times b_i$, counting the area of this rectangle means that

- we first count the length a_i; this requires $a_i - 1$ counting steps;
- then, we count the width b_i; this requires $b_i - 1$ counting steps;
- then, we need one multiplication step: to multiply a_i by b_i.

Thus, overall, we need $(a_i - 1) + (b_i - 1) + 1 = a_i + b_i - 1$ steps.

Once we have s such sets (parallelepipeds and rectangles), and we have computed the number of elements n_i in each of them, we need $s - 1$ additions to compute the sum $n_1 + n_2 + \ldots + n_s$. Thus, we arrive at the following definitions.

Definition 10.1 By a *parallelepiped*, we mean a triple $P = (a, b, c)$ of positive integers. For each parallelepiped $P = (a, b, c)$,

- the product $a \cdot b \cdot c$ is called its *volume* and denoted by $N(P)$;
- the expression $a + b + c - 1$ is called its *number of computational steps* and denoted by $C(P)$.

Definition 10.2 By a *rectangle*, we mean a pair $R = (a, b)$ of positive integers. For each rectangle $R = (a, b)$,

- the product $a \cdot b$ is called its *area* and denoted by $N(R)$;
- the expression $a + b - 1$ is called its *number of computational steps* and denoted by $C(R)$.

Definition 10.3 By an arrangement $S = (S_1, \ldots, S_s)$, we mean a finite list consisting of parallelepipeds and rectangles. For each arrangement $S = (S_1, \ldots, S_s)$,

- the sum $\sum_{i=1}^{s} N(S_i)$ is called its *number of elements* and denoted by $N(S)$;
- the value $\sum_{i=1}^{s} C(S_i) + s - 1$ is called its *number of computational steps* and denoted by $C(S)$.

Definition 10.4 For a given positive integer n, we say that an arrangement $S = (S_1, \ldots, S_s)$ is *optimal* if it has n elements $N(S) = n$, and among all the arrangements with n elements, it has the smallest number of computational steps $C(S)$. The number of computational steps in the optimal arrangement will be denoted by $C(n)$.

Formulation of the computational problem. Given an integer n, find the optimal arrangement and the corresponding value $C(n)$.

In principle, one can use exhaustive search, but this is not practical. In principle, for each n, we can try all possible arrangements – there are only n elements to arrange, so there is a finite number of possible arrangements with n elements. However, for large n, there are too many possible arrangements to be practically possible to enumerate them all. Thus, we need a more efficient algorithm. Such an algorithm will be presented in this chapter.

Idea. Before we target the problem of computing $C(n)$ for general arrangements, let us first solve the simpler problems in which arrangements are limited. We will then use the solutions to these simpler problems to solve the more general problem.

Simple counting. For a simple counting, we need $C_0(n) = n - 1$ computational steps.

First simplified problem: a single rectangle. Let us first consider the simplest possible case, when we are only allowing one set, and this set has to be a rectangle. For each n, all such possible rectangles correspond to values a and b for which

$a \cdot b = n$, i.e., to values a that divide n: $a|n$; the value b is equal to n/a and is, thus, uniquely determined by the value a. The total number of computational steps in this arrangement is equal to $a + (n/a) - 1$. Thus, for each n, the smallest possible number $C_1(n)$ of computational steps in such an arrangement is equal to

$$C_1(n) = \min_{a:a|n} \left(a + \frac{n}{a} - 1 \right). \tag{10.1}$$

By enumerating all possible divisors of each integer n, we can therefore compute $C_1(n)$. For example, the integer $n = 10$ has 4 divisors $a = 1$, $a = 2$, $a = 5$, and $a = 10$, so we have

$$C_1(10) = \min(1 + 10 - 1, 2 + 5 - 1, 5 + 2 - 1, 10 + 1 - 1) = \min(10, 6, 6, 10) = 6,$$

with the optimal rectangle 2×5.

The values $C_1(n)$ corresponding to the first 30 integers are as follows:

n	1	2	3	4	5	6	7	8	9	10
$C_1(n)$	1	2	3	3	5	4	7	5	5	6

n	11	12	13	14	15	16	17	18	19	20
$C_1(n)$	11	6	13	8	7	7	17	8	19	8

n	21	22	23	24	25	26	27	28	29	30
$C_1(n)$	9	12	23	9	9	14	11	10	29	10

The corresponding optimal rectangles are:

1	2	3	4	5
1×1	1×2	1×3	2×2	1×5

6	7	8	9	10
2×3	1×7	2×4	3×3	2×5

11	12	13	14	15
1×11	3×4	1×13	2×7	3×5

16	17	18	19	20
4×4	1×17	3×6	1×19	4×5

21	22	23	24	25
3×7	2×11	1×23	4×6	5×5

26	27	28	29	30
2×13	3×9	4×7	1×29	5×6

Comment. One can easily check, by taking the derivative and checking its sign, that the function $a + (n/a) - 1$ decreases for $a \le \sqrt{n}$ and increases for $a \ge \sqrt{n}$. Thus, to find its smallest possible value, it is not necessary to check all divisors a of n: it is sufficient to only check the divisors which are the closest to $\sqrt{n}$. The divisor which is the closest to $\sqrt{n}$ from the right is equal to n divided by the closest divisor from the left, so they lead to the exact same rectangle.

Thus, it is sufficient to take, as a, the largest divisor which is still smaller than or equal to $\sqrt{n}$. For example, for $n = 1$, we have $\sqrt{10} = 3....$, so out of 4 possible divisors $a = 1$, $a = 2$, $a = 5$, and $a = 10$, we should take $a = 2$.

Second simplified problem: a single parallelepiped. Let us now assume that we allow a single set, and this set should be a parallelepiped. For each n, all such possible parallelepipeds correspond to values a, b, and c for which $(a \cdot b) \cdot c = n$. The corresponding values c have to divide n. The product $a \cdot b$ is then equal to n/c and is, thus, uniquely determined by the value c. The total number of computational steps in this arrangement is equal to $c + (a + b - 1)$.

Once c is fixed, and thus, the product $n' = a \cdot b$ is fixed (as $n' = n/c$), the smallest possible number of computational steps is attained when the value $a + b - 1$ is the smallest among all pairs (a, b) for which $a \cdot b = n'$.

We already know, from the previous problem, that this smallest number of computational steps is equal to $C_1(n') = C_1(n/c)$. Thus, for each n, the smallest possible number $C_2(n)$ of computational steps in such an arrangement is equal to

$$C_2(n) = \min_{c:c|n} \left(c + C_1 \left(\frac{n}{c} \right) \right). \tag{10.2}$$

By enumerating all possible divisors c of each integer n, we can therefore compute $C_2(n)$. For example, the integer $n = 24$ has 8 divisors $c = 1$, $c = 2$, $c = 3$, $c = 4$, $c = 6$, $c = 8$, $c = 12$, and $c = 24$, so we have

$$C_2(24) = \min(1 + C_1(24), 2 + C_1(12), 3 + C_1(8), 4 + C_1(6), 6 + C_1(4),$$

$$8 + C_1(3), 12 + C_1(2), 24 + C_1(1)) =$$

$$\min(1 + 9, 2 + 6, 3 + 5, 4 + 4, 6 + 3, 8 + 3, 12 + 2, 24 + 1) =$$

$$\min(10, 8, 8, 8, 9, 11, 14, 25) = 8,$$

exactly as we observed earlier.

Here, the optimal value is attained for $c = 2$, so the optimal parallelepiped has $c = 2$ and $n' = n/c = 12$. For $n' = 12$, the optimal rectangle is 3×4, so for $n = 24$, the optimal parallelepiped is $2 \times 3 \times 4$.

The values $C_2(n)$ corresponding to the first 30 integers are as follows:

n	1	2	3	4	5	6	7	8	9	10
$C_2(n)$	2	3	4	4	6	5	8	5	6	7

n	11	12	13	14	15	16	17	18	19	20
$C_2(n)$	12	6	14	9	8	7	18	7	20	8

n	21	22	23	24	25	26	27	28	29	30
$C_2(n)$	10	13	24	8	10	15	8	10	30	9

The corresponding optimal parallelepipeds are:

1	2	3	4	5
$1 \times 1 \times 1$	$1 \times 1 \times 2$	$1 \times 1 \times 3$	$1 \times 2 \times 2$	$1 \times 1 \times 5$

6	7	8	9	10
$1 \times 2 \times 3$	$1 \times 1 \times 7$	$2 \times 2 \times 2$	$1 \times 3 \times 3$	$1 \times 2 \times 5$

11	12	13	14	15
$1 \times 1 \times 11$	$2 \times 2 \times 3$	$1 \times 1 \times 13$	$1 \times 2 \times 7$	$1 \times 3 \times 5$

16	17	18	19	20
$2 \times 2 \times 4$	$1 \times 1 \times 17$	$2 \times 3 \times 3$	$1 \times 1 \times 19$	$2 \times 2 \times 5$

21	22	23	24	25
$1 \times 3 \times 7$	$1 \times 2 \times 11$	$1 \times 1 \times 23$	$2 \times 3 \times 4$	$1 \times 5 \times 5$

26	27	28	29	30
$1 \times 2 \times 13$	$3 \times 3 \times 3$	$2 \times 2 \times 7$	$1 \times 1 \times 29$	$2 \times 3 \times 5$

Case when we allow simple counting, a single rectangle, or a single parallelepiped. In this case, we can select the option with the smallest number of computational steps, i.e., we get

$$C'(n) = \min(C_0(n), C_1(n), C_2(n)). \qquad (10.3)$$

The values $C'(n)$ corresponding to the first 30 integers are as follows:

n	1	2	3	4	5	6	7	8	9	10
$C'(n)$	0	1	2	3	4	4	6	5	5	6

n	11	12	13	14	15	16	17	18	19	20
$C'(n)$	10	6	12	8	7	7	16	7	18	8

n	21	22	23	24	25	26	27	28	29	30
$C'(n)$	9	12	22	8	9	14	8	10	28	9

The corresponding optimal arrangements are (where a single number means a simple counting):

1	2	3	4	5
1	2	3	2×2	5

6	7	8	9	10
2×3	7	$2 \times 2 \times 2$	3×3	$1 \times 2 \times 5$

11	12	13	14	15
11	$2 \times 2 \times 3$	13	2×7	3×5

16	17	18	19	20
$2 \times 2 \times 4$	17	$2 \times 3 \times 3$	19	$2 \times 2 \times 5$

21	22	23	24	25
3×7	2×11	23	$2 \times 3 \times 4$	5×5

26	27	28	29	30
2×13	$3 \times 3 \times 3$	$2 \times 2 \times 7$	29	$2 \times 3 \times 5$

How to solve the original problem. Our objective is to find the number $C(n)$ of computational steps that correspond to the optimal arrangement of n elements. For every positive integer n, the optimal arrangement corresponds:

- either to simple counting or a single rectangle or parallelepiped, in which case we need $C'(n)$ computational steps,
- or to several rectangles or parallelepipeds, in which case we need $C'(n')$ for the first of them, an optimal arrangement for the remaining $n - n'$ elements – leading to $C(n - n')$ steps – plus 1 additional step to add up these numbers.

The optimal arrangement corresponds to the alternative with the smallest number of computational steps:

$$C(n) = \min \left(C'(n), \min_{n':0<n'<n} (C'(n') + C(n - n') + 1) \right). \qquad (10.4)$$

We can use this formula to sequentially compute the values $C(1)$, $C(2)$, ...
   The values $C(n)$ corresponding to the first 30 integers are as follows:

n	1	2	3	4	5	6	7	8	9	10
$C(n)$	0	1	2	3	4	4	5	5	5	6

n	11	12	13	14	15	16	17	18	19	20
$C(n)$	7	6	7	8	7	7	8	7	8	8

n	21	22	23	24	25	26	27	28	29	30
$C(n)$	9	10	11	8	9	10	8	9	10	9

The corresponding optimal arrangements are:

1	2	3	4	5
1	2	3	2×2	5
6	7	8	9	10
2×3	$1 + 2 \times 3$	$2 \times 2 \times 2$	3×3	$1 \times 2 \times 5$
11	12	13	14	15
$1 + 2 \times 5$	$2 \times 2 \times 3$	$1 + 2 \times 2 \times 3$	2×7	3×5
16	17	18	19	20
4×4	$1 + 4 \times 4$	$2 \times 3 \times 3$	$1 + 2 \times 3 \times 3$	$2 \times 2 \times 5$
21	22	23	24	25
3×7	$1 + 3 \times 7$	$2 + 3 \times 7$	$2 \times 3 \times 4$	5×5
26	27	28	29	30
$1 + 5 \times 5$	$3 \times 3 \times 3$	$1 + 3 \times 3 \times 3$	$2 + 3 \times 3 \times 3$	$2 \times 3 \times 5$

Comment. For some integers n, there are several optimal arrangements; in this case, we list one of them. For example, for $n = 21$, in addition to the arrangement 3×7 that we listed in the above table, the same number of computational steps can also be achieved if we use a different arrangement $1 + 4 \times 5$.

Computational complexity of computing $C(n)$. For each n, to compute $C_1(n)$, we need to check all numbers $a \leq \sqrt{n}$: whether they divide n. This requires $\sqrt{n}$ computational steps. Thus, to compute the values $C_1(n)$ for all n from 1 to N, we need $\sum_{n=1}^{N} \sqrt{n} \sim N^{3/2}$ computational steps.

To compute $C_2(n)$ for each n, we also need to check all the numbers $c \leq n -$ whether they divide n. Thus, to compute the values $C_2(n)$ for all n from 1 to N, we need $1 + 2 + \ldots + N \sim N^2$ computational steps.

Computing $C'(n)$ for all $n \leq N$ requires $O(N)$ steps.

Finally, computing each value $C(n)$ according to the formula (10.4) requires n steps, to the total of $1 + 2 + \ldots + N = O(N^2)$ steps.

All these procedures require $O(N^{3/2}) + O(N^2) + O(N) + O(N^2) = O(N^2)$ steps, so we can compute $C(N)$ in time quadratic in N.

Comment. Some objects have such a shape that we cannot place one on top of the others, so we cannot make a parallelepiped out of them. In this case, we have to limit ourselves to rectangles. For such objects, we can repeat the above computations, with the only difference that there is no need to compute $C_2(n)$ and $C'(n)$ is now computed according to a new formula $C'(n) = \min(C_0(n), C_1(n))$.

What if we do not know how many elements we have. In the above text, we assumed that the customer knows how many objects she picked, and the question was how to arrange these objects so as to check that the number of objects is correct.

But what if we do not know how many objects n we have picked? If we placed them into a parallelepiped of sizes a, b, and c, then among all the values a, b, and

c for which $a \cdot b \cdot c = n$, we must select the triple for which $a + b + c - 1$ is the smallest possible. To get an approximate solution to this optimization problem, let us consider a simplified version of this problem, in which the values a, bb, and c can be arbitrary real numbers (not necessarily integers). This simplified problem can be easily solved, and its solution is $a = b = c = \sqrt[3]{n}$, i.e., a cube.

Thus, it is reasonable to stack the objects in such a way that they form a cube. First, we form a cube of linear size 2, then we add elements on all sides to get a cube of linear size 3, etc.

References

1. O. Kosheleva, Geometric approach to error-less counting. Appl. Math. Sci. **4**(63), 3161–3170 (2010)
2. J.A. Van de Walle, *Elementary and Middle School Mathematics: Teaching Developmentally* (Allyn and Bacon, Boston, 2006)
3. M. Burns, *Math by All Means: Multiplication, Grade 3* (Math Solutions Publications, Sausalito, 1995)
4. L.J. Morrow (ed.), *The Teaching and Learning of Algorithms in School Mathematics* (National Council of Teachers of Mathematics, Reston, 1998)
5. J. Pallotta. R. Bolster, *Hershey's Milk Chocolate Multiplication Book* (Cartwheel Publ., 2002)
6. J.–W. Son, A comparison of how textbooks teach multiplication of fractions and division of fractions in Korea and in the U.S., in *Proceedings of the 29th Conference of the International Group for the Psychology of Mathematics Education PME*, ed. by H.L. Chick, J.L. Vincent. PME Publ., Melbourne, 2005, vol. 4, pp. 201–208

Chapter 11
How to Enhance Student Motivations by Borrowing from Modern Practices: Can We Learn Algorithms from People Who Compute Fast

In the previous chapter, we showed how to use *usual* modern practices when teaching math. In this chapter, we focus on *unusual* modern practices, namely, on the ability of some people to perform calculations unusually fast. In the past, mathematicians actively used this ability. With the advent of computers, there is no longer need for human calculators – even fast ones. However, recently, it was discovered that there exist, e.g., multiplication algorithms which are much faster than standard multiplication. Because of this discovery, it is possible than even faster algorithms will be discovered. It is therefore natural to ask: did fast human calculators of the past use faster algorithms – in which case we can learn from their experience – or they simply performed all operations within a standard algorithm much faster? This question is difficult to answer directly, because the fast human calculators' self-description of their algorithm is very fuzzy. In this chapter, we use an indirect analysis to argue that fast human calculators most probably used the standard algorithm.

The results from this chapter first appeared in [1, 2].

People who computed fast: a historical phenomenon. In history, several people have been known for their extraordinary ability to compute fast. Before the 20th century invention of computers, their computational abilities were actively used.

For example, in the 19th century, Johann Martin Zaharias Dase (1824–1861) from Vienna performed computations so much faster than everyone else that professional mathematicians hired him to help with their calculations; see, e.g., [3–6]. Dase:

- computed π (in his head!) by using a formula

$$\pi/4 = \arctan\left(\frac{1}{2}\right) + \arctan\left(\frac{1}{5}\right) + \arctan\left(\frac{1}{8}\right), \qquad (11.1)$$

with a record-breaking accuracy of 200 digits, during the period of under two months in 1844 – while it took several centuries for researchers to compute π with a previously known accuracy of 140 digits;

© Springer-Verlag GmbH Germany 2018
O. Kosheleva and K. Villaverde, *How Interval and Fuzzy Techniques Can Improve Teaching*, Studies in Computational Intelligence 750,
https://doi.org/10.1007/978-3-662-55993-2_11

- calculated the logarithms table with 7 digit accuracy – this table included logarithms of all the numbers from 1 to a million, and
- performed many other computational tasks.

Karl Friedrich Gauss himself, the famous *Princeps mathematicorum* ("The Prince of Mathematicians"), recommended that Dase be paid by the Hamburg Academy of Science to perform calculations – this was historically the first time when a person was paid for performing mathematical calculations.

People who computed fast: how they computed? Calculations have been (and are) important in many practical problems. Because of this practical importance, people have therefore always been trying to speed up computations. One natural way to speed up computations is to learn from the people who can do computations fast.

People who computed fast: their self-explanations were fuzzy and unclear. In spite of numerous attempts to interview the fast calculators and to learn from them how they compute, researchers could not extract a coherent algorithm. One of the main reasons is that most of the fast calculators were *idiots savants*: their intellectual abilities outside computations were below average. Their explanations of their algorithms were always fuzzy and imprecise, formulated in terms of words of natural language rather than in precise mathematical terms.

To us familiar with fuzzy logic and its applications this fuzziness should not be surprising. It is normal that people in general – and not necessarily idiots savants – cannot precisely describe

- how they drive,
- how they operate machines,
- how they walk,
- how they translate from one language to another,
- how they recognize faces,
- how they control different situations, etc.

– this is why fuzzy control and other fuzzy techniques, techniques for translating this fuzzy description into a precise strategy, have so many practical applications; see, e.g., [7].

What may be somewhat unusual about fast computations is that while the self-description of fast calculators was fuzzy and imprecise, the results of the computations were always correct and precise.

With the appearance of computers, the interest in fast human calculators waned. With the appearance of computers, the need for human calculators disappeared, and the interest in their skills waned. Such folks may provide a good entertainment, they may be of interest to psychologists who study how we reason and how we perform menial tasks – but mathematicians were no longer interested.

Yes, fast human calculators can perform calculations faster than an average human being – but electronic computers can perform the same computations much faster. From this viewpoint, fast human calculators remained a curiosity.

Why interest waned: implicit assumptions. One of the main reasons why in the 1940s and 1950s the interest in fast human calculators waned is that it was implicitly assumed that the standard techniques of addition and multiplication are the best.

For example, it was assumed that the fastest way to add the two n-digit numbers is to add them digit by digit, which requires $O(n)$ operations with digits.

It was also implicitly assumed that the fastest way to multiply two n-digit numbers is the standard way to multiply the first number by each of the digits of the second numbers, and then to add the results. Each multiplication by a digit and each addition requires $O(n)$ steps, thus multiplication by all n digits and the addition of all n results require $n \cdot O(n) = O(n^2)$ computational steps.

From this viewpoint, the only difference between a normal human calculator, a fast human calculator, and an electronic computer is in the speed with which we can perform operations with digits. From this viewpoint, the only thing we can learn from fast human calculators is how to perform operations with digits faster. Once the electronic computers became faster than fast human calculators, the need to learn from the fast human calculators disappeared.

A surprising 1960s discovery of fast multiplication algorithms. The above implicit assumption about arithmetic operations was not questioned until a surprising sequence of discoveries was made in the 1960s; see, e.g., [8].

These discoveries started with the discovery of the Fast Fourier Transform algorithm, an algorithm that enables us to compute the Fourier Transform

$$\hat{f}(\omega) = \frac{1}{\sqrt{2\pi}} \cdot \int f(t) \cdot \exp(\mathrm{i} \cdot \omega \cdot t) \, dt, \tag{11.2}$$

where $\mathrm{i} = \sqrt{-1}$, in time $O(n \cdot \log(n))$ – instead of the $n \cdot O(n) = O(n^2)$ time needed for a straightforward computation of each of n values of $\hat{f}(\omega)$ as an integral (i.e., in effect, a sum) over n different values $f(t)$.

The ability to compute Fourier transform fast lead to the ability to speed the computation of the convolution of two functions:

$$h(t) \stackrel{\text{def}}{=} \int f(s) \cdot g(t-s) \, ds. \tag{11.3}$$

A straightforward computation of the convolution requires that for each of the n values $h(t)$, we compute the integral (sum) of n different products $f(s) \cdot g(t-s)$ corresponding to n different values s. Thus, the straightforward computation requires $O(n^2)$ computational steps.

Indeed, it is known that the Fourier transform of the convolution is equal to the product of Fourier transforms. Thus, to compute convolution, we can do the following:

- first, we compute Fourier transforms $\hat{f}(\omega)$ and $\hat{g}(\omega)$;
- then, we compute the Fourier transform of h as

$$\hat{h}(\omega) = \hat{f}(\omega) \cdot \hat{g}(\omega); \tag{11.4}$$

- finally, we apply the inverse Fourier transform to the function $\hat{h}(\omega)$ and compute the desired convolution $h(t)$.

What is the computation time of this algorithm?

- Both Fourier transform and inverse Fourier transform can be computed in time $O(n \cdot \log(n))$.
- The point-by-point multiplication $\hat{h}(\omega) = \hat{f}(\omega) \cdot \hat{g}(\omega)$ requires n computational steps.

Thus, the overall computation time requires

$$O(n \cdot \log(n)) + n + O(n \cdot \log(n)) = O(n \cdot \log(n)) \tag{11.5}$$

steps, which is much faster than $O(n^2)$.

V. Strassen was the first to notice, in 1968, than this idea can lead to fast multiplication of long integers. Indeed, an integer x in a number system with base b can be represented as a sum

$$x = \sum_{i=1}^{n} x_i \cdot b^i. \tag{11.6}$$

In these notations, the product $z = x \cdot y$ of two integers $x = \sum_{i=1}^{n} x_i \cdot b^i$ and $y = \sum_{i=1}^{n} y_i \cdot b^i$ can be represented as

$$z = \left(\sum_{i=1}^{n} x_i \cdot b^i \right) \cdot \left(\sum_{j=1}^{n} y_j \cdot b^j \right) =$$

$$\sum_{i=1}^{n} \sum_{j=1}^{n} x_i \cdot y_j \cdot b^i \cdot b^j = \sum_{i=1}^{n} \sum_{j=1}^{n} x_i \cdot y_j \cdot b^{i+j}. \tag{11.7}$$

By combining terms at different values b^k, we conclude that

$$z = \sum_{k=1}^{n} z_k \cdot b^k, \tag{11.8}$$

where

$$z_k = \sum_{i} x_i \cdot y_{k-i}. \tag{11.9}$$

This is convolution, and we know that convolution can be computed in time

$$O(n \cdot \log(n)).$$

The values z_k are not exactly the digits of the desired number z, since the sum (11.9) can exceed the base b. Thus, some further computations are needed. However, even with these further computations, we can multiply two numbers in almost the same time

$$O(n \cdot \log(n) \cdot \log(\log(n))). \qquad (11.10)$$

The corresponding algorithm, first proposed by A. Schönhage and V. Strassen in their 1971 paper [9], remains the fastest known – and it is much faster than the standard $O(n^2)$ algorithm.

Fast multiplication: open problems. Fast algorithms drastically reduced the computation time. Fast Fourier transform is one of the main tools for signal processing. These successes has led to the need to find faster and faster algorithms. From this viewpoint, it is desirable to look for even faster algorithms for multiplying numbers.

The fact that researchers succeeded in discovering algorithms which are much faster than traditional multiplication gives us hope that even faster algorithms can be found.

Interest in fast human calculators revived. Where to look for these algorithms? One natural source is folks who did compute fast. From this viewpoint, the fact that they were unable to clearly explain what algorithm they used becomes an advantage: maybe the algorithm that they actually used is some fast multiplication algorithm? This possibility revived an interest in fast human calculators.

So, the question is:

- did fast human calculators use fast multiplication algorithm(s), or
- they used the standard algorithm but simply performed operations with digits faster?

Some of the human calculators do use special tricks to speed up their computations; see, e.g., [4, 5, 10] and a website [11] hosted by Oleg Stepanov from St. Petersburg, Russia, one of the fastest living calculators. These tricks help humans compute faster – usually because they avoid the typical human limitations of memory and of computations with many digits. But computers do not have these limitations, so the known tricks of the fast human calculators cannot help them compute faster.

Are there other tricks that can help computers too?

Direct analysis is impossible. The most well-known fast human calculator, Johann Martin Zacharias Dase, died almost 150 years ago. Even when he was alive, his self-descriptions were not sufficient to find out how exactly he performed the computations.

We know, from experience, that self-description of people with such phenomenal abilities are often fuzzy. For example, according to detailed self-explanations provided in [12], to different symbols, we associate different colors and/or tones and then find the result by an informal process of combining and matching these colors and/or tones.

There might have been hope that our knowledge of fast multiplication algorithms can help in this understanding, but it did not work out. In other words, the direct analysis of Dase's behavior has been impossible – and it is still impossible. We therefore need to perform an *indirect* analysis.

Indirect analysis: main idea. A natural way to check which algorithm is used by a computational device – be it a human calculator or an electronic computer – is to find out how the computation time changes with the size n (= number of digits) of the numbers that we are multiplying.

- If this computation time grows with n as n^2, then it is reasonable to conclude that the standard algorithm is used – since for this algorithm, the computation time grows as n^2.
- On the other hand, if the computation time grows with n as $\approx n \cdot \log(n)$ (or even slower), then it is reasonable to conclude that a fast multiplication algorithm is used.

  – It may be a Strassen-type algorithm, for which the computation time grows as

  $$n \cdot \log(n).$$

  – It may be a (yet unknown) faster algorithm in which case the computation time grows even slower than $n \cdot \log(n)$.

Data that we can use. Interestingly, there is a data on the time that Dase needed to perform multiplication of numbers of different size. This data comes from fact that Dase's performance was analyzed and tested by several prominent mathematicians of his time – including Gauss himself. Specifically:

- Dase multiplied two 8-digit numbers in 54 sec;
- he multiplied two 20-digit numbers in 6 min;
- he multiplied two 40-digit numbers in 40 min; and
- he multiplied two 100-digit numbers in 8 h and 45 min.

Analysis. For the standard multiplication algorithm, the number of computational steps grows with the numbers size n as n^2. Thus, for the standard multiplication algorithm, the computation time for performing the computation also grows as n^2:

$$t(n) = C \cdot n^2. \tag{11.11}$$

So, for this algorithm, two different number sizes $n_1 < n_2$, we would have $t(n_1) = C \cdot n_1^2$ and $t(n_2) = C \cdot n_2^2$ and thus,

$$\frac{t(n_2)}{t(n_1)} = \frac{n_2^2}{n_1^2} = \left(\frac{n_2}{n_1}\right)^2. \tag{11.12}$$

For a faster algorithm, e.g., for the algorithm that requires $O(n \cdot \log(n))$ running time, the corresponding ratio will be smaller:

$$\frac{t(n_2)}{t(n_1)} = \frac{n_2 \cdot \log(n_2)}{n_1 \cdot \log(n_1))} = \frac{n_2}{n_1} \cdot \frac{\log(n_2)}{\log(n_1)}. \tag{11.13}$$

Thus, to check whether a human calculators uses the standard algorithm or a faster one it is sufficient to compare the corresponding time ratio $\dfrac{t(n_2)}{t(n_1)}$ with the square $\left(\dfrac{n_2}{n_1}\right)^2$:

- If the time ratio is smaller than the square, this means that the human calculator used an algorithm which is much faster than the standard one.
- On the other hand, if the time ratio is approximately the same as the square, thus means that the human calculators most probably used the standard algorithm – or at least some modification of it that does not drastically speed up the computations.
- If it turns out that the time ratio is larger than the square, this would mean, in effect, that a human calculator used an algorithm which is asymptotically even slower than the standard one – this can happen, e.g., if there is an overhead needed to store values etc.

According to the above data, we have:

- $t(8) = 0.9 \,\text{min}$,
- $t(20) = 6 \,\text{min}$,
- $t(40) = 40 \,\text{min}$, and
- $t(100) = 8 \cdot 60 + 45 = 525 \,\text{min}$.

Here, for $n_1 = 8$ and $n_1 = 20$, we have:

$$\frac{t(n_2)}{t(n_1)} = \frac{t(20)}{t(8)} = \frac{6}{0.9} \approx 6.7 > \left(\frac{n_2}{n_1}\right)^2 = \left(\frac{20}{8}\right)^2 = 2.5^2 = 6.25. \tag{11.14}$$

For $n_1 = 8$ and $n_2 = 40$, we have

$$\frac{t(n_2)}{t(n_1)} = \frac{t(40)}{t(8)} = \frac{40}{0.9} \approx 44 > \left(\frac{n_2}{n_1}\right)^2 = \left(\frac{40}{8}\right)^2 = 5^2 = 25. \tag{11.15}$$

Finally, for $n_1 = 8$ and $n_2 = 100$, we have

$$\frac{t(n_2)}{t(n_1)} = \frac{t(100)}{t(8)} = \frac{525}{0.9} \approx 583 > \left(\frac{n_2}{n_1}\right)^2 = \left(\frac{100}{8}\right)^2 = 12.5^2 \approx 156. \tag{11.16}$$

In all these cases, the computation time of a human calculator grows faster then n^2 corresponding to the standard algorithm.

The same conclusion can be made if instead of comparing each value n with the smallest value n_1, we compare these values with each other. For $n_1 = 20$ and $n_2 = 40$, we have

$$\frac{t(n_2)}{t(n_1)} = \frac{t(40)}{t(20)} = \frac{40}{6} \approx 7 > \left(\frac{n_2}{n_1}\right)^2 = \left(\frac{40}{20}\right)^2 = 2^2 = 4. \tag{11.17}$$

For $n_1 = 20$ and $n_2 = 100$, we have

$$\frac{t(n_2)}{t(n_1)} = \frac{t(100)}{t(20)} = \frac{525}{6} \approx 88 > \left(\frac{n_2}{n_1}\right)^2 = \left(\frac{100}{20}\right)^2 = 5^2 = 25. \tag{11.18}$$

Finally, for $n_1 = 40$ and $n_2 = 100$, we have

$$\frac{t(n_2)}{t(n_1)} = \frac{t(100)}{t(40)} = \frac{525}{40} \approx 13 > \left(\frac{n_2}{n_1}\right)^2 = \left(\frac{100}{40}\right)^2 = 2.5^2 = 6.25. \tag{11.19}$$

Thus, it is reasonable to conclude that the fast human calculators did not use any algorithm which is faster than the standard one.

Possible future work. People who perform computations fast appear once in a while. It may be a good idea to record and analyze their computation time – and maybe record their fuzzy explanations and try to make sense of them.

References

1. O. Kosheleva, Can we learn algorithms from people who compute fast: an indirect analysis in the presence of fuzzy descriptions, in *Proceedings of the 2009 World Congress of the International Fuzzy Systems Association IFSA'2009*, Lisbon, Portugal, 20–24 July 2009. pp. 1394–1397
2. O. Kosheleva, V. Kreinovich, Can we learn algorithms from people who compute fast, in *Soft Computing in Humanities and Social Sciences*, ed. by R. Seising, V. Sanz (Springer, Heidelberg, 2011), pp. 267–275
3. P. Beckmann, *A History of π* (Barnes and Noble, New York, 1991)
4. M. d'Ocagne, *Le Calcul Simplifié par les Procédés Mecaniques et graphiques* (Gauthier-Villars, Paris, 1905)
5. M. d'Ocagne, *Le Calcul Simplifié: Graphical and Mechanical Methods for Simplifying Calculation* (MIT Press, Cambridge, 1986)
6. D.R. Hofstadter, *Godel, Escher, Bach: an Eternal Golden Braid* (Basic Books, New York, 1999)
7. G.J. Klir, B. Yuan, *Fuzzy Sets and Fuzzy Logic: Theory and Applications* (Prentice-Hall, Upper Saddle River, 1995)
8. Th.H. Cormen, C.E. Leiserson, R.L. Rivest, C. Stein, *Introduction to Algorithms* (MIT Press, Cambridge, 2001)
9. A. Schönhage, V. Strassen, Schnelle Multiplikation grosser Zahlen. Computing **7**, 281–292 (1971)
10. R. Tocquet, *The Magic of Numbers* (Fawcett Publications, Robbinsdale, 1965)

11. O. Stepanov, http://stepanov.lk.net/
12. A.R. Luria, *The Mind of a Mnemonist: a Little Book About a Vast Memory* (Hardard University Press, Cambridge, 1987)

Chapter 12
How to Enhance a General Student Motivation to Study: Asymmetric Paternalism

In the previous chapters, we provided examples of how uncertainty-related examples and ideas help explain, to the students, why a *specific* material is useful and interesting. In addition to this, we also need to make sure that the students are excited about studying *in general*, that their levels of interest and commitment remain high. If a math instructor convinces the engineering students that they need to learn math, this should not lead to them getting less interested in studying engineering disciplines, ideally they should be excited about all the topics that they study. In this chapter, we therefore analyze how to increase this *general* level of interest. It turns out that properly taking uncertainty into account can help with this task as well.

This idea is based on the fact that while in general, human beings are rational decision makers, in many situations, they exhibit unexplained "inertia", reluctance to switch to a better decision. We show that this seemingly irrational behavior can be explained if we take uncertainty into account, and we explain how this phenomenon can be utilized in education.

The results from this chapter first appeared in [1].

Traditional approach to human decision making: a brief reminder. In the traditional approach to decision making (see, e.g., [2, 3]), the decision maker's preferences $A_1, \ldots, A_n$ can be characterized by their "utility values" $u(A_1), \ldots, u(A_n)$, so that an alternative A_i is preferable to the alternative A_j if and only if $u(A_i) > u(A_j)$.

Empirical testing of the traditional approach to decision making is not easy. The traditional approach to decision making is a theoretical description of human behavior. Since its appearance, researchers have been testing to what extent this approach adequately describes the actual behavior of human decision makers.

Such a testing is not easy, since the traditional approach relates an empirically "testable" behavior (such as preferring one alternative A_i to another alternative A_j) with the difficult-to-test comparison between the (usually unknown) utility values.

© Springer-Verlag GmbH Germany 2018
O. Kosheleva and K. Villaverde, *How Interval and Fuzzy Techniques Can Improve Teaching*, Studies in Computational Intelligence 750,
https://doi.org/10.1007/978-3-662-55993-2_12

A testable consequence of the traditional approach to decision making. Although a direct test of the traditional approach to decision making is not easy, some testable consequences of the traditional approach can be derived.

For example, in the traditional approach, unless the two alternatives A_i and A_j have the exact same utility value $u(A_i) = u(A_j)$, we have two possibilities:

- either $u(A_i) > u(A_j)$, i.e., the alternative A_i is better,
- or $u(A_j) > u(A_i)$, i.e., the alternative A_j is better.

In the first case,

- if we originally only had an alternative A_i, and then we are adding the alternative A_j, then we stick with A_i;
- on the other hand, if we originally only had an alternative A_j, and then we are adding the alternative A_i, then we switch our choice to A_i.

Similarly, in the second case,

- if we originally only had an alternative A_j, and then we are adding the alternative A_i, then we stick with A_j;
- on the other hand, if we originally only had an alternative A_i, and then we are adding the alternative A_j, then we switch our choice to A_j.

These two cases can be summarized in the following 2-stage experiment: In the first stage,

- first, we present the decision maker with only one alternative A_i; the decision maker does not have a choice, so he or she selects this alternative A_i;
- after that, we provide the decision maker with another possible alternative A_j; so now the decision maker has two alternatives A_i and A_j.

We then record the user's choice in this first stage.

After some amount of time, we start with the second stage of our experiment:

- first, we present the decision maker with only one alternative A_j; the decision maker does not have a choice, so he or she selects this alternative A_j;
- after that, we provide the user with another possible alternative A_i, so now the decision maker has two alternatives A_i and A_j.

We then record the user's choice in this second stage.

According to the above-described traditional approach to decision making, on both stages, the decision maker should make the same choice:

- if the alternative A_i has a larger utility value, then on both stages, the decision maker should chose A_i;
- on the other hand, if the alternative A_j has a larger utility value, then on both stages, the decision maker should chose A_j.

The above testable consequence is in perfect agreement with common sense. The above behavior not only follows from the mathematics of the traditional decision making, it is also in perfect agreement with common sense.

Indeed, unless the two alternatives A_i and A_j are absolutely equivalent for the decision maker (which happens very rarely), either the first one is better or the second one is better.

So, on both stages of the above experiment, a rational decision maker should make the same choice:

- if to this decision maker, the alternative A_i is preferable to the alternative A_j, then on both stages, the decision maker should chose A_i;
- on the other hand, if to this decision maker, the alternative A_j is preferable to the alternative A_i, then on both stages, the decision maker should chose A_j.

For close alternatives, decision makers do not behave in this rational fashion. Interestingly, in the actual tests of the above experiment, human decision makers do not follow this seemingly rational behavior; see, e.g., [4–7]. Specifically, they exhibit "inertia", the desire not to change an alternative.

Namely, if the alternatives are close in value, then the decision makers exhibit the following behavior in our two-stage experiment. In the first stage,

- first, we present the decision maker with only one alternative A_i; the decision maker does not have a choice, so he or she selects this alternative A_i;
- after that, we provide the decision maker with another possible alternative A_j; so now the decision maker has two alternatives A_i and A_j.

On this stage, most decision makers continue to stick to the original choice A_i.

After some amount of time, we perform the second stage of our experiment:

- first, we present the decision maker with only one alternative A_j; the decision maker does not have a choice, so he or she selects this alternative A_j;
- after that, we provide the user with another possible alternative A_i, so now the decision maker has two alternatives A_i and A_j.

On this stage, most decision makers continue to stick to the original choice A_j.

An example where such seemingly irrational behavior occurred is the employees' choice between two financial retirement plans A_i and A_j.

- Originally, the employees had only one option: retirement plan A_i.
- After this, an additional option A_j is introduced.
- In spite of this new option, most employees decided to keep the old option A_i.

This, by itself, is not inconsistent with rational behavior: we can simply conclude that for most employees, the original option A_i is better than the new option A_j.

However, at the same time, a very similar group of employees was presented with a different scenario:

- Originally, the employees had only one option: retirement plan A_j.
- After this, an additional option A_i is introduced.
- In spite of this new option, most employees decided to keep the old option A_j.

Since we concluded that for most employees, the option A_i is better than the option A_j, we would expect most employees from this second group to switch from the original option A_j to the new option A_i. However, in reality, most employees from this second group stayed with their original option A_j.

In behavioral economics, this "inertial" behavior is called *present-biased preferences*: whatever options we have selected at present biases our future choices.

Maybe human behavior is irrational? How can we explain this seemingly irrational behavior? One possible explanation is that many people do often make bad (irrational) decisions: waste money on gambling, waste one's health on alcohol and drugs, etc.

However, the above inertial behavior occurs not only among decision makers who exhibit self-destructive irrational behavior, it is a common phenomenon which occurs among the most successful people as well.

It is therefore reasonable to look for an explanation of this seemingly irrational behavior. It turns out that we can come up with such an explanation if we take into account uncertainty related to decision making.

How to take into account uncertainty in decision making situations. Each alternative decision can lead to different possible situations. For example, a decision about selecting a financial retirement plan can lead to different amounts available to the decision maker by the moment of his or her retirement:

- A more conservative retirement plan – e.g., investing all the retirement money in the government-guaranteed bonds – will not lead to a large increase of the invested amount, but, on the positive side, has a smaller probability of losing the retirement money.
- On the other hand, a riskier retirement plan – e.g., investing all the retirement money into stocks – has a larger probability of failing, but it can also lead to a much larger amount of money available for retirement.

In the traditional approach to decision making, we first estimate the utilities $U_1, \ldots, U_m$ of different possible consequences $c_1, \ldots, c_m$ of our actions – e.g., the utilities of having different amounts of money by the time of the retirement – and then estimate the utility of each alternative decision as the expected value of this utility: $u(A) = p(c_1 \mid A) \cdot U_1 + \ldots + p(c_m \mid A) \cdot U_m$, where $p(c_k \mid A)$ is the conditional probability of the consequence c_k under the condition that we select an alternative A.

In real life, we do not know the exact values of these probabilities $p(c_k \mid A)$, we only know them with uncertainty. Usually, we only know the approximate estimates of these probabilities. In some cases, we have bounds

$$\underline{p}(c_k \mid A) \leq p(c_k \mid A) \leq \overline{p}(c_k \mid A)$$

on these probabilities, i.e., we know intervals $[\underline{p}(c_k \mid A), \overline{p}(c_k \mid A)]$ that contain the (unknown) probabilities $p(c_k \mid A)$. In such situations, for each alternative A, instead of a single value $u(A)$, we get an interval of possible values:

$$[\underline{u}(A), \overline{u}(A)] =$$

$$[\underline{p}(c_1 \mid A), \overline{p}(c_1 \mid A)] \cdot U_1 + \ldots + [\underline{p}(c_m \mid A), \overline{p}(c_m \mid A)] \cdot U_m.$$

In other situations, we have expert estimates of the unknown probabilities $p(c_k \mid A)$. Such expert estimates can be naturally described by fuzzy numbers. In this case, the resulting utility estimate $u(A)$ is also a fuzzy number.

Comment. In effect, fuzziness means that instead of single pair of bounds for the unknown probability (and, as a result, for the utility $u(A)$), we can provide different bounds which are valid with different degrees of confidence.

In other words, for each such quantity (probability or utility), instead of a single interval, we get a nested family of confidence intervals corresponding to different levels of uncertainty. Nested families are, in effect, equivalent to fuzzy numbers; see, e.g., [8–11], so this natural idea of representing uncertainty is indeed mathematically equivalent to using fuzzy numbers.

Uncertainty explains present-biased preferences. When the approximate estimates $\widetilde{u}_i$ and $\widetilde{u}_j$ for the (unknown) utilities $u(A_i)$ and $u(A_j)$ are close, this, due to the uncertainty, means that it is quite possible that the actual values $u(A_i)$ and $u(A_j)$ are equal. It is also possible that $u(A_i) > u(A_j)$ and it is also possible that $u(A_j) > u(A_i)$.

Let us illustrate these possibilities on a simple example where every estimate has the exact same accuracy ε. In other words, we know that $|u(A_i) - \widetilde{u}_i| \le \varepsilon$ and that $|u(A_j) - \widetilde{u}_j| \le \varepsilon$. In this case, based on the estimate $\widetilde{u}_i$ for $u(A_i)$, the only information that we have about the actual (unknown) value of the utility $u(A_i)$ is that this value belongs to the interval $\mathbf{u}_i \stackrel{\text{def}}{=} [\widetilde{u}_i - \varepsilon, \widetilde{u}_i + \varepsilon]$. Similarly, based on the estimate $\widetilde{u}_j$ for $u(A_j)$, the only information that we have about the actual (unknown) value of the utility $u(A_j)$ is that this value belongs to the interval $\mathbf{u}_j \stackrel{\text{def}}{=} [\widetilde{u}_j - \varepsilon, \widetilde{u}_j + \varepsilon]$.

When the estimates $\widetilde{u}_i$ and $\widetilde{u}_j$ are close – to be more precise, when $|\widetilde{u}_i - \widetilde{u}_j| < 2\varepsilon$ – the intervals $\mathbf{u}_i$ and $\mathbf{u}_j$ intersect.

These intervals represent the set of possible values for, correspondingly, $u(A_i)$ and $u(A_j)$. Thus, the fact that these intervals intersect means that it is possible that the same real number is a value of both $u(A_i)$ and of $u(A_j)$ – i.e., that the corresponding utilities may be equal. Similarly, we can show that in this case:

- there are values $u_i \in \mathbf{u}_i$ and $u_j \in \mathbf{u}_j$ for which $u_i < u_j$, and
- there are also values $u'_i \in \mathbf{u}_i$ and $u'_j \in \mathbf{u}_j$ for which $u'_i > u'_j$.

In other words, when the estimates $\widetilde{u}_i$ and $\widetilde{u}_j$ are close, all three situations are possible:

- it is possible that the alternatives A_i and A_j have exactly the same utility to the decision maker;
- it is possible that for this decision maker, the alternative A_i leads to better results than A_j;
- it is also possible that for this decision maker, the alternative A_j leads to better results than A_i.

Switching to a different alternative usually has a cost, a small but still a cost. For example, in the case of a financial retirement plan, there is a trader's charge for selling stocks and for buying government bonds (and vice versa). Thus, it only makes sense to perform this switch if we are reasonably sure that switching will indeed lead to a better alternative – i.e., in utility terms, to the larger value of utility.

If the utility estimates are close, i.e., if $|\widetilde{u}_i - \widetilde{u}_j| < 2\varepsilon$, we have no guarantee that the new alternative is indeed better than the previously selected one. In this case, it is prudent to stick to the original choice – exactly as actual decision makers are doing.

Comment. On the other hand, if one of the estimates is much larger than the other, it makes sense to switch.

Specifically, if $\widetilde{u}_i > \widetilde{u}_j + 2\varepsilon$, then every value $u(A_i) \in \mathbf{u}_i = [\widetilde{u}_i - \varepsilon, \widetilde{u}_i + \varepsilon]$ is larger than every value $u(A_j) \in \mathbf{u}_j = [\widetilde{u}_j - \varepsilon, \widetilde{u}_j + \varepsilon]$. In this case, we are guaranteed that $u(A_i) > u(A_j)$. Thus, if our original choice was the worse alternative A_j, it makes sense to switch to a better alternative A_i.

Analogy with interval-valued control of a mobile robot. The rationality of inertia under uncertainty can be illustrated on the example of a similar situation: how an intelligent mobile robot makes decisions about its motion.

In the traditional control, we make decisions based on the current values of the quantities. For example, when controlling a mobile robot, we make decisions about changing its trajectory based on the moment-by-moment measurements of this robot's location and/or velocity. Measurements are never 100% accurate; the resulting measurement noise leads to random deviations of the robot from the ideal trajectory – shaking and "wobbling". Since each change in direction requires that energy from the robot's battery go to the robot's motor, this wobbling drains the batteries and slows down the robot's motion.

A natural way to avoid this wobbling is to change a direction only if it is absolutely clear (beyond the measurement uncertainty) that this change will improve the robot's performance. The idea was one of the several interval-related ideas that in 1997, led our university robotic team to the 1st place in the robotic competitions organized by the American Association for Artificial Intelligence (AAAI); see, e.g., [12]. A similar idea also improves the motion of interval-valued fuzzy control; see, e.g., [13–15].

Asymmetric paternalism: practical application of present-biased preferences. At first glance, one may think that the above explanation is of purely theoretical value: OK, we explained how people actually make decisions. How does that help in practice?

The reason why we got interested in coming up with this explanation is that the phenomenon of present-biases preferences is actually actively used in practice. This use is called *asymmetric paternalism*; see, e.g., [5, 16, 17]. Let us explain how this application works.

Suppose that we have two types of behavior, one slightly worse for an individual, and one slightly better. For example, when thirsty, a kid can drink either a healthy fruit juice or a soda drink which has no health value. Our intent is to enforce the healthy alternative.

Traditional paternalism literally enforces the healthy choice by prohibiting all other choices. Alas, practice has shown that in many cases, this literal enforcement does not work.

It turns out that much better results can be achieved if we at first provide only the desired alternative – and then gradually introduce all the other alternatives. For example, we have only healthy drinks for the first few weeks of a school orientation, but then we allow all the choices. Due to the present-biased preferences, kids will tend to stick to their original healthier choice without the need for strict (and non-working) enforcement. Experience shows that this approach really works; see, e.g., [16, 17].

The same strategy works even better for adults – whom we cannot legally enforce into healthy choices.

How does our explanation help? If this approach works, do we need any explanation to make it work? Well, sometimes this approach works, and sometimes it does not. As of now, empirical attempts were the only way to check whether this approach will work for given alternatives A_i and A_j.

Suppose that we want to enforce A_i as opposed to A_j by introducing A_i first. How can we tell beforehand whether the decision maker will stick to A_i and not switch to A_j?

Our explanation provides an answer to this question: if $[\underline{u}_i, \overline{u}_i]$ is the interval of possible values of $u(A_i)$ and $[\underline{u}_j, \overline{u}_j]$ is the interval of possible values of $u(A_j)$, then the decision maker sticks with the original choice of A_i if and only if there is no guarantee that the new choice A_j is better. In mathematical terms, this means that the smallest possible value $\underline{u}_j$ corresponding to the new choice does not exceed the largest possible value $\overline{u}_i$ corresponding to the original choice: $\underline{u}_j \leq \overline{u}_i$. For fuzzy numbers, we can get a similar answer for "not switching with a given confidence", if we similarly compare the intervals (α-cuts) for $u(A_i)$ and $u(A_j)$ corresponding to this given confidence level.

Potential applications to education. As of now, the asymmetric paternalism techniques have been used in economic and medical situations [5, 16, 17]. In our opinion, this phenomenon can also be efficiently applied to education.

For example, it is well known that when the students just come to class from recess or from home, it is difficult to get their attention. On the other hand, once they get engaged in the class material, it is difficult for them to stop when the bell rings. To take advantage of this phenomenon, it is desirable to start a class with engaging fun material; once the students got into the studying state A_i they will (hopefully) remain in this state even when a somewhat less fun necessary material is presented (and which provides them with a possibility to switch to a passive state A_j).

References

1. O. Kosheleva, F. Modave, Asymmetric paternalism: description of the phenomenon, explanation based on decisions under uncertainty, and possible applications to education, in *Proceedings of the 27th International Conference of the North American Fuzzy Information Processing Society NAFIPS'2008*, New York, 19–22 May 2008
2. R.L. Keeney, H. Raiffa, *Decisions with Multiple Objectives* (Wiley, New York, 1976)
3. H. Raiffa, *Decision Analysis* (Addison-Wesley, Reading, 1970)

4. E.J. Johnson, J. Hershey, J. Meszaros, H. Kunreuther, Framing, probability distortions, and insurance decisions. J. Risk Uncertain. **7**, 35–53 (1993)
5. G. Loewenstein, T. Brennan, K.G. Volpp, Asymmetric paternalism to improve health behavior. J. Am. Med. Assoc. (JAMA) **298**, 2415–2417 (2007)
6. B.C. Madrian, D.F. Shea, The power of suggestion: inertia in 401(k) participation and savings behavior. Q. J. Econ. **116**(4), 1149–1187 (2001)
7. T. O'Donoghue, M. Rabin, Doing it now or later. Am. Econ. Rev. **97**, 31–46 (2005)
8. D. Dubois, H. Prade, Operations on fuzzy numbers. Int. J. Syst. Sci. **9**, 613–626 (1978)
9. G.J. Klir, B. Yuan, *Fuzzy Sets and Fuzzy Logic: Theory and Applications* (Prentice-Hall, Upper Saddle River, 1995)
10. H.T. Nguyen, O. Kosheleva, V. Kreinovich, Is the success of fuzzy logic really paradoxical? or: towards the actual logic behind expert systems. Int. J. Intell. Syst. **11**, 295–326 (1996)
11. H.T. Nguyen, E.A. Walker, *A First Course in Fuzzy Logic* (CRC Press, Boca Raton, 2006)
12. D. Morales, T.C. Son, Interval methods in robot navigation. Reliab. Comput. **4**(1), 55–61 (1998)
13. H.T. Nguyen, V. Kreinovich, Q. Zuo, Interval-valued degrees of belief: applications of interval computations to expert systems and intelligent control. Int. J. Uncertainty, Fuzziness Knowl. Based Syst. (IJUFKS) **5**(3), 317–358 (1997)
14. K.C. Wu, A robot must be better than a human driver: an application of fuzzy intervals, in *NAFIPS/IFIS/NASA'94, Proceedings of the First International Joint Conference of The North American Fuzzy Information Processing Society Biannual Conference*, eds. by L. Hall, H. Ying, R. Langari, J. Yen. *The Industrial Fuzzy Control and Intelligent Systems Conference, and The NASA Joint Technology Workshop on Neural Networks and Fuzzy Logic*, San Antonio, 18–21 December 1994. pp. 171–174
15. K.C. Wu, Fuzzy interval control of mobile robots. Comput. Electr. Eng. **22**(3), 211–219 (1996)
16. C. Cammerer, S. Issacharoff, G. Loewenstein, T. O'Donoghue, M. Rabin, Regulation for conservatives: behavioral economics and the case for 'asymmetric paternalism'. Univ. Pa. Law Rev. **151**(3), 1211–1254 (2003)
17. R.H. Thaler, C.R. Sunstein, Libertarian paternalism. Am. Econ. Rev. **93**(2), 175–179 (2003)

Chapter 13
Financial Motivation: How to Incentivize Students to Graduate Faster

At prime research universities, students study full-time and receive their Bachelors's degree in four years. In contrast, at urban universities, many students study only part-time, and take a longer time to graduate. The sooner such a student graduates, the sooner will the society start benefitting from his or her newly acquired skills – and the sooner the student will start earning more money. It is therefore desirable to incentivize students to graduate faster. In the present chapter, we propose a first-approximation solution to the problem of how to distribute a given amount of resources so as to maximally speed up students graduation.

The results from this chapter first appeared in [1].

Fact: students at urban universities often take longer to graduate. At prime research universities, students study full-time and receive their Bachelor's degrees in four years. In contrast, at urban universities, many students study only part-time. As a result, these students take longer to graduate.

Speeding up graduation is a win-win idea. From the viewpoint of the student, the sooner he or she graduates, the sooner will his or her salary increase reflecting the newly acquired skills.

From the viewpoint of the society as a whole, the sooner a student graduates, the sooner will the society start benefitting from his or her newly acquired skills. In other words, speeding up graduation is a win-win idea.

How can we speed up graduation. Among the main reasons why some students at urban universities only study part-time are financial reasons. So, to speed up student graduation, it is desirable to provide financial incentives.

Towards a corresponding optimization problem. In the ideal world, we should be able to fully support every student. In real life, however, our resources are limited. So, the question is: what is the best way to distribute these resources so that we can maximally speed up student graduation – or, equivalently, maximally increase the number of classes n that a student takes every semester.

© Springer-Verlag GmbH Germany 2018
O. Kosheleva and K. Villaverde, *How Interval and Fuzzy Techniques Can Improve Teaching*, Studies in Computational Intelligence 750,
https://doi.org/10.1007/978-3-662-55993-2_13

This is a problem that we will be solving in this chapter.

Modeling student's decisions. In order to solve the above problem, let us first formulate it in precise terms. According to the general decision making theory (see, e.g., [2–4]), every agent selects a decision that maximizes his or her utility u. So, to understand the student's behavior, we need to understand how this utility u depends on the number n of classes per semester.

In general, it is reasonable to assume that this dependence $u(n)$ is smooth – even analytical, so the dependence can be well approximated by a Taylor series $u(n) = u_0 + u_1 \cdot n + u_2 \cdot n^2 + \ldots$ The number of hours n does not differ that much between different students, so the range of n is small, and on a small range, a few first terms in the Taylor expansion are usually sufficient to reasonably accurately describe the dependence. Let us see how many terms we need for our problem.

The 0-th order term u_0 can be interpreted as a utility of simply being at a university. Since this term does not depend on the number of classes n that a student is taking, it does not affect the student's decision. Therefore, we can safely ignore this term and assume that $u_0 = 0$.

The next term $u_1 \cdot n$ represents the gain in knowledge (minus effort) per class. While there are minor differences in how much material students learn, in the first approximation, it is reasonable to assume that this amount is approximately the same for all the students. Since maximizing the function $u(n) = u_1 \cdot n + \ldots$ and maximizing the function $\dfrac{u(n)}{u_1} = n + \ldots$ are equivalent tasks, we can safely assume that $u_1 = 1$, i.e., that the linear term has the form $u(n) = n$.

If we only had this linear term, then the more classes the student would take, the larger this student's utility. In other words, in this approximation, a student would take as many classes as there are available. This is clearly not what we observe. This means that in order to explain the actual student behavior, it is not sufficient to only consider linear terms in the dependence $u(n)$, we need to consider at least the terms of the next order – i.e., quadratic terms. Thus, we arrive at the utility expression $u(n) = n + u_1 \cdot n^2$.

If $u_1 > 0$, then this expression increases with n and so, we face the same problem as before. So, to explain the actual student behavior, we need to assume that $u_1 < 0$. In this case, the utility function has a clear maximum: when $\dfrac{du}{dn} = 1 + 2u_1 \cdot n = 0$, i.e., when $n = \dfrac{1}{2|u_1|}$.

For each student, we observe the actual number of classes n_a that this student takes, so we can conclude that for this student, $u_1 = -\dfrac{1}{2n_a}$ and thus, the student's utility function has the form

$$u(n) = n - \frac{1}{2n_a} \cdot n^2.$$

Modeling student population. In the above first approximation model, decisions by each student are characterized by a single parameter n_a – the number of classes that this student takes. Thus, to describe a student population, it is sufficient to describe the distribution of this parameter. This distribution can be described, e.g., by the probability density $\rho(n)$ which is defined, as usual, as the ratio $\dfrac{P(n \leq n_a \leq n + \Delta n)}{\Delta n}$ of the proportion $P(n \leq n_a \leq n + \Delta n)$ of students for whom the actual number of classes n_a is between n and $n + \Delta$ and the width Δn of the corresponding interval $[n, n + \Delta n]$.

Adding an incentive. A natural incentive is to give a discount for each course above a certain threshold n_0. This incentive adds, to the original utility, a new term

$$k \cdot (n - n_0),$$

where $k > 0$ is the per-course value of this discount.

Once we select a threshold n_0, we can determine the per-course discount value b by equating the total discount the total amount of offered discounts to the available amount A, i.e., from the condition that

$$k \cdot \int_{n_0} \rho(x) \cdot (n - n_0) \, dn = A.$$

From this condition, we can describe the value k as follows:

$$k = \frac{A}{\displaystyle\int_{n_0} \rho(x) \cdot (n - n_0) \, dn}.$$

Decision making in the presence of this incentive. Once we add the incentive, for $n < n_0$, we get the same utility as before, but for $n > n_0$, we get a new utility expression

$$u_i(n) = n - \frac{1}{2n_a} \cdot n^2 + k \cdot (n - n_0).$$

As a result, a student who previously selected n_a courses will now optimize a new objective function $u_i(a)$ and get a new number of courses n_i.

Our objective. We want to select a threshold n_0 in such a way that the average increase in the number of courses is the largest possible, i.e., that the value

$$\int_{n_0} \rho(n_a) \cdot (n_i(n_a) - n_a) \, dn_a$$

is the largest possible.

Now, the problem has been reformulated in precise terms, so we can start solving it.

Towards a solution to the problem. Differentiating the new utility function $u_i(n)$ with respect to n and equating the derivative to 0, we conclude that for the value n_i at which this objective function attains its maximum, we get

$$1 - \frac{n_i}{n_a} + k = 0,$$

hence $n_i = (1+k) \cdot n_a$.

Thus, $n_i(n_a) - n_a = k \cdot n_a$, and so, the objective function that we use to select a threshold n_0 takes the form

$$k \cdot \int_{n_0} \rho(n) \cdot n \, dn.$$

Substituting the above expression for k, we conclude that we need to maximize the following expression:

$$A \cdot \frac{\displaystyle\int_{n_0} \rho(n) \cdot n \, dn}{\displaystyle\int_{n_0} \rho(n) \cdot (n - n_0) \, dn}.$$

Dividing the objective function by a constant does not change the value at which this function attains its maximum. So, the above maximization problem is equivalent to the problem of maximizing the following ratio:

$$\frac{\displaystyle\int_{n_0} \rho(n) \cdot n \, dn}{\displaystyle\int_{n_0} \rho(n) \cdot (n - n_0) \, dn}.$$

Maximizing an expression E is equivalent to minimizing its reciprocal $\frac{1}{E}$. Thus, maximizing the above ratio is equivalent to minimizing the reciprocal ratio

$$\frac{\displaystyle\int_{n_0} \rho(n) \cdot (n - n_0) \, dn}{\displaystyle\int_{n_0} \rho(n) \cdot n \, dn}.$$

Here,

$$\int_{n_0} \rho(n) \cdot (n - n_0) \, dn = \int_{n_0} \rho(n) \cdot n \, dn - n_0 \cdot \int_{n_0} \rho(n) \, dn,$$

and therefore, the above ratio takes the form

$$1 - \frac{n_0 \cdot \int_{n_0} \rho(n)\, dn}{\int_{n_0} \rho(n) \cdot n\, dn}.$$

Minimizing this expression $1 - r$ is equivalent to maximizing r, i.e., to minimizing the ratio

$$\frac{n_0 \cdot \int_{n_0} \rho(n)\, dn}{\int_{n_0} \rho(n) \cdot n\, dn}.$$

This minimization, in its turn, is equivalent to maximizing the reciprocal ratio

$$\frac{1}{n_0} \cdot \frac{\int_{n_0} \rho(n) \cdot n\, dn}{\int_{n_0} \rho(n)\, dn}.$$

One can easily check that the ratio

$$\frac{\int_{n_0} \rho(n) \cdot n\, dn}{\int_{n_0} \rho(n)\, dn}$$

is, by definition, equal to the conditional mean of the variable n under the condition that $n \geq n_0$:

$$\frac{\int_{n_0} \rho(n) \cdot n\, dn}{\int_{n_0} \rho(n)\, dn} = E[n \mid n \geq n_0].$$

Thus, we arrive to the following conclusion.

Solution to the optimization problem. To maximize the effect of the incentive, we should select a threshold n_0 for which the following ratio is the largest possible:

$$\frac{E[n \mid n \geq n_0]}{n_0}.$$

Discussion. For distributions with "light" tails – similar to the normal distribution – the above ratio decreases with n_0. Thus, for such distributions, to achieve the largest effect, we should select the smallest possible threshold n_0.

For heavy-tailed distributions [5, 6] – e.g., for the Pareto distribution, when $\rho(n) = C \cdot n^{-\alpha}$ for all $n \geq N$ for some small N – the situation is different. For example, for the Pareto distribution, the above ratio does not depend on the threshold n_0; therefore, to decide which threshold to select, it is not sufficient to use the above first approximation: we must consider the next approximation as well.

References

1. O. Kosheleva, How to incentivize students to graduate faster. Int. J. Innov. Manag. Inf. Prod. (IJIMIP) **3**(4), 31–35 (2012)
2. P.C. Fishburn, *Utility Theory for Decision Making* (Wiley, New York, 1969)
3. R.D. Luce, H. Raiffa, *Games and Decisions: Introduction and Critical Survey* (Dover, New York, 1989)
4. H. Raiffa, *Decision Analysis: Introductory Lectures on Choices Under Uncertainty* (McGraw Hill, New York, 1997)
5. B. Mandelbrot, *The Fractal Geometry of Nature* (Freeman, San Francisco, 1983)
6. S.I. Resnick, *Heavy-Tail Phenomena: Probabilistic and Statistical Modeling* (Springer, New York, 2007)

Part II
In What Order to Present the Material

Chapter 14
In What Order to Present the Material: An Overview of Part II

In Part 1 of this book, we described how uncertainty ideas can help motivated students to study. Once the students are motivated and teaching starts, we need to decide *in what order* we should present the material. Some courses first provide the basic ideas of all the topics, and only after all the basics are described, provide the technical details; other courses first deal with one topic, then go to another topic, etc. Which approach is better? When is each of these approaches better? Within each of these approaches, which topic should be taught first and which next? These are all important questions which can often seriously impact the course results; see, e.g., [1–8].

In this part of the book, we show that interval and fuzzy techniques can help us decide in what order to present the material. Specifically, we show that these techniques are helpful on all levels, from the most general level of designing a curriculum to the most detailed level of teaching an individual topic to the highest level of designing an inter-disciplinary curriculum.

- First, in Chap. 15, we discuss the general ideas behind the curriculum design; specifically, we provide a justification for an empirically successful idea of a "spiral" curriculum, in which we repeatedly revisit each topic.
- In Chaps. 16 and 17, we analyze how much time we need to allocate to each topic within the curriculum. This is related to the question of whether it is necessary to first achieve a perfect knowledge of one topic before moving to another one. This question turns out to be related to the pedagogical controversy about teaching to the test.
- In Chap. 18, we analyze what is the best order to topics.
- In Chap. 19, we discuss specifics of an inter-disciplinary curriculum; the corresponding results turn out to be useful not only for students, but also for professionals who want to engage in inter-disciplinary research.
- Finally, in Chap. 20, we deal with the teaching of each individual topic: should we start with concrete examples and present more abstract ideas later on, or should we introduce abstract ideas as early as possible.

© Springer-Verlag GmbH Germany 2018
O. Kosheleva and K. Villaverde, *How Interval and Fuzzy Techniques Can Improve Teaching*, Studies in Computational Intelligence 750,
https://doi.org/10.1007/978-3-662-55993-2_14

References

1. V. Davydov, *Types of Generalizations in Instruction: Logical and Psychological Problems in the Structuring of Schhol Curricular* (NCTM, Reston, 1990)
2. J.A. Kaminski, V.M. Sloutsky, A.F. Heckler, Do children need concrete instantiations to learn an abstract concept?, eds.by R. Sun, N. Miyakein. *Proceedings of the 27th Annual Conference of the Cognitive Science Society, Vancouver, BC, 26–29 July 2006* (Lawrence Erlbaum, Mahwah, NJ, 2006), pp. 411–416
3. J.A. Kaminski, V.M. Sloutsky, A.F. Heckler, The advantage of abstract examples in learning math. Science **320**, 454–455 (2008)
4. L.M. Lesser, M.A. Tchoshanov, The effect of representation and representational sequence on students' understanding, eds. by G.M. Lloyd, M. Wilson, J.M.L. Wilkins, S.L. Behm in *Proceedings of the 27th Annual Meeting of the North American Chapter of the International Group for the Psychology of Mathematics Education PME–NA'2005*, Roanoke, Virginia, 20–23 October 2005
5. L.M. Lesser, M.A. Tchoshanov, Selecting representations. Tex. Math. Teach. **53**(2), 20–26 (2006)
6. S. Paper, M.A. Tchoshanov, The role of representation(s) in developing mathematical understanding. Theory Pract. **40**(2), 118–127 (2001)
7. M.A. Tchoshanov, *Visual Mathematics* (ABAK Publ, Kazan, 1997). (in Russian)
8. J. Van Patten, C. Chao, C. Reigeluth, A review of strategies for sequencing and synthesizing instruction. Rev. Educ. Res. **56**(4), 437–471 (1986)

Chapter 15
Spiral Curriculum: Towards Mathematical Foundations

One of the fundamental ideas of modern education is the idea of spiral curriculum, when students repeatedly revisit the same sequence of topics at the increasing levels of depth, detail, and sophistication. In this chapter, we show that under reasonable assumptions, the optimal sequence of presenting the material should indeed follow a spiral pattern.

The results from this chapter first appeared in [1].

What is a spiral curriculum. The concept of a spiral curriculum was developed by Harvard's professor Jerome S. Bruner, a one-time President of the American Psychological Association, and one of the major modern thinkers on education (see, e.g., [2]). This notion was first developed in his classic monograph [3].

The main idea of the spiral curriculum is that students repeatedly revisit the same sequence of topics at the increasing levels of depth, detail, and sophistication. This idea has been further developed in his following books [4–9].

This approach has been successfully used at all levels of education, from the kindergarten level to the level of university education.

What we do in this chapter. In this chapter, we:

- describe the problem of selecting the optimal sequence of presenting the material,
- formalize this problem as a precise optimization problem, and
- prove that the resulting optimal sequence is indeed a spiral.

Thus, we provide an additional mathematical justification of the spiral curriculum approach.

What techniques we use. In this chapter, we use techniques based on symmetries and symmetry groups.

These techniques have been successfully used in physics. In particular, in [10, 11], we used symmetry techniques to describe optimal trajectories and optimal shapes – e.g., including optimal shapes of celestial bodies; see also [12].

© Springer-Verlag GmbH Germany 2018
O. Kosheleva and K. Villaverde, *How Interval and Fuzzy Techniques Can Improve Teaching*, Studies in Computational Intelligence 750,
https://doi.org/10.1007/978-3-662-55993-2_15

In [13, 14], we extended these techniques to education problems. In this chapter, we use these techniques to explain the optimality of spiral curriculum.

It is reasonable to restrict ourselves to finite-parametric descriptions. Let us start with an appropriate description of the set of all the material that needs to be learned.

The ultimate objective of our mathematical analysis is to help the educators with selecting the best sequence of presenting the material. For this purpose, we must be able to present different parts of the material in the computer.

Inside any given computer, we can only store finitely many bits, and therefore, we can represent the information only about finitely many parameters. So, it is reasonable to restrict ourselves to *finite-dimensional descriptions*.

In other words, different parts of the material can be described by the values finitely many parameters $x_1, \ldots, x_n$. So, each part of the material can be represented as a point in a n-dimensional space R^n.

We need a curve, not a trajectory. Ideally, we should describe which part of the material should be presented at different moments of time t. In other words, we need to know the optimal dependence $x(t)$ of the material presented at moment t on this moment of time. So, from the mathematical viewpoint, it is desirable to describe a *trajectory*, i.e., a mapping from the set of real numbers (representing time) to the n-dimensional space R^n.

From the practical viewpoint, however, there are two aspects to this choice:

- first, we must select the order in which we present the material;
- second, we must select the speed with which the material is presented.

Selecting the proper order is difficult. Once the order is selected, selecting the speed of presentation is much easier: we just need to keep track of how fast students learn, and make sure that they learn the material to the desired level before moving to the next part of the material.

In view of this fact, in the following text, we will concentrate on the most difficult part of the problem: selecting the proper order. From the mathematical viewpoint, the order is represented by the corresponding *curve*, i.e., by a 1-dimensional set describing the covered material. In these terms, we would be looking for the optimal curve.

We need a family of curves, not a single curve. The optimal sequence of presenting the material depends on what exactly we want: e.g., as we have shown in [13, 14], this sequence depends on

- whether we want the students to be able to flawlessly apply this knowledge right away or
- whether we allow for a follow-up training period (like internship for medical doctors) in which they work under proper supervision.

As a result, we do not expect to find a *single* optimal curve, we expect to find a *family* of curves which are optimal under different optimality criteria.

In the computer, we can only use finite-parametric families of curves. As we have mentioned earlier, the ultimate objective of our mathematical analysis is to help the educators with selecting the best sequence of presenting the material. For this purpose, we must be able to present different curves in the computer.

From the purely mathematical viewpoint, we can have families characterized by *infinite* number of parameters: e.g., the family of all possible curves. However, inside any given computer, we can only store finitely many bits, and therefore, we can represent the information only about *finitely many* parameters. So, it is reasonable to restrict ourselves to *finite-dimensional family of curves*.

Main problem: which families of curves should we choose. In principle, different families of curves can be used. Therefore, it is important to choose the right family.

Currently, this choice is mainly made *ad hoc*, at best, by testing a few possible families and choosing the one that performs the best on a few benchmarks. Since only a few families are analyzed, we are not sure that we did not miss the real good approximating family. (And since only a few benchmarks are used for comparison, we are not sure that the chosen family is indeed the best one.) It is, therefore, desirable to find the *optimal* family of curves.

What is "optimality criterion". Our objective is to find an optimal family of curves. When we say "optimal", we mean optimal w.r.t. to some *optimality criterion*. When we say that some *optimality criterion* is given, we mean that, given two different families of approximating sets, we can decide whether the first one is better, or that the second one is better, or that these families are of the same quality w.r.t. the given criterion. In mathematical terms, this means that we have a *pre-ordering relation* $\preceq$ on the set of all possible finite-dimensional families of sets.

We want to solve an ambitious problem: enumerate all finite-dimensional families of curves that are optimal relative to some natural criteria. One way to approach the problem of choosing the "best" family of curves is to select *one* optimality criterion, and to find a family of curves that is the best with respect to this criterion. The main drawback of this approach is that there can be different optimality criteria, and they can lead to different optimal solutions.

It is, therefore, desirable not only to describe a family of curves that is optimal relative to *some* criterion, but to describe *all* families of curves that can be optimal relative to different natural criteria. In this chapter, we are planning to undertake exactly this more ambitious task.

Numerical optimality criteria. Pre-ordering is the general formulation of optimization problems in general, not only of the problem of choosing a family of sets. In general optimization theory, in which we are comparing arbitrary *alternatives* $A, B, \ldots$, from a given set $\mathscr{A}$, the most frequent case of such a pre-ordering is when a *numerical criterion* is used, i.e., when a function $J : \mathscr{A} \to R$ is given for which $A \preceq B$ if and only if $J(A) \leq J(B)$.

For example, we may want to select a family for which the speed of learning the given material is the largest, or the percentage of material retained under a certain period of time is the largest. For both criteria, the speed and the percentage depend on the student body. We can therefore consider worst-case criteria – by selecting the curve under which the guaranteed learning time if the smallest or the guaranteed retention percentage is the largest. Alternative, we can consider average-case criteria – by selecting the curve under which the average learning time is the smallest or the average retention percentage is the largest.

Non-numerical optimality criteria naturally appear. For "worst-case" optimality criteria, it often happens that there are several different alternatives that perform equally well in the worst case, but whose performance differ drastically in the average cases. In this case, it makes sense, among all the alternatives with the optimal *worst-case* behavior, to choose the one for which the *average* behavior is the best possible. This very natural idea leads to the optimality criterion that is *not* described by a numerical optimality criterion $J(A)$: in this case, we need *two* functions: $J_1(A)$ describes the worst-case behavior, $J_2(A)$ describes the average-case behavior, and $A \preceq B$ if and only if:

- either $J_1(A) < J_2(B)$,
- or $J_1(A) = J_1(B)$ and $J_2(A) \leq J_2(B)$.

We could further specify the described optimality criterion and end up with *a* natural criterion. However, as we have already mentioned, the goal of this chapter is not to find *a* family of curves that is optimal relative to some criterion, but to describe *all* families of curves that are optimal relative to some natural optimality criteria. In view of this goal, in the following text, we will not specify the criterion, but, vice versa, we will describe a very general class of *natural* optimality criteria.

So, let us formulate what "natural" means.

The optimality criterion must be invariant with respect to reasonable symmetries. Problems related to geometric sets often have natural *symmetries*. For example, in our presentation of knowledge, we may want to rotate the original coordinates $x_1, \ldots, x_d$ and consider new ones. The choice of a *rotated* coordinate system is equivalent to rotating all the points: $x \to R(x)$. As a result, in the new coordinates, each curve $X \in A$ from a family of curves A will be described by a "rotated" curve $R(X) = \{R(x) \mid x \in X\}$, and the original family A turns into a "rotated" family $R(A) = \{R(X) \mid X \in A\}$. It is reasonable to require that the relative quality of the two families of sets do not change under this rotation, if A is better than B, then $R(A)$ is better than $R(B)$.

Another reasonable symmetry is re-scaling. Usually, the choice of units to describe the parameters x_i is rather arbitrary. If we replace the original unit with a new unit which is λ times smaller, then the new values of the parameter are increased by a factor of λ, i.e., $x \to \lambda \cdot x$. It is, therefore, natural to require that the desired optimality criterion be invariant with respect to such rescalings.

The criterion must be final. If the criterion does not select any family as an optimal one, i.e., if, according to this criterion, none of the families is better than the others, then this criterion is of no use in selection.

If the criterion considers several different families equally good, then we can always use some other criterion to help select between these "equally good" ones, thus designing a two-step criterion. If this new criterion still does not select a unique family, we can continue this process until we arrive at a combination multi-step criterion for which there is only one optimal family.

Therefore, we can always assume that our criterion is *final* in the sense that it selects one and only one optimal family.

Definitions and the general result. Our goal is to choose the best finite-parametric family of curves. To formulate this problem precisely, we must formalize what a finite-parametric family is and what it means for a family to be optimal. In accordance with our informal description, both formalizations will use natural symmetries.

So, we will first formulate how symmetries can be defined for families of sets, then what it means for a family of sets to be finite-dimensional, and finally, how to describe an optimality criterion. Curves will be then introduced as particular cases of sets.

Definition 15.1 Let $g : M \to M$ be a 1-1-transformation of a set M, and let A be a family of subsets of M. For each set $X \in A$, we define the result $g(X)$ of applying this transformation g to the set X as $\{g(x) \mid x \in X\}$, and we define the result $g(A)$ of applying the transformation g to the family A as the family $\{g(X) \mid X \in A\}$.

Definition 15.2 Let M be a smooth manifold. A group G of transformations $M \to M$ is called a *Lie transformation group*, if G is endowed with a structure of a smooth manifold for which the mapping $g, a \to g(a)$ from $G \times M$ to M is smooth.

We want to define r-parametric families sets in such a way that symmetries from G would be computable based on parameters. Formally:

Definition 15.3 Let M and N be smooth manifolds.

- By a *multi-valued function $F : M \to N$* we mean a function that maps each $m \in M$ into a discrete set $F(m) \subseteq N$.
- We say that a multi-valued function is *smooth* if for every point $m_0 \in M$ and for every value $f_0 \in F(m)$, there exists an open neighborhood U of m_0 and a smooth function $f : U \to N$ for which $f(m_0) = f_0$ and for every $m \in U$, $f(m) \subseteq F(m)$.

Definition 15.4 Let G be a Lie transformation group on a smooth manifold M.

- We say that a class A of closed subsets of M is *G-invariant* if for every set $X \in A$, and for every transformation $g \in G$, the set $g(X)$ also belongs to the class.
- If A is a G-invariant class, then we say that A is a *finitely parametric family of sets* if there exist:

  - a (finite-dimensional) smooth manifold V;
  - a mapping s that maps each element $v \in V$ into a set $s(v) \subseteq M$; and
  - a smooth multi-valued function $\Pi : G \times V \to V$

such that:

- the class of all sets $s(v)$ that corresponds to different $v \in V$ coincides with A, and
- for every $v \in V$, for every transformation $g \in G$, and for every $\pi \in \Pi(g, v)$, the set $s(\pi)$ (that corresponds to π) is equal to the result $g(s(v))$ of applying the transformation g to the set $s(v)$ (that corresponds to v).

- Let $r > 0$ be an integer. We say that a class of sets B is a r-*parametric class* of sets if there exists a finite-dimensional family of sets A defined by a triple (V, s, Π) for which B consists of all the sets $s(v)$ with v from some r-dimensional sub-manifold $W \subseteq V$.

Definition 15.5 Let $\mathscr{A}$ be a set, and let G be a group of transformations defined on $\mathscr{A}$.

- By an *optimality criterion*, we mean a *pre-ordering* (i.e., a transitive reflexive relation) $\preceq$ on the set $\mathscr{A}$.
- An optimality criterion is called G-*invariant* if for all $g \in G$, and for all $A, B \in \mathscr{A}$, $A \preceq B$ implies $g(A) \preceq g(B)$.
- An optimality criterion is called *final* if there exists one and only one element $A \in \mathscr{A}$ that is preferable to all the others, i.e., for which $B \preceq A$ for all $B \neq A$.
- An optimality criterion is called G-*natural* if it is G-invariant and final.

Definition 15.6 Let M be a smooth manifold. A set $X \subset M$ *is called a* curve if it is an image of a smooth mapping from a real line to M.

Theorem 15.1 *Let M be a manifold, let G be a d-dimensional Lie transformation group on M, and let $r < d$ be a positive integer. Let $\mathscr{A}$ denote the class of all r-parametric families of sets from M, and let $\preceq$ be a G-natural optimality criterion on the class $\mathscr{A}$. Then:*

- *the $\preceq$-optimal family A_{opt} is G-invariant; and*
- *each set X from the $\preceq$-optimal family A_{opt} is a union of orbits of $\geq (d - r)$-dimensional subgroups of the group G.*

Proof Since the criterion $\preceq$ is final, there exists one and only one optimal family of sets. Let us denote this family by A_{opt}.

1. Let us first show that this family A_{opt} is indeed G-invariant, i.e., that $g(A_{\text{opt}}) = A_{\text{opt}}$ for every transformation $g \in G$.

Indeed, let $g \in G$. From the optimality of A_{opt}, we conclude that for every family $B \in \mathscr{A}$, $g^{-1}(B) \preceq A_{\text{opt}}$. From the G-invariance of the optimality criterion, we can now conclude that $B \preceq g(A_{\text{opt}})$. This is true for all $B \in \mathscr{A}$ and therefore, the family $g(A_{\text{opt}})$ is optimal. But since the criterion is final, there is only one optimal family; hence, $g(A_{\text{opt}}) = A_{\text{opt}}$. So, A_{opt} is indeed invariant.

2. Let us now show an arbitrary set X_0 from the optimal family A_{opt} consists of orbits of $\geq (d - r)$-dimensional subgroups of the group G.

Indeed, the fact that A_{opt} is G-invariant means, in particular, that for every $g \in G$, the set $g(X_0)$ also belongs to A_{opt}. Thus, we have a (smooth) mapping $g \to g(X_0)$ from the d-dimensional manifold G into the $\leq r$-dimensional set $G(X_0) = \{g(X_0) \mid g \in G\} \subseteq A_{opt}$. In the following, we will denote this mapping by g_0.

Since $r < d$, this mapping cannot be 1-1, i.e., for some sets $X = g'(X_0) \in G(X_0)$, the pre-image $g_0^{-1}(X) = \{g \mid g(X_0) = g'(X_0)\}$ consists of one than one point. By definition of $g(X)$, we can conclude that $g(X_0) = g'(X_0)$ iff $(g')^{-1}g(X_0) = X_0$. Thus, this pre-image is equal to $\{g \mid (g')^{-1}g(X_0) = X_0\}$. If we denote $(g')^{-1}g$ by $\tilde{g}$, we conclude that $g = g'\tilde{g}$ and that the pre-image $g_0^{-1}(X) = g_0^{-1}(g'(X_0))$ is equal to $\{g'\tilde{g} \mid \tilde{g}(X_0) = X_0\}$, i.e., to the result of applying g' to $\{\tilde{g} \mid \tilde{g}(X_0) = X_0\} = g_0^{-1}(X_0)$. Thus, each pre-image $(g_0^{-1}(X) = g_0^{-1}(g'(X_0)))$ can be obtained from one of these pre-images (namely, from $g_0^{-1}(X_0)$) by a smooth invertible transformation g'. Thus, all pre-images have the same dimension D.

We thus have a *stratification* (fiber bundle) of a d-dimensional manifold G into D-dimensional strata, with the dimension D_f of the factor-space being $\leq r$. Thus, $d = D + D_f$, and from $D_f \leq r$, we conclude that $D = d - D_f \geq n - r$.

So, for every set $X_0 \in A_{opt}$, we have a $D \geq (n - r)$-dimensional subset $G_0 \subseteq G$ that leaves X_0 invariant (i.e., for which $g(X_0) = X_0$ for all $g \in G_0$). It is easy to check that if $g, g' \in G_0$, then $gg' \in G_0$ and $g^{-1} \in G_0$, i.e., that G_0 is a *subgroup* of the group G. From the definition of G_0 as $\{g \mid g(X_0) = X_0\}$ and the fact that $g(X_0)$ is defined by a smooth transformation, we conclude that G_0 is a smooth sub-manifold of G, i.e., a $\geq (n - r)$-dimensional subgroup of G.

To complete our proof, we must show that the set X_0 is a union of orbits of the group G_0. Indeed, the fact that $g(X_0) = X_0$ means that for every $x \in X_0$, and for every $g \in G_0$, the element $g(x)$ also belongs to X_0. Thus, for every element x of the set X_0, its entire orbit $\{g(x) \mid g \in G_0\}$ is contained in X_0. Thus, X_0 is indeed the union of orbits of G_0. $\square$

We need to describe appropriate symmetries. The above general result was formulated in terms of general symmetry groups. So to apply this general result to our education problem, it is necessary to find appropriate symmetries.

In geometric situations described in [10, 11], we have natural geometric symmetries like rotation. In the education problems, there are no natural symmetries, but we must find the appropriate symmetries in some indirect way.

To be able to describe symmetries, we first describe closeness. To describe symmetries, we will describe "closeness" between different parts of the material, and define symmetries as transformations which preserves this closeness – so that if x and y are ρ-close they either remain ρ-close after this transformation, or at least become $f(\rho)$-close for an appropriate function $f(\rho)$.

Closeness means that for each value of a distance ρ and for each point x, we have a set of all the points whose distance to x does not exceed x. We have different sets for different closeness criteria, for different points x, and for different value ρ. Let us describe the family of all such sets.

Linear transformations. It is reasonable to restrict ourselves to *linear* transformations $R^n \to R^n$, i.e., of all transformations of the type $x_i \to a_i + \sum_j a_{ij} x_j$ with an invertible matrix a_{ij}.

Main result. We will show that the ellipsoids are the *simplest* optimal family, i.e., that of all possible optimal finite-parametric families that correspond to different G_e-invariant optimality criteria, ellipsoids have the smallest number of parameters.

Definition 15.7 By a closed domain, we mean a closed set that is equal to the closure of the set of its interior points.

Theorem 15.2 *Let $n > 0$ be an integer, $M = R^n$, G_e be the group of all affine transformations, and $\preceq$ be a natural (i.e., G_e-invariant and final) optimality criterion on the class $\mathscr{A}$ of all r-parametric families of connected bounded closed domains from R^n. Then:*

- $r \geq n(n+3)/2$;
- *if $r = n(n+3)/2$, then the optimal family coincides either with the family of all ellipsoids, or, for some $\lambda \in (0, 1)$, with the family of all regions obtained from ellipsoids by subtracting λ times smaller homothetic ellipsoids.*

Comment. If we restrict ourselves to *convex* sets (or only to simply connected sets), we get ellipsoids only.

Proof Due to Theorem 15.1, the optimal family A_{opt} is affine invariant, i.e., for every $X \in A_{\text{opt}}$, and for every transformation $g \in G_e$, the set $g(X)$ also belongs to A_{opt}.

1. Let us first show that $r \geq n(n+3)/2$. Indeed, it is known (see, e.g., [15]) that for every open bounded set X, among all ellipsoids that contain X, there exists a unique ellipsoid E of the smallest volume. We will say that this ellipsoid E *corresponds* to the set X. Let us consider the set of ellipsoids $\mathscr{E}_c$ that correspond (in this sense) to all possible sets $X \in A_{\text{opt}}$.

Let us fix a set $X_0 \in A_{\text{opt}}$, and let E_0 denote an ellipsoid that corresponds to X_0.

An arbitrary ellipsoid E can be obtained from any other ellipsoid (in particular, from E_0) by an appropriate affine transformation g: $E = g(E_0)$. The ratio of volumes is preserved under arbitrary linear transformations g; hence, since the ellipsoid E_0 is the smallest volume ellipsoid that contains X_0, the ellipsoid $E = g(E_0)$ is the smallest volume ellipsoid that contains $g(X_0) = X$.

Hence, an arbitrary ellipsoid $E = g(E_0)$ corresponds to some set $g(X_0) \in A_{\text{opt}}$. Thus, the family $\mathscr{E}_c$ of all ellipsoids that correspond to sets from A_{opt} is simply equal to the set $\mathscr{E}$ of all ellipsoids. Thus, we have a (locally smooth) mapping from an r-dimensional set A_{opt} onto the $\dfrac{n(n+3)}{2}$-dimensional set of all ellipsoids. Hence,

$$r \geq n(n+3)/2.$$

2. Let us now show that for $r = n(n+3)/2$, the only G_e-invariant families A are ellipsoids and "ellipsoid layers" (described in Theorem 15.2).

Indeed, let X_0 be an arbitrary set from the invariant family, and let E_0 be the corresponding ellipsoid. Let $g_0 \in G_e$ be an affine transformation that transform E_0 into a ball $E_1 = g(E_0)$. This ball then contains the set $X_1 = g_0(E_0) \in A_{\mathrm{opt}}$.

Let us show, by reduction to a contradiction, that the set X_1 is invariant w.r.t. arbitrary rotations around the center of the ball E_1. Indeed, if it is not invariant, then the set R of all rotations that leave X_1 invariant is different from the set of all rotations $SO(n)$. Hence, R is a proper closed subgroup of $SO(n)$. From the structure of $SO(n)$, it follows that there exists a 1-parametric subgroup R_1 of $SO(n)$ that intersects with R only in the identity transformation 1. This means that if $g \in R_1$ and $g \neq 1$, we have $g \notin R$, i.e., $g(X_1) \neq X_1$.

If $g(X_1) = g'(X_1)$ for some $g, g' \in R_1$, then we have $g^{-1}g'(X_1) = X_1$, where $g^{-1}g' \in R_1$. But such an equality is only possible for $g^{-1}g' = 1$, i.e., for $g = g'$. Thus, if $g, g' \in R_1$ and $g \neq g'$, then the sets $g(X_1)$ and $g'(X_1)$ are different. In other words, all the sets $g(X_1)$, $g \in R_1$, are different.

Since the family A is G_e-invariant, all the sets $g(X_1)$ for all $g \in R_1 \subseteq G_e$ also belong to A. For all these sets, the corresponding ellipsoid is $g(E_1)$, the result of rotating the ball E_1, i.e., the same ball $g(E_1) = E_1$. Hence, we have a 1-parametric family of sets contained in the ball E_1.

By applying appropriate affine transformations, we will get 1-parametric families of sets from A in an arbitrary ellipsoid. So, we have an $n(n+3)/2$-dimensional family of ellipsoids, and inside each ellipsoid, we have a 1-dimensional family of sets from A. Thus, A would contain a $\left(\dfrac{n(n+3)}{2} + 1 \right)$-parametric family of sets, which contradicts to our assumption that the dimension r of the family A is exactly

$$\frac{n(n+3)}{2}.$$

This contradiction shows that our initial assumption was false, and for

$$r = \frac{n(n+3)}{2},$$

the set X_1 is invariant w.r.t. rotations. Hence, with an arbitrary point x, the set X_1 contains all the points that can be obtained from x by arbitrary rotations, i.e., the entire sphere that contains x. Since X_1 is connected, X_1 is either a ball, or a ball from which a smaller ball was deleted.

The original set $X_0 = g_0^{-1}(X_1)$ is an affine image of this set X_1, and therefore, X_0 is either an ellipsoid, or an ellipsoid with an ellipsoidal hole inside. $\square$

Resulting symmetries. The reason we decided to consider closeness is that we wanted to describe symmetries as (linear) transformations which preserve closeness.

From the previous section, we know that closeness is described by ellipsoids. By an appropriate selection of coordinates, every ellipsoid can be described as a sphere (it is sufficient to take eigenvectors of the corresponding quadratic form as coordinates).

Thus, in the appropriate coordinates, closeness is described by Euclidean distance $\rho(x, y) = \sqrt{(x_1 - y_1)^2 + \ldots + (x_n - y_n)^2}$. Transformations preserving Euclidean distance are rotations and shifts. Since we are also allowing transformations which change the distance – but keep equal distances equal – we must also allow dilations $x \to \lambda \cdot x$.

So, the natural group of symmetries G_a is generated by shifts, rotations, and dilations.

Main result: mathematical justification of the spiral curriculum method. We have already argued that for the above "optimal education" problem, the natural group of symmetries G_a is generated by shifts, rotations, and dilations. In accordance with Theorem 15.1, to find the optimal curves, we must therefore describe all 1-dimensional orbits of subgroups of this group G_a.

How we can solve the corresponding mathematical problem. A 1-D orbit is an orbit of a 1-D subgroup. This subgroup is uniquely determined by its "infinitesimal" element, i.e., by the corresponding element of the Lie algebra of the group G. This Lie algebra if easy to describe. For each of its elements, the corresponding differential equation (that describes the orbit) is reasonably straightforward to solve.

General result. One can see that in general, the resulting curve is indeed a spiral – although some limit cases are also possible, and a limit of a spiral may have as different shape.

3-D case. Let us illustrate all the possibilities in the 3-D case. A generic 1-dimensional orbit of the corresponding group G_a is a *conic spiral* that is described (in cylindrical coordinates) by the equations $z = k\rho$ and $\rho = R_0 \exp(c\varphi)$. Its limit cases are:

- a *logarithmic* (Archimedean) *spiral*: a planar curve ($z = 0$) that is described (in polar coordinates) by the equation $\rho = R_0 \exp(c\varphi)$.
- a *cylindrical spiral*, that is described (in appropriate coordinates) by the equations $z = k\phi$, $\rho = R_0$.
- a *circle* ($z = 0$, $\rho = R_0$);
- a *semi-line* (*ray*); and
- a *straight line*.

2-D case. In the 2-D case, the general case if a logarithmic spiral, and the limit cases are a circle, a semi-line, and a straight line.

Comment about the 3-D case. In the 3-D case, there is an alternative (slightly more geometric) way of describing 1-D orbits: by taking into consideration that an orbit, just like any other curve in a 3-D space, is uniquely determined by its curvature $\kappa_1(s)$ and torsion $\kappa_2(s)$, where s is the arc length measured from some fixed point.

The fact that this curve is an orbit of a 1-D group means that for every two points x and x' on this curve, there exists a transformation $g \in G$ that maps x into x'. Shifts and rotations do not change κ_i, they may only shift s (to $s + s_0$); dilations also change s to $s \to \lambda \cdot s$ and change the numerical values of κ_i. So, for every s, there exist $\lambda(s)$ and $s_0(s)$ such that the corresponding transformation turns a point corresponding. to $s = 0$ into a point corresponding to s.

As a result, we get functional equations that combine the two functions $\kappa_i(s)$ and these two functions $\lambda(s)$ and $s_0(s)$. Taking an infinitesimal value s in these functional equations, we get differential equations, whose solution leads to the desired 1-D orbits.

Conclusion. We start with the problem of describing an optimal curve, i.e., an optimal order of presenting the material. As a result of our analysis, we show that for every reasonable optimality criterion, all the curves from the optimal family of curves are spirals (or limits of spirals).

Thus, we have indeed provided a mathematical justification for the spiral curriculum.

References

1. O. Kosheleva, Spiral curriculum: towards mathematical foundations, in *Proceedings of the International Conference on Information Technology InTech'07*, Sydney, Australia, 12–14 December 2007. pp. 152–157
2. H. Gardner, Jerome S. Bruner, in *Fifty Modern Thinkers on Education. From Piaget to the present*, ed. by J.A. Palmer (Routledge, London, 2001)
3. J. Bruner, *The Process of Education* (Harvard University Press, Cambridge, 1960)
4. J.S. Bruner, *Toward a Theory of Instruction* (Belkapp Press, Cambridge, 1966)
5. J.S. Bruner, *The Relevance of Education* (Norton, New York, 1971)
6. J. Bruner, *Child's Talk: Learning to Use Language* (Norton, New York, 1983)
7. J. Bruner, *Actual Minds, Possible Worlds* (Harvard University Press, Cambridge, 1986)
8. J. Bruner, *The Culture of Education* (Harvard University Press, Cambridge, 1996)
9. J. Bruner, J. Goodnow, A. Austin, *A Study of Thinking* (Wiley, New York, 1956)
10. A. Finkelstein, O. Kosheleva, V. Kreinovich, Astrogeometry, error estimation, and other applications of set-valued analysis. ACM SIGNUM Newsl. **31**(4), 3–25 (1996)
11. A. Finkelstein, O. Kosheleva, V. Kreinovich, Astrogeometry: towards mathematical foundations. Int. J. Theor. Phys. **36**(4), 1009–1020 (1997)
12. S. Li, Y. Ogura, V. Kreinovich, *Limit Theorems and Applications of Set Valued and Fuzzy Valued Random Variables* (Kluwer Academic Publishers, Dordrecht, 2002)
13. R. Aló, O. M. Kosheleva, Optimization techniques under uncertain criteria, and their possible use in computerized education, in *Proceedings of the 25th International Conference of the North American Fuzzy Information Processing Society NAFIPS'2006*, Montreal, Quebec, Canada, IEEE Press, Piscataway, New Jersey, 3–6 June 2006
14. O.M. Kosheleva, R. Aló, Towards economics of education: optimization under uncertainty, in *Proceedings of the Second International Conference on Fuzzy Sets and Soft Computing in Economics and Finance FSSCEF'2006*, St. Petersburg, Russia, June 28– July 1, 2006, pp. 63–70
15. H. Busemann, *The Geometry of Geodesics* (Academic Press, New York, 1955)

Chapter 16
How Much Time to Allocate to Each Topic?

Since we cannot spend as much time as we would like to on teaching all the topics, it is necessary to optimally distribute the limited amount of time between different topics. In this chapter, we explain how general techniques of optimization under uncertainty can be used in education.

The results from this chapter first appeared in [1, 2].

Basic assumptions about training. In order to find the optimal training schedule, let us make some (simplifying but realistic) assumptions about training.

In principle, there are several different types of items that we want a student to learn; for example:

- when we teach typing, we want the student to acquire the motor skills of typing all the symbols on the keyboard and all pairs of consequent symbols;
- when we teach words from a foreign language, we want the student to learn all these words;
- when we teach cross-country driving, we want the student to develop motor skills corresponding to different types of terrain: flat surface, rugged terrain, uphill, downhill, narrow bridge, etc.

Let us denote the total number of types to learn by T. For simplicity, we assume that acquiring skills necessary for each of these types takes the same number of training situations s. So, to learn all necessary types, a students needs at least $T \cdot s$ repetitions. If we denote, by T_0, the time necessary for handling each repetition, then the total time for training a student for all necessary types is equal to $T_0 \cdot T \cdot s$.

In many learning situations, the total number T of necessary types is large, so the above time of total training is unrealistically large. Therefore, we cannot expect every single student to be 100% skilled in every possible situation type.

Since we cannot train a student to be skilled in every possible situation, it is therefore necessary to train a student in such a way that the student will be able to handle *the largest possible number* of these types.

© Springer-Verlag GmbH Germany 2018
O. Kosheleva and K. Villaverde, *How Interval and Fuzzy Techniques Can Improve Teaching*, Studies in Computational Intelligence 750,
https://doi.org/10.1007/978-3-662-55993-2_16

In future applications, some of these types are more frequent, some are less frequent. So, if we know that a student can only learn, say, t different words, and we have to choose which of these words the student will learn perfectly well, we should choose t most frequent ones.

A *skill* of a student can be thus characterized by the number t of the types of items in which this student is well skilled.

To estimate frequencies of different types, we can use a general (semi-empirical) law discovered by G. K. Zipf (see, e.g., [3, 4]), according to which, if we order types from the most frequent to the least frequent one, then the frequency f_i of ith type is proportional to $1/i$: $f_i = c/i$ for some constant c. The value of this constant can be determined from the fact that the sum of all these frequencies should be equal to 1: $f_1 + \cdots + f_T = 1$. Since $1 + 1/2 + \cdots + 1/T \approx \ln(T)$, we thus conclude that $c \cdot \ln(T) = 1$, $c = 1/\ln(T)$, and

$$f_i = \frac{1}{\ln(T) \cdot i}. \tag{16.1}$$

Traditional learning. In traditional learning of a language, a student in trained on texts from real language. In traditional typing lessons, student learn to type by typing real-life texts. In all these cases, a student is trained on a real-life flow of items.

Let us denote by I the time allocated for training. Since handling each repetition takes time T_0, during this training time, the trainee will see $N = I/T_0$ repetitions. According to our assumption about the training time, the student will be trained only in those types i for which he or she has seen at least s repetitions. Out of the total of N repetitions, the student will see $N \cdot f_i$ repetitions of ith type; so, the student will be trained in all the types for which $N \cdot f_i \geq s$. Substituting Zipf's expression (1) for f_i, we conclude that the student will learn all the types i for which $\dfrac{I}{T_0} \cdot \dfrac{1}{\ln(T) \cdot i} \geq s$, i.e., for which $i \leq \dfrac{I}{T_0 \cdot \ln(T) \cdot s}$. Therefore, the resulting student's skill level t (i.e., the total number of types in which this student will be skilled), will be equal to

$$t = \frac{I}{T_0 \cdot \ln(T) \cdot s}. \tag{16.2}$$

This formula describes the skill level acquired during a given training time I.

We can also consider the inverse problem: we want a student to be trained for a certain skill level t, and we need to know the time I required for this training. From the formula 16.2, we can conclude that

$$I = t \cdot T_0 \cdot \ln(T) \cdot s. \tag{16.3}$$

Optimal training. We can generate repetitions in arbitrary order, not necessarily with real-life frequencies. If we want a student to be trained on t different types, then we need to generate exactly s repetitions of this type.

If we fix the total training time I, then during this time, we can generate $N = I/T_0$ repetitions. Since learning each type requires s repetitions, the total amount of different types in which a student can get skilled is equal to $t = N/s = (I/T_0) \cdot s$. Thus, after this training, the student will acquire the skill level

$$t = \frac{I}{T_0 \cdot s}. \tag{16.4}$$

This formula describes the skill level acquired during a given training time I.

We can also consider the inverse problem: we want a student to be trained for a certain skill level t, and we need to know the time I required for this training. From the formula 16.4, we can conclude that

$$I = t \cdot T_0 \cdot s. \tag{16.5}$$

Conclusion: optimal training is faster and better. By comparing the formulas 16.2 and 16.4, we conclude that during the same training time, the skill level acquired during the automated training can be much higher ($\ln(T)$ times higher) that the skill level acquired in traditional training.

Similarly, by comparing the formulas 16.3 and 16.5, we conclude that the training time necessary to acquire a given skill can be much shorter ($\ln(T)$ times shorter) for the automated training than for traditional training.

How to optimally combine classroom and field training. Usually, after classroom training, a students goes into realistic situations (field training) which solidify his training. How can we best organize this combined training?

Let us denote the time that we can allocate for classroom training by I_{au}, and the training time for the follow-up field training by I_{tr}. During the follow-up training, the student encounters $N_{tr} = I_{tr}/T_0$ repetitions. Of these repetitions, $N_{tr} \cdot f_i$ are of type i.

If this number of repetitions is $\geq s$, then for this type, the student acquires necessary skills during the follow-up training, so there is no need to simulate patients of this type during the automated training. Thus, we get all types from 1 to

$$t_{tr} = \frac{I_{tr}}{T_0 \cdot \ln(T) \cdot s} \tag{16.6}$$

covered.

For each type $i > t_{tr}$, we get $f_i \cdot N_{tr} = \dfrac{I_{tr}}{T_0 \cdot \ln(T) \cdot i} < s$ repetitions covered during traditional training. So, if we want the student to get the necessary skills, we must generate the remaining number of repetitions

$$n_i = s - \frac{I_{tr}}{T_0 \cdot \ln(T) \cdot i} \tag{16.7}$$

during the automated training.

We want to learn as many new types as possible. How many situation types can we thus learn? During the time I_{au}, we can only generate $N_{au} = I_{au}/T_0$ repetitions. Since learning type i requires n_i repetitions, the skill level t acquired by a student can be determined by the formula $\dfrac{I_{au}}{T_0} = N_{au} = \sum_{i=t_{tr}}^{t} n_i$. Substituting the above expression for n_i, we conclude that

$$\frac{I_{au}}{T_0} = s \cdot (t - t_{tr}) - \frac{I_{tr}}{T_0 \cdot \ln(T)} \cdot \sum_{i=t_{tr}}^{t} \frac{1}{i}.$$

Since $1 + 1/2 + \cdots + 1/i \approx \ln(i)$, we can rewrite this equation as

$$\frac{I_{au}}{T_0} = s \cdot (t - t_{tr}) - \frac{I_{tr} \cdot (\ln(t) - \ln(t_{tr}))}{T_0 \cdot \ln(T)}. \tag{16.8}$$

So, we can make two conclusions:

• If the training times I_{au} and I_{tr} are given, then the resulting acquired skill t can be determined from the equation 16.8, where t_{tr} is determined from the equation 16.6.
• Vice versa, if we know the training time I_{au} for the classroom training, and the required skill level t, then we must find t_{tr} for the equation 16.8, and then use the formula 16.6 to determine the necessary traditional training period as $I_{tr} = t_{tr} \cdot T_0 \cdot \ln(T) \cdot s$.

In both cases, the number of repetitions of different types $i = t_{tr}, t_{tr} + 1, \ldots, t$ generated during the classroom training is determined by the formula 16.7.

Other applications. In [5], we used a similar idea to optimize the types of virtual patients used by doctors during medical training – specifically, during a training of surgeons for spinal cord stimulation procedures; see, e.g., [6].

References

1. R. Aló, O.M. Kosheleva, Optimization techniques under uncertain criteria, and their possible use in computerized education, in *Proceedings of the 25th International Conference of the North American Fuzzy Information Processing Society NAFIPS'2006, Montreal, Quebec, Canada, 3–6 June 2006* (2006) (IEEE Press, Piscataway)
2. O.M. Kosheleva, R. Aló, Towards economics of education: optimization under uncertainty, in *Proceedings of the Second International Conference on Fuzzy Sets and Soft Computing in Economics and Finance FSSCEF'2006*, St. Petersburg, Russia, 28 June – 1 July (2006), pp. 63–70
3. B. Mandelbrot, *The Fractal Geometry of Nature* (Freeman, San Francisco, 1983)
4. G.K. Zipf, *Human Behavior and the Principle of Least-Effort* (Addison-Wesley, Cambridge, 1949)

5. R. Aló, K. Aló, V. Kreinovich, Towards intelligent virtual environment for training medical doctors in surgical pain relief, in *Proceedings of The Eighth International Fuzzy Systems Association World Congress IFSA'99*, Taipei, Taiwan, 17–20 Aug 1999 (1999), pp. 260–264
6. S. Horsch, L. Clayes, *Spinal Cord Stimulation II* (Steinkopff, Darmstadt, 1995)

Chapter 17
What Is Wrong with Teaching to the Test: Uncertainty Techniques Help in Understanding the Controversy

In the USA, in the last decade, standards have been adapted for each grade level. These standards are annually checked by state-wide tests. The results of these tests often determine the school's funding and even the school's future existence. Due to this importance, a large amount of time is spent on teaching to the tests.

Most teachers believe that this testing approach is detrimental to student education. This belief seems to be empirically supported by the fact that so far, the testing approach has not led to spectacular improvements promised by its proponents. While this empirical evidence is reasonably convincing, the teacher community has not yet fully succeeded in clearly explaining their position to the general public – because the opposing argument (of the need for accountability) also seems to be reasonably convincing.

In this chapter, we show that the situation becomes much clearer if we take uncertainty into account – and that, hopefully, a proper use of uncertainty can help in resolving this situation.

The results from this chapter first appeared in [1].

What is "teaching to the test"? In the last few decades, in the US school education, state-wide tests have been developed for testing the mathematical knowledge of students at the end of each grade. Student performance on these state tests (and on similar tests in other disciplines) has become the most important criterion of how the performance of schools and teachers are gauged:

- Funding of individual schools is largely determined by the test results.
- In some cases, schools are disbanded and teachers are fired if the test results are unsatisfactory several years in a row.

Because of the importance of the test results, schools are understandably paying a large amount of attention to making sure that the students pass these tests. In other words, instead of spending most of the time teaching the material – as it was in the past – teachers now spend a significant amount of time teaching "to the test".

© Springer-Verlag GmbH Germany 2018
O. Kosheleva and K. Villaverde, *How Interval and Fuzzy Techniques Can Improve Teaching*, Studies in Computational Intelligence 750,
https://doi.org/10.1007/978-3-662-55993-2_17

The results of teaching to the test are not as spectacular as the proposers hoped. The main idea behind the tests sounds reasonable:

- if we do not gauge how well students are doing in a way that will enable us to compare different schools and different teachers,
- then how will we know which schools are doing better and which schools need improvement?

The authors of this idea expected that with well-established testing, the students' knowledge will drastically improve. Alas, these expectations turned out to be too optimistic:

- In some states and some school districts, there has been some improvement.
- However, overall, this program has not been a spectacular success as its proponents hoped.
- Moreover, there is anecdotal evidence that in many cases, with the introduction of state-wide testing, the students' knowledge actually decreased.

Teaching to the test: a current controversy. There is a current controversy, both in the media and in the scholarly publications (see, e.g., [2–4]) about the state-wide tests and teaching to the test.

In a simplified form, the controversy is fought along the following lines:

- on the one hand, many politicians believe that tests are a good idea (while they acknowledge the limitations and drawbacks of the existing tests, and the need to improve the tests);
- on the other hand, most teachers strongly believe that the entire approach of teaching to the tests is flawed and detrimental to education.

In the media, this controversy gets personal and nasty:

- Some politicians accuse the teacher community and the teacher unions of defending weak under-performing teachers at the expense of the student success.
- Some teachers accuse the politicians of ignorance-motivated over-simplified populist interference with a complex teaching process.

While there may be some anecdotal evidence in support of such accusations, the situation is definitely more complex than the simplified picture one may get from the media coverage:

- on the one hand, several knowledgable politicians, with successful teaching experience, are in favor of the state-wide tests,
- on the other hand, many very good teachers, teachers who are extremely successful in teaching, are strongly against the current emphasis on these tests.

Population is somewhat confused. One of the frustrating aspects of the current controversy is that the general population is somewhat confused about it:

- on the one hand, it is reasonable to require accountability, and this accountability logic naturally leads to the current testing program;

- on the other hand, respected teachers are against this program, and empirical evidence also shows that it has not led to spectacular successes – contrary to natural expectations motivated by accountability.

What we do in this chapter. In this chapter, we argue that the confusion – and, to some extent, the controversy itself – is largely due to the simplification of the complex pedagogical process. Specifically, we argue that if we properly take uncertainty into account, then the situation becomes much clearer.

In this explanation, we follow up on our previous research [5–13] on taking into account uncertainty (in particular, fuzzy uncertainty; see, e.g., [14, 15]) when gauging the pedagogical process.

The background of our main idea. In general, it is assumed that learning comes from repetitions:

- once a student has repeated a certain procedure certain number of times,
- the student has mastered it.

This is why an important part of learning each idea of high school mathematics is practice. For example:

- unless students do a lot of exercises where they have to add fractions,
- they will not master this skill well enough to be able to easily add two fractions, and
- this will hinder their progress in the following mathematical topics like dealing with polynomials (where the ability to add fractions is already assumed).

In general:

- the only way to learn to write is to practice writing,
- the only way to learn to spell is to practice spelling,
- the only way to learn a foreign language is to practice it, and
- the only way to learn the multiplication table is to practice multiplication.

The required number of repetitions depends:

- on the complexity of the topic,
- on the match between this particular topic and the student's individual interests and prior skills,
- etc.

However, the fact remains:

- for every topic and for every student,
- there is a number of iterations after which the student will master this topic.

From this viewpoint, let us analyze both the traditional teaching process and the new situation of teaching to the test.

Analysis of the traditional teaching process. The main objective of a school mathematics program is that after graduation, students should have certain skills. These skills often build on each other, so that one skill requires another one.

For example:

- In order to be able to solve quadratic equations, we need to know how to add, how to subtract, how to multiply, etc.
- In order to be able to handle polynomials with rational (fractional) coefficients, a student needs to be able to perform arithmetic operations with fractions, etc.

To illustrate our main idea, let us consider a simple sequence of two skills A and B, for which the skill B requires that the student also has learned skill A.

For example, we can take,

- as A, the ability to add, subtract, multiply, and divide fractions, and
- as B, the ability to process polynomials with rational coefficients.

For simplicity, let us assume that

- skill A is learned in one year, and
- skill B is learned in the following year.

At the end of the A-B sequence, a student should have mastered both skills A and B. Let us assume that the student needs

- n_A iterations to master skill A, and
- n_B iterations to master skill B.

This means that by the end of the school education, a student should have done

- n_A iterations of skill A and
- n_B iterations of skill B.

Since the skill B is only taught in Year 2, we should have all n_B iterations in Year 2. However, since practicing the skill B often involves practicing skill A, a student practices the skill A also when she learns skill B.

Not every problem related to learning skill B necessarily involves skill A. For example, when the students learn that $(a + b) \cdot (a - b) = a^2 - b^2$ or that $(a + b)^2 = a^2 + 2 \cdot a \cdot b + b^2$, these polynomial problems do not include any operations with fractions.

Let us denote by r the proportion of problems of type B that involve using skill A. Then, during n_B exercises needed to master skill B, the student, in effect, performs $r \cdot n_B$ exercises in which she practices skill A as well.

In the traditional teaching approach, without annual checks, where the only objective is mastery at the end, all we therefore need is to be have a total of n_A exercises in skill A by the end of Year 2.

- In Year 2 we, in effect, have $r \cdot n_B$ exercises in this skill.
- This means that it is sufficient to have $n_A - r \cdot n_B$ exercises in skill A in Year 1.

Yes, this number $n_A - r \cdot n_B$ is smaller than n_A, so by the end of Year 1, the students have not yet fully mastered skill A, but this is normal in education – the skills come with practice.

How situation changes when we teach to the test. What happens when we teach to the test?

- According to the school program, Year 1 is devoted to teaching skill A.
- So, to test how well the students learned after this year, it is reasonable to design a test with questions about skill A.

Once we give this test, the results are far from perfect – because, as we have mentioned, by the end of Year 1, the students only had $n_A - r \cdot n_B < n_A$ exercises, and they have not yet mastered this skill. The argument "Is this how much we want our graduates to know about A?" sounds convincing, so a pressure is placed on schools to improve the score on the test at the end of Year 1.

The only way to do it is to increase the number of skill-A-related exercises in Year 1 to n_A.

Teaching to the test: a seemingly positive result. Now, the test grades for Year 1 go up – because:

- in the past, the students did not have enough exercises to master skill A, while
- now, they have enough exercises, so they do master skill A at the end of Year 1.

The progress is visible, results are good. But are they?

Teaching to the test: while there is a drastic improvement in test scores, there is no significant improvement of the school graduates knowledge. Let us see how teaching to the test affects the main school objective – to make sure that the graduates learn both skills A and B at the end of both years. Let us show that with respect to this criterion, we should not expect any significant improvement. Indeed:

- in the past, we had a total of n_A exercises in skill A:
- now, the students have $n_A + r \cdot n_B$ exercises in skill A.

In both cases, we have enough exercises to master skill A. So, in both cases, we should have the same reasonably positive result.

Teaching to the test: a serious problem. At first glance, it may seem like teaching to the test may not be as spectacularly successful as its proponents claim, but overall, its results are positive – at least the test grades improved.

However, there *is* a serious problem. The problem is that school time is limited. The time that now schools use to practice additional $r \cdot n_B$ repetitions of skill A in Year 1 has to come at the expense of something else. Clearly, it comes at the expense of other topics that are not explicitly included in the statewide test.

As a result,

- while students' knowledge of the topics included in the test (like skills A and B) does not decrease (and many even increase),
- the students' mastery of some other skills will necessarily drastically decrease.

This is what teachers object to when they object to "teaching to the test".

We clarified the problem – but what is a solution? In the above text, we explained the problem. This explanation also helps to find a solution to this problem.

In order to compare different schools, to compare different teachers, we need to have some objective way of gauging the student success. But we need to make sure that this comparison does not hinder the students' progress.

In the ideal world, we should design better tests – this is one of the few things with which everyone agrees. However, even with the existing tests, we can drastically improve the situation if we *no longer require* that

- at the end of each school year,
- students should have a perfect knowledge of all the topics that they learned during this year.

This requirements comes from the "crisp" thinking, thinking that does not take uncertainty into account – a student either mastered the skill or did not. In reality, after a few exercises of the skill A, a student usually achieves mastery *to a degree*.

As a result, in the traditional approach,

- the student will have an imperfect score on the test for skill A at the end of Year 1;
- this is OK, as long as this score is what we should expect after $n_A - r \cdot n_B$ exercises, so that we will be sure that

  – after additional $r \cdot n_B$ exercises involving skill A in Year 2
  – the student will achieve the true mastery of skill A.

This should be the guideline to developing required satisfaction level on annual exams. Any increase of this satisfaction level should be *discouraged* because

- it would indicate that the teachers are over-emphasizing skill A in Year 1, while
- they could use fewer exercises of this particular skill and spend the remaining time teaching the students some other useful skills.

How fuzzy logic can help. Fuzzy logic has been explicitly designed to handle situations in which some property is true to a degree – so fuzzy logic seems to be a perfect tool for this analysis.

Our idea is more general than teaching-to-the-test controversy. Our main objective is to help in understanding and resolving the "teaching to the test" controversy. However, the same idea can be applied to all levels of education as well. We should not aim for perfect knowledge on intermediate classes.

For example, college students taking a computer science sequence may be somewhat shaky about programming at the end of the first class, but their basic skills are reinforced in the following classes.

We used this idea in [5, 9] to plan an optimal teaching schedule, and it worked. We hope that this idea can lead to further improvement of the teaching process.

References

1. O. Kosheleva, What is wrong with teaching to the test: uncertainty techniques help in understanding (and hopefully resolving) the controversy, in *Proceedings of the 2011 Annual Conference of the North American Fuzzy Information Processing Society NAFIPS'2011, El Paso, Texas, 18–20 March 2011* (2011)
2. A. Bell, Teaching for the test, in *Teaching Mathematics*, ed. by M. Selinger (The Open University Press, London, 1999), pp. 41–46
3. U. Boser, Teaching to the test? *Education Week 7 June 2000* (2000)
4. W.J. Popham, Teaching to the test? Educational Leadership **58**(6), 16–20 (2001)
5. R. Aló and O.M. Kosheleva, Optimization techniques under uncertain criteria, and their possible use in computerized education, in *Proceedings of the 25th International Conference of the North American Fuzzy Information Processing Society NAFIPS'2006, Montreal, Quebec, Canada, 3–6 June 2006* (IEEE Press, Piscataway, 2006)
6. O. Castillo, P. Melin, E. Gamez, V. Kreinovich, O. Kosheleva, Intelligence techniques are needed to further enhance the advantage of groups with diversity in problem solving, in *Proceedings of the 2009 IEEE Workshop on Hybrid Intelligent Models and Applications HIMA'2009, Nashville, Tennessee, 30 March–2 April 2009* (2009) pp. 48–55
7. O. Kosheleva, Potential application of fuzzy techniques to math education: emphasizing paradoxes as a (seemingly paradoxical) way to enhance the learning of (strict) mathematics, in *Proceedings of the 27th International Conference of the North American Fuzzy Information Processing Society NAFIPS'2008, New York, 19–22 May 2008* (2008)
8. O. Kosheleva, Early start can inhibit learning: a geometric explanation, in *Proceedings of the 2009 World Congress of the International Fuzzy Systems Association IFSA'2009, Lisbon, Portugal, 20–24 July 2009* (2009) pp. 438–442
9. O.M. Kosheleva and R. Aló, Towards economics of education: optimization under uncertainty, *Proceedings of the Second International Conference on Fuzzy Sets and Soft Computing in Economics and Finance FSSCEF'2006, St. Petersburg, Russia, 28 June–1 July 2006* (2006) pp. 63–70
10. O.M. Kosheleva, M. Ceberio, Processing educational data: from traditional statistical techniques to an appropriate combination of probabilistic, interval, and fuzzy approaches, in *Proceedings of the International Conference on Fuzzy Systems, Neural Networks, and Genetic Algorithms FNG'05, Tijuana, Mexico, 13–14 October 2005* (2005) pp. 39–48
11. O. Kosheleva, F. Modave, Asymmetric paternalism: description of the phenomenon, explanation based on decisions under uncertainty, and possible applications to education, in *Proceedings of the 27th International Conference of the North American Fuzzy Information Processing Society NAFIPS'2008, New York, 19–22 May 2008* (2008)
12. O. Kosheleva, V. Kreinovich, Towards Optimal Effort Distribution in Process Design under Uncertainty, with Application to Education, in *Proceedings of the 4th International Workshop on Reliable Engineering Computing REC'2010, Singapore, 3–5 March 2010*(2010) pp. 509–525
13. P. Pinheiro da Silva, A. Velasco, O. Kosheleva, Degree-based ideas and technique can facilitate inter-disciplinary collaboration and education, in *Proceedings of NAFIPS'2010, Toronto, Canada, 12–14 July 2010* (2010) pp. 388–393
14. G.J. Klir, B. Yuan, *Fuzzy Sets and Fuzzy Logic: Theory and Applications* (Prentice-Hall, Upper Saddle River, 1995)
15. H.T. Nguyen, E.A. Walker, *A First Course in Fuzzy Logic* (CRC Press, Boca Raton, 2006)

Chapter 18
In What Order to Present the Material: Fractal Approach

In the previous chapters, we described the optimal frequencies with which we repeat each of the items that a student has to learn. Once we know the number of repetitions of each item, the next natural question is: in what order should we present these repetitions? Should we first present all the repetitions of item 1, then all the repetitions of item 2, etc., or should we randomly mix these repetitions?

Towards mathematical formulation of the corresponding optimization problem. Each item is characterized by several (n) numerical characteristics, so we can geometrically represent each item as a point in the corresponding n-dimensional space.

- Similar items have close values of these characteristics, so the distance between the points corresponding to similar items is small.
- Vice versa, when the items are different, they at least some of these characteristics have different values on these items, so the resulting distance is large.

Thus, the distance between the corresponding points in a multi-D space can be viewed as a measure of similarity between the items.

In terms of multi-D space, an order in which we present repetitions is described as a function $x(t)$, where x is a multi-D point corresponding to the item presented at moment $t = k \cdot \Delta t$, where Δt is the time between repetitions.

As we have mentioned, when we have a few items to learn, we can easily learn them all, so there is no need for sophisticated optimization. Optimization becomes necessary when there are many items – and thus, many repetitions. In this case, similar to the way we simplify the physical problems if we approximate a collection of atoms by a continuous medium, we can approximate the discrete dependence $x(t)$ on discrete time t by a continuous function $x(t)$ of continuous time t.

What is the optimal trajectory $x(t)$? The experience of learning shows that often, presenting the items in random order is beneficial. To allow for this possibility,

© Springer-Verlag GmbH Germany 2018
O. Kosheleva and K. Villaverde, *How Interval and Fuzzy Techniques Can Improve Teaching*, Studies in Computational Intelligence 750,
https://doi.org/10.1007/978-3-662-55993-2_18

instead of looking for a deterministic function $x(t)$, we look for *random* processes $x(t)$. Since a deterministic function is a particular case of a random process, we are thus not restricting ourselves.

Let us consider Gaussian random processes. A Gaussian random process can be uniquely characterized by its mean $m(t) \stackrel{\text{def}}{=} E[x(t)]$ and autocorrelation function $A(t, s) \stackrel{\text{def}}{=} E[(x(t) - x(s))^2]$.

Students come with different levels of preparation. Therefore, a good learning strategy should work not only for a student that comes from 0, but also for a student that comes at moment t_0 with the knowledge that other students have already acquired by this time. From this viewpoint, a student's education starts at the moment t_0. It is therefore natural to require that the random process should look the same whether we start with a point $t = 0$ or with some later point t_0. Hence, the characteristics of the process should be the same, i.e., $m(t) = m(t + t_0)$ and $A(t, s) = A(t + t_0, s + t_0)$ for every t, s, and t_0.

From the first condition, we conclude that $m(t) = \text{const}$. Thus, by changing the origin of the coordinate system, we can safely assume that $m(t) = 0$.

From the second condition, for $t_0 = -s$, we conclude that $A(t, s) = A(t - s, 0)$, i.e., that the autocorrelation function depends only on the difference between the times: $A(t, s) = a(t - s)$, where we denoted $a(t) \stackrel{\text{def}}{=} A(t, 0)$. In other words, the random process must be *stationary*.

The final question is: what autocorrelation function $a(t)$ should we use?

We must choose a family of functions, not a single function. The function $a(t)$ depends on how intensely we train. In more intensive training, we present the material faster, and thus, within the same time interval t, we can cover more diverse topics. More diverse topics means that the average change $a(t)$ can be larger. A natural way to describe this increase is by proportionally enlarging all the distances, which leads from $a(t)$ to $C \cdot a(t)$. In other words, if $a(t)$ is a reasonable function for some training, then a new function $C \cdot a(t)$ should also be reasonable.

We can say that the functions $a(t)$ and $C \cdot a(t)$ describe exactly the same learning strategy, but with different intensities. Since intensity can be different, we cannot select a unique function $a(t)$ and claim it to be the best, because for every function $a(t)$, the function $C \cdot a(t)$ describes exactly the same learning strategy. In view of this, instead of formulating a problem of choosing the best autocorrelation *function*, it is more natural to formulate a problem of choosing the best *family* $\{C \cdot a(t)\}_C$ of autocorrelation functions.

Which family is the best? We may need non-numerical optimality criteria. Among all the families $\{C \cdot a(t)\}_C$, we want to choose the best one.

In mathematical optimization problems, numerical criteria are most frequently used, when to every alternative (in our case, to each family) we assign some value expressing its performance, and we choose an alternative (in our case, a family) for which this value is the largest. In our problem, as such a numerical criterion, we can select, e.g., the average grade on some standardized test A.

However, it is not necessary to restrict ourselves to such numerical criteria only. For example, if we have several different families that have the same average average grade A, we can choose between them the one that has the minimal level of uncomfortableness U. In this case, the actual criterion that we use to compare two families is not numerical, but more complicated: *a family F_1 is better than the family F_2 if and only if either $A(F_1) < A(F_2)$, or $A(F_1) = A(F_2)$ and $U(F_1) < U(F_2)$.* A criterion can be even more complicated. What a criterion *must* do is to allow us, for every pair of families, to tell whether the first family is better with respect to this criterion (we'll denote it by $F_1 \succ F_2$), or the second is better ($F_1 \prec F_2$), or these families have the same quality in the sense of this criterion (we'll denote it by $F_1 \sim F_2$). Of course, it is necessary to demand that these choices be consistent, e.g., if $F_1 \prec F_2$ and $F_2 \prec F_3$ then $F_1 \prec F_3$.

The optimality criterion must select a unique optimal family. Another natural demand is that this criterion must choose a *unique* optimal family (i.e., a family that is better with respect to this criterion than any other family). The reason for this demand is very simple.

If a criterion does not choose a family at all, then it is of no use.

If several different families are "the best" according to this criterion, then we still have a problem to choose among those "best". Therefore, we need some additional criterion for that choice. For example, if several families turn out to have the same average grade, we can choose among them a with the minimal uncomfortableness.

So what we actually do in this case is abandon that criterion for which there were several "best" families, and consider a new "composite" criterion instead: F_1 is better than F_2 according to this new criterion if either it was better according to the old criterion or according to the old criterion they had the same quality and F_1 is better than F_2 according to the additional criterion.

In other words, if a criterion does not allow us to choose a unique best family, it means that this criterion is not final. We have to modify it until we come to a final criterion that will have that property.

The optimality criterion must be scale-invariant. The next natural condition that the criterion must satisfy is connected with the fact that the numerical value of the time t depends on the choice of the unit for measuring time.

If we replace the original unit of time by a new unit which is λ times larger (i.e., replace minutes by hours), then numerical values change from t to $\tilde{t} = t/\lambda$. The autocorrelation function that in the old units is described by a family $\{C \cdot a(t)\}$, in the new units, has a new form $\{C \cdot a(\lambda \cdot t)\}$,

Since this change is simply a change in a unit of time, it is reasonable to require that going from $a(t)$ from $a(\lambda \cdot t)$ should not change the *relative* quality of the autocorrelation functions, i.e., if a family $\{C \cdot a(t)\}_C$ is better that the family $\{C \cdot a'(t)\}_C$, then for every $\lambda > 0$, the family $\{C \cdot a(\lambda \cdot t)\}_C$ must be still better than the family $\{C \cdot a'(\lambda \cdot t)\}_C$.

Definition 18.1

- By an *autocorrelation function* we mean a monotonically non-strictly decreasing function from non-negative real numbers to non-negative real numbers.
- By a *family* of functions we mean the family $\{C \cdot a(t)\}_C$, where $a(t)$ is a given autocorrelation function and C runs over arbitrary positive real numbers.
- A pair of relations $(\prec, \sim)$ is called *consistent* [1] if it satisfies the following conditions:

(1) if $a \prec b$ and $b \prec c$ then $a \prec c$;
(2) $a \sim a$;
(3) if $a \sim b$ then $b \sim a$;
(4) if $a \sim b$ and $b \sim c$ then $a \sim c$;
(5) if $a \prec b$ and $b \sim c$ then $a \prec c$;
(6) if $a \sim b$ and $b \prec c$ then $a \prec c$;
(7) if $a \prec b$, then $b \prec a$ or $a \sim b$ are impossible.

Definition 18.2

- Assume a set A is given. Its elements will be called *alternatives*. By an *optimality criterion* we mean a consistent pair $(\prec, \sim)$ of relations on the set A of all alternatives. If $b \prec a$, we say that a is *better* than b; if $a \sim b$, we say that the alternatives a and b are *equivalent* with respect to this criterion.
- We say that an alternative a is *optimal* (or *best*) with respect to a criterion $(\prec, \sim)$ if for every other alternative b either $b \prec a$ or $a \sim b$.
- We say that a criterion is *final* if there exists an optimal alternative, and this optimal alternative is unique.
- Let $\lambda > 0$ be a real number. By the λ-*rescaling* $R_\lambda(\rho)$ of a function $a(t)$ we mean a function $(R_\lambda a)(t) \overset{\text{def}}{=} a(\lambda \cdot t)$.
- By the λ-*rescaling* $R_\lambda(F)$ of a family F, we mean the set of the functions that are obtained from $f \in F$ by λ-rescaling.

In this chapter, we consider optimality criteria on the set $\mathscr{F}$ of all families.

Definition 18.3 We say that an optimality criterion on F is *scale-invariant* if for every two families F and G and for every number $\lambda > 0$, the following two conditions are true:

(i) if F is better than G in the sense of this criterion (i.e., $G \prec F$), then

$$R_\lambda(G) \prec R_\lambda(F);$$

(ii) if F is equivalent to G in the sense of this criterion (i.e., $F \sim G$), then

$$R_\lambda(F) \sim R_\lambda(G).$$

As we have already remarked, the demands that the optimality criterion is final and scale-invariant are quite reasonable. The only problem with them is that at first glance they may seem rather weak. However, they are not, as the following theorem shows:

Theorem 18.1 ([1]) *If a family F is optimal in the sense of some optimality criterion that is final and scale-invariant, then every function $a(t)$ from this optimal family F which has the form $a(t) = A \cdot t^\alpha$ for some real numbers A and α.*

In other words, the optimal configuration is a fractal random process.

- When $\alpha = 2$, we have a straightforward trajectory, without any randomness.
- The value $\alpha = 0$ means that values of $x(t)$ and $x(s)$ for $t \neq s$ are completely uncorrelated, i.e., that we have a white noise.
- Intermediate values $\alpha \in (0.2)$ correspond to different levels of fractal randomness.

Our experience showed that such *fractal* order indeed leads to improvement in learning. The exact value of the parameter α – corresponding to the fractal dimension of the corresponding trajectories – should be adjusted to the learning style of the students.

Reference

1. H.T. Nguyen, V. Kreinovich, *Applications of Continuous Mathematics to Computer Science* (Kluwer, Dordrecht, 1997)

Chapter 19
How AI-Type Uncertainty Ideas Can Improve Inter-Disciplinary Education and Collaboration: Lessons from a Case Study

In many application areas, there is a need for inter-disciplinary collaboration and education. However, such education and collaboration are not easy. On the example of our participation in a cyberinfrastructure project, we show that many obstacles on the path to successful collaboration and education can be overcome if we take into account that each person's knowledge of a statement is often a matter of *degree* – and that we can therefore use appropriate degree-based ideas and techniques.

The results from this chapter first appeared in [1, 2].

Practical problem: need to combine geographically separate computational resources. Before we start explaining the problems of inter-disciplinary communication and our proposed solution to this problem, let us first briefly describe the context of cyberinfrastructure in which our inter-disciplinary communication took place. In different knowledge domains in science and engineering, there is a large amount of data stored in different locations, and there are many software tools for processing this data, also implemented at different locations. Users may be interested in different information about this domain.

Sometimes, the information required by the user is already stored in *one of* the *databases*. For example, if we want to know the geological structure of a certain region in Texas, we can get this information from the geological map stored in Austin. In this case, all we need to do to get an appropriate response from the query is to get this data from the corresponding database.

In other cases, different pieces of the information requested by the user are *stored at different locations*. For example, if we are interested in the geological structure of the Rio Grande Region, then we need to combine data from the geological maps of Texas, New Mexico, and the Mexican state of Chihuahua. In such situations, a correct response to the user's query requires that we access these pieces of information from different databases located at different geographic locations.

© Springer-Verlag GmbH Germany 2018
O. Kosheleva and K. Villaverde, *How Interval and Fuzzy Techniques Can Improve Teaching*, Studies in Computational Intelligence 750,
https://doi.org/10.1007/978-3-662-55993-2_19

In many other situations, the appropriate answer to the user's request requires that we not only collect the relevant data $x_1, \ldots, x_n$, but that we also use some *data processing* algorithms $f(x_1, \ldots, x_n)$ to process this data. For example, if we are interested in the large-scale geological structure of a geographical region, we may also use the gravity measurements from the gravity databases. For that, we need special algorithms to transform the values of gravity at different locations into a map that describes how the density changes with location. The corresponding data processing programs often require a lot of computational resources; as a result, many such programs reside on computers located at supercomputer centers, i.e., on computers which are physically separated from the places where the data is stored.

The need to combine computational resources (data and programs) located at different geographic locations seriously complicates research.

Centralization of computational resources – traditional approach to combining computational resources; its advantages and limitations. Traditionally, a widely used way to make these computational resources more accessible was to move all these resources to a *central location*. For example, in the geosciences, the US Geological Survey (USGS) was trying to become a central repository of all relevant geophysical data. However, this centralization requires a large amount of efforts: data is presented in different formats, the existing programs use specific formats, etc. To make the central data repository efficient, it is necessary:

- to reformat all the data,
- to rewrite all the data processing programs – so that they become fully compatible with the selected formats and with each other, etc.

The amount of work that is needed for this reformatting and rewriting is so large that none of these central repositories really succeeded in becoming an easy-to-use centralized database.

Cyberinfrastructure – a more efficient approach to combining computational resources. Cyberinfrastructure technique is a new approach that provides the users with the efficient way to submit requests without worrying about the geographic locations of different computational resources – and at the same time avoid centralization with its excessive workloads. The main idea behind this approach is that *we keep all (or at least most) the computational resources*

- *at their current locations,*
- *in their current formats.*

To expedite the use of these resources:

- we supplement the local computational resources with the "metadata", i.e., with the information about the formats, algorithms, etc.,
- we "wrap up" the programs and databases with auxiliary programs that provide data compatibility into *web services*,

and, in general, we provide a cyberinfrastructure that uses the metadata to automatically combine different computational resources.

For example, if a user is interested in using the gravity data to uncover the geological structure of the Rio Grande region, then the system should automatically:

- get the gravity data from the UTEP and USGS gravity databases,
- convert them to a single format (if necessary),
- forward this data to the program located at San Diego Supercomputer Center, and
- move the results back to the user.

This example is exactly what we have been designing under the NSF-sponsored Cyberinfrastructure for the Geosciences (GEON) project; see, e.g., [3–14], and what we are currently doing under the NSF-sponsored Cyber-Share project. This is similar to what other cyberinfrastructure projects are trying to achieve.

Technical advantages of cyberinfrastructure: a brief summary. In different knowledge domains, there is a large amount of data stored in different locations; algorithms for processing this data are also implemented at different locations. Web services – and, more generally, cyberinfrastructure – provide the users with an efficient way to submit requests without worrying about the geographic locations of different computational resources (databases and programs) – and avoid centralization with its excessive workloads [15]. Web services enable the user to receive the desired data $x_1, \ldots, x_n$ and the results $y = f(x_1, \ldots, x_n)$ of processing this data.

Main advantage of cyberinfrastructure: the official NSF viewpoint. Up to now, we concentrated on the technical advantages of cyberinfrastructure. However, its advantages (real and potential) go beyond technical. According to the final report of the National Science Foundation (NSF) Blue Ribbon Advisory Panel on Cyberinfrastructure, "a new age has dawned in scientific and engineering research, pushed by continuing progress in computing, information, and communication technology, and pulled by the expanding complexity, scope, and scale of today's challenges. The capacity of this technology has crossed thresholds that now make possible a comprehensive 'cyberinfrastructure' on which to build new types of scientific and engineering knowledge environments and organizations and to pursue research in new ways and with increased efficacy.

Such environments and organizations, enabled by cyberinfrastructure, are increasingly required to address national and global priorities, such as understanding global climate change, protecting our natural environment, applying genomics-proteomics to human health, maintaining national security, mastering the world of nanotechnology, and predicting and protecting against natural and human disasters, as well as to address some of our most fundamental intellectual questions such as the formation of the universe and the fundamental character of matter."

Main advantage of cyberinfrastructure: in short. Cyberinfrastructure greatly enhances the ability of scientists to discover, reuse and combine a large number of resources, including data and services.

Need for inter-disciplinary collaboration. A successful cyberinfrastructure requires an intensive collaboration between

- domain scientists – who provide the necessary information and metadata, and
- computer scientists who provide the corresponding cyberinfrastructure.

Moreover, since we combine data obtained by different subdomains, we also need collaboration between representatives of these subdomains.

Need for inter-disciplinary education. For the collaboration between researchers from different disciplines (even different sub-disciplines) to be successful, we need to *educate* collaborating researchers in the basics of each others' disciplines.

Inter-disciplinary collaboration and education: a typical communication situation. Let us give an example of a typical problem that we encountered when we started collaboration within our Cyber-ShARE Center.

- Suppose that a computer science has an interesting idea on how to better organize the geosciences' data and/or metadata. This is a typical *collaboration* problem.
- Alternatively, a computer scientist may simply want to teach, to a geosciences colleague, a few existing computer science ideas on how to organize data and/or metadata. This is a typical *education* problem.

How to convey a computer science idea to a geoscientist?

First possibility: just convey this idea. One possibility is simply to describe this idea in Computer Science terms.

Alas, many of these terms are usually very specific. Even many computer scientists – those whose research is unrelated to cyberinfrastructure – are not very familiar with these terms and with the ideas behind them.

The only serious way for a geoscientist to understand and learn these terms, notions, ideas is to learn the material of several relevant computer science courses – i.e., in effect, to get a second degree in computer science. A few heroes may end up doing this, but it is unrealistic to expect such deep immersion in a normal inter-disciplinary collaboration.

Second possibility: try to illustrate this idea in the domain science terms. Alternatively, to make it clearer, a computer scientist can try to explain his or her ideas on the example of a toy geosciences problem.

The limitation of this approach is that the computer scientist is usually not a specialist in the domain science (in our case, in geosciences). As a result, his or her description of the toy problem is, inevitably, flawed: e.g., oversimplified. Hence, the problem that the new idea is trying to solve in this example is often not meaningful to a geoscientist – and since the motivation is missing, it is difficult to understand the idea.

Conveying a problem: a similar situation. A similar situation occurs when instead of communicating an *idea*, we try to communicate a *problem*. Specifically, suppose that a geoscientist (or, more generally, a domain scientist) has a real problem in which, he believes, cyberinfrastructure can help.

Comment. There are such problems – otherwise, the geoscientist would not seek collaboration with a computer scientist.

First possibility: just convey this problem. One possibility is simply to describe this idea in the geosciences terms.

Alas, these terms are usually very specific: even many geoscientists – those whose research is unrelated to the specific sub-domain – may be not very familiar with these terms and with the ideas behind them.

The only serious way for a computer scientist to understand and learn these terms, notions, ideas is to learn the material of several relevant geosciences courses – i.e., in effect, to get a second degree in the domain science. A few heroes may end up doing this, but, as we have mentioned earlier, it is unrealistic to expect such deep immersion in a normal inter-disciplinary collaboration.

Second possibility: try to illustrate this idea in terms understandable to a computer scientist. Alternatively, to make it clearer, a geoscientist can try to explain his or her problem by using terms understandable to a computer scientist.

A limitation of this approach is that the geoscientist is usually not a specialist in computer science. As a result, his or her description of the problem is, inevitably, flawed: e.g., oversimplified. Hence, the problem is difficult to understand.

Consequences. As a result of the above problems, our weekly meetings – in which we tried to understand domain science problems and explain possible solutions – were, for a while, not very productive. For a while, they turned into what we called "fight club", when

- a geoscientist would find (and explain) flaws in a toy geosciences model that a computer scientist uses to describe his or her ideas, while
- a computer scientist would find (and explain) flaws in the way a geoscientist would describe his or her problem.

And then we succeeded. And then we – serendipitously – found a solution to our struggles. After we found this solution, we started thinking why it worked – and discovered an explanation – via the matter-of-degree ideology.

Our solution may be known, but our explanation seems to be new. While our approach is probably known – at least there exist other successful inter-disciplinary collaborations, so some solutions have been found – we could not find a theoretical explanation for its success. To the best of our knowledge, our explanation is new.

What we would like to do: our main objective. Our main goal is to explain to others how this problem can be solved – and thus, make other inter-disciplinary collaborations more productive. This is what we plan to do in this paper.

The fact that we are not simply proposing an empirical solution, that we have a theoretical justification for our successful strategy, hopefully makes our case more convincing – so we hope that others will follow our strategy.

An additional objective. Since we have a theoretical explanation in terms of degrees (numbers), we hope to transform our original *qualitative* idea into a more precise *quantitative* strategy.

This is mostly the subject of future work: since it is desirable not only to propose formulas, but also to show that these formulas – based on an inevitably simplified description of the communication situation – really work. To convincingly test whether this idea works or not, we need to have numerous examples of using this idea – and to have these examples, we must first convince others to use our strategy.

Thus, the convincing is still the main objective of this paper.

Main idea: let us use examples of successful cyberinfrastructure collaborations. The above problems may sound unsolvable if we restrict ourselves to a specific domain science. However, the very fact that we are not starting from scratch, that there are already examples of successful inter-disciplinary collaborations – shows that these communication problems are solvable.

Describing these successful examples is a way to convince scientists that collaboration is possible and potentially beneficial.

It turns out that, moreover, these outside examples themselves helped us to solve our communication problem.

What we did. Instead of trying to describe his ideas in purely computer science terms or on a toy geosciences example, a computer scientist described these ideas on the example of his applying similar ideas to a complete different area: solar astronomy.

What happened. This description was inevitably less technical – since none of us is a specialist in solar astronomy – and therefore, much more understandable.

Positive results. As a result, we got a much better understanding of the original computer science idea.

Recommendation. When a communication problem occurs because of the different areas of expertise of the describer and the respondent, try to convey the message on the example on a *different* domain, a domain in which both the describer and the respondent have a similar level of sophistication.

Idea of degrees. Every person has different degrees of knowledge in different areas.

There are many potential ways to measure these degrees. A natural way is to gauge the degree of expertise the way we gauge the student's knowledge: by counting the proportion of correct answers on some test describing the knowledge. In this case, the degree is a number between 0 and 1, with 0 representing no knowledge at all and 1 meaning perfect knowledge.

Comment. This may sound like a probabilistic definition, but it is important to notice that when the knowledge is imperfect, the resulting knowledge is not a random selection. Usually, in every discipline, we have:

- the simplest facts that practically everyone knows,
- somewhat more sophisticated facts and results that fewer people know,
- etc.,
- all the way to subtle technical details that only true experts know.

An imperfect knowledge usually means that a person knows all the facts and results of limited sophistication level: from very basic when this knowledge is small to very deep when the knowledge is greater.

For example, when a person has a basic level of understanding, this person knows the basic facts, but lacks knowledge about more sophisticated details – so this person's idea of these details will be most probably wrong.

In general, the body of knowledge contains statements of different degree of sophistication. Our definition of the degree of expertise as simply a proportion means, crudely speaking, that this body of knowledge contains an equal number of statements at different levels of sophistication.

Similarly, it is reasonable to conclude that an individual body of knowledge of a person in a certain area is equally distributed between different levels – from the simplest to the level of sophistication of this person.

Let us use these degrees in our communication problem. Let us re-formulate the above communication situations in terms of the corresponding degrees.

First situation: a specialist conveys an idea or a problem in the terms of his/her discipline. Let us start with a situation in which a computer scientist describes his/her ideas in computer science terms – or a geoscientist describes his or her problem in geoscience terms.

The first person, the person who describes the idea or the problem is an expert in his or her area, so this person's degree d_1 is close to 1: $d_1 \approx 1$. The original idea (or problem) is therefore described on this persons' level of expertise.

The second person, the person to whom this idea (or this problem) needs to be conveyed is not a specialist in the corresponding terms, so his or her degree of expertise d_2 in the describer's domain is much smaller: $d_2 \approx 0$.

By definition of the degree of expertise, this means that only the d_2th part of the original idea – the part corresponding to the sophistication level below this person's degree of expertise – will be properly understood. So, the result degree of understanding d is equal to the respondent's degree of expertise $d_2 \approx 0$ in the describer's domain.

Second situation: a specialist translates an idea or a problem into the terms of the other discipline. Let us now describe a situation in which a computer scientist tries to describe his/her ideas in geoscience terms – or a geoscientist tries to describe his or her problem in computer science terms.

The first person, the person who describes the idea or the problem, is not an expert in the domain in which this person is trying to describe, so this person's degree of expertise d_1 in this domain is close to 0: $d_1 \approx 0$. So, when this person translates his or her idea (problem) into this new domain, this translation is absolutely correct only at the sophistication level d_1.

In other words, while the main idea may be correct, most technical details will be wrong – since the describer is not an expert in the new domain.

The second person, the person to whom this idea (or this problem) needs to be conveyed is a specialist in the corresponding terms, so he or she will see all the errors

– and thus, will be unable to understand all the details beyond the very basic, at the level d_1.

Thus, the resulting degree of understanding is equal to the describer's degree of expertise $d_1 \approx 0$ in the respondent's domain.

General case. Let us now consider a general case, when the describer translates his or her idea (or problem) into a domain in which he or she has a degree of expertise d_1 and the respondent has a degree of expertise d_2.

In general, similar to the above two situations, there are two problems that prevent us from perfect understanding:

- first, the describer's level may be too low, so his or her presentation has a lot of inaccuracies that prevent understanding;
- second, the describer's level may be too high, so his or her presentation may be too sophisticated for the responder to understand – which also prevents understanding.

Because of these two possible problems, let us consider two subcases corresponding to the above two situations:

- when the new domain is closer to the describer's area of expertise, i.e., when $d_2 \leq d_1$, and
- when the new domain is closer to the respondent's area of expertise, i.e., when $d_1 \leq d_2$.

Case when $d_2 \leq d_1$. In this case, the describer's degree of sophistication in the new domain is higher than the respondent's, so the respondent will not be able to detect inaccuracies in the describer's presentation. The only problem here is that since the describer's level of sophistication d_1 may be higher than the respondent's level d_2, the corresponding part of the presentation will not be clear to the respondent.

From all the knowledge corresponding to the levels of sophistication from 0 to d_1, only the parts corresponding to levels from 0 to $d_2 \leq d_1$ will be properly understood. We have argued above that the knowledge is more or less uniformly distributed across different levels of sophistication. Out of d_1 different levels, only d_2 levels lead to understanding.

As a result, the proportion d of properly understood message is approximately equal to the ratio d_2/d_1.

Case when $d_1 \leq d_2$. In this case, the describer's degree of sophistication in the new domain is smaller than the respondent's, so the respondent will be able to understand all the terms that the describer is using. However, because of the possible difference of the levels of expertise, the respondent will be able to detect inaccuracies in all the levels of sophistication beyond d_1.

Thus, from all the knowledge corresponding to the levels of sophistication from 0 to d_2, only the parts corresponding to levels from 0 to $d_1 \leq d_2$ will be properly understood. We have argued above that the knowledge is more or less uniformly distributed across different levels of sophistication. Out of d_2 different levels on which the recipient receives information, only d_1 levels lead to understanding.

As a result, the proportion d of property understood message is approximately equal to the ratio d_2/d_1.

General formula. In both cases, the degree of understanding d can be obtained by dividing the smallest of the degrees d_1 and d_2 by the largest of these two degrees:

$$d = \frac{\min(d_1, d_2)}{\max(d_1, d_2)}. \tag{19.1}$$

This indeed explain the success of our empirical strategy. When the describer formulates his or message either in his or her own domain terms or in terms of the respondent's domain, we have $\min(d_1, d_2) \approx 0$ and $\max(d_1, d_2) \approx 1$, so $d \approx 0$.

When the describer instead formulates his or her own message in the language of the third domain, in which $d_1 \approx d_2$, we have $\min(d_1, d_2) \approx \max(d_1, d_2)$ and therefore, $d \approx 1$.

What will be the ideal case. According to the above formula (19.1), the degree of understanding d attains the largest possible value 1 when $\min(d_1, d_2) = \max(d_1, d_2)$, i.e., when both degrees of expertise coincide: $d_1 = d_2$. So, to maximize the degree of understanding, we must find a common domain in which both the describer and the respondent have the same level of expertise.

Dependence on the domain. In general, the further the area a from the person's main area of expertise a_0, the smaller this person's degree of sophistication $d(a)$ in this area a.

Let $\rho(a, a_0)$ describe the "distance" between different domains. Thus, the degree $d(a)$ should be a decreasing function of this distance: $d(a) = f(\rho(a, a_0))$ for some decreasing function $f(\rho)$.

Dependence on the expert. This function $f(\rho)$ is, in general, different for different experts:

- Some experts are more "narrow", for them this decrease is more steep: for such experts, even for reasonably close areas $a \approx a_0$, the level of expertise $d(a)$ is very low.
- Some experts are more "broad", they retain some level of expertise even in sufficiently distant domains a.

To take this difference into account, let us describe the expert's "radius" of possible expertise by r – we can define it, e.g., as the distance $\rho(a, a_0)$ at which the corresponding degree of expertise $d(a)$ drops to a certain threshold d_0. The broader the expert, the larger this radius.

Resulting formula. We can then reasonably conjecture that for all experts, we have

$$d(a) = f_0\left(\frac{\rho(a, a_0)}{r}\right) \tag{19.2}$$

for some universal monotonically decreasing function $f_0(\rho)$.

Towards recommendations. We want to select a domain a for which $d_1(a) = d_2$. Due to formula (19.2), this means that

$$f_0\left(\frac{\rho(a, a_{01})}{r_1}\right) = f_0\left(\frac{\rho(a, a_{02})}{r_2}\right). \tag{19.3}$$

Since the function $f(\rho)$ is monotonic, this means that

$$\frac{\rho(a, a_{01})}{r_1} = \frac{\rho(a, a_{02})}{r_2}. \tag{19.4}$$

Resulting recommendation. Select an area a for which the equality (19.4) holds.

Example Let us consider the case when the domains are represented by points in a usual Euclidean space, with a standard metric $\rho(a, a_0)$. In this case, it is reasonable to look for the location of a on the straight line

$$\{\alpha \cdot a_{01} + (1 - \alpha) \cdot a_{02} : 0 \leq \alpha \leq 1\} \tag{19.5}$$

connecting the describer's and the respondent's areas of expertise a_{01} and a_{02}. For a point

$$a = \alpha \cdot a_{01} + (1 - \alpha) \cdot a_{02} \tag{19.6}$$

on this line, the formula (19.3) takes the form

$$\frac{\alpha \cdot \rho(a_{01}, a_{02})}{r_1} = \frac{(1 - \alpha) \cdot \rho(a_{01}, a_{02})}{r_2}. \tag{19.7}$$

Dividing both sides of this equality by the distance $\rho(a_{01}, a_{02})$ and multiplying both sides by $r_1 \cdot r_2$, we conclude that

$$\alpha \cdot r_2 = (1 - \alpha) \cdot r_1, \tag{19.8}$$

hence

$$\alpha = \frac{r_1}{r_1 + r_2} \tag{19.9}$$

and

$$a = \frac{r_1}{r_1 + r_2} \cdot a_{01} + \frac{r_2}{r_1 + r_2} \cdot a_{02}. \tag{19.10}$$

An alternative recommendation. An alternative recommendation is to use an "interpreter", i.e., a person who has a reasonable (although not perfect) understanding in both fields.

Here, a describer first use the terms of his or her domain to convey the idea (or problem) to the interpreter. In this transaction, because of the interpreter's knowledge, the degree of understanding $d = \dfrac{\min(d_1, d_2)}{\max(d_1, d_2)}$ is reasonably high.

The interpreter then translates the message into the respondent's domain and conveys thus translated message to the respondent. Here, also, the degree of understanding is reasonably high.

Comment. This strategy, by the way, works well too. We hope that the above formulas will help to optimize this approach as well.

References

1. P. Pinheiro da Silva, A. Velasco, O. Kosheleva, Degree-based ideas and technique can facilitate inter-disciplinary collaboration and education, in *Proceedings of the NAFIPS'2010, Toronto, Canada, July 12–14, 2010* (2010), pp. 388–393
2. P. Pinheiro da Silva, A. Velasco, O. Kosheleva, V. Kreinovich, How AI-type uncertainty ideas can improve inter-disciplinary collaboration and education: lessons from a case study. J. Adv. Comput. Intell. Intell. Inf. JACIII **14**(6), 700–707 (2010)
3. M.S. Aguiar, G.P. Dimuro, A.C.R. Costa, R.K.S. Silva, F.A. Costa, V. Kreinovich, The multi-layered interval categorizer tesselation-based model, in *IFIP WG2.6 Proceedings of the 6th Brazilian Symposium on Geoinformatics Geoinfo'2004, Campos do Jordão, Brazil, November 22–24, 2004*, ed. by C. Iochpe, G. Câmara (2004), pp. 437–454
4. R. Aldouri, G.R. Keller, A.Q. Gates, J. Rasillo, L. Salayandia, V. Kreinovich, J. Seeley, P. Taylor, S. Holloway, GEON: Geophysical data add the 3rd dimension in geospatial studies, in *Proceedings of the ESRI International User Conference 2004, San Diego, California, August 9–13, 2004* (2004)
5. M.G. Averill, K.C. Miller, G.R. Keller, V. Kreinovich, R. Araiza, S.A. Starks, Using expert knowledge in solving the seismic inverse problem, in *Proceedings of the 24nd International Conference of the North American Fuzzy Information Processing Society NAFIPS'2005, Ann Arbor, Michigan, June 22–25, 2005* (2005), pp. 310–314
6. M. Ceberio, S. Ferson, V. Kreinovich, S. Chopra, G. Xiang, A. Murguia, J. Santillan, How to take into account dependence between the inputs: from interval computations to constraint–related set computations, with potential applications to nuclear safety, bio– and geosciences, in *Proceedings of the Second International Workshop on Reliable Engineering Computing REC'2006, Savannah, Georgia, February 22–24, 2006* (2006), pp. 127–154
7. M. Ceberio, V. Kreinovich, S. Chopra, B. Ludäscher, Taylor model–type techniques for handling uncertainty in expert systems, with potential applications to geoinformatics, in *Proceedings of the 17th World Congress of the International Association for Mathematics and Computers in Simulation IMACS'2005, Paris, France, July 11–15, 2005* (2005)
8. G.R. Keller, T.G. Hildenbrand, R. Kucks, M. Webring, A. Briesacher, K. Rujawitz, A.M. Hittleman, D.J. Roman, D. Winester, R. Aldouri, J. Seeley, J. Rasillo, T. Torres, W.J. Hinze, A. Gates, V. Kreinovich, L. Salayandia, A community effort to construct a gravity database for the United States and an associated web portal', in *Geoinformatics: Data to Knowledge, 2006, Geological Society of America Publ., Boulder, Colorado*, ed. by A.K. Sinha (2006), pp. 21–34
9. E. Platon, K. Tupelly, V. Kreinovich, S. A. Starks, and K. Villaverde, "Exact bounds for interval and fuzzy functions under monotonicity constraints, with potential applications to biostratigraphy", In: *Proceedings of the 2005 IEEE International Conference on Fuzzy Systems FUZZ–IEEE'2005*, Reno, Nevada, May 22–25, 2005, pp. 891–896

10. C.G. Schiek, R. Araiza, J.M. Hurtado, A.A. Velasco, V. Kreinovich, V. Sinyansky, Images with uncertainty: efficient algorithms for shift, rotation, scaling, and registration, and their applications to geosciences, in *Soft Computing in Image Processing: Recent Advances*, ed. by M. Nachtegael, D. Van der Weken, E.E. Kerre, W. Philips (Springer, Berlin, 2005), pp. 35–64
11. A.K. Sinha (ed.), *Geoinformatics: Data to Knowledge* (Geological Society of America Publisher, Boulder, 2006)
12. R. Torres, G.R. Keller, V. Kreinovich, L. Longpré, S.A. Starks, Eliminating duplicates under interval and fuzzy uncertainty: an asymptotically optimal algorithm and its geospatial applications. Reliab. Comput. **10**(5), 401–422 (2004)
13. Q. Wen, A.Q. Gates, J. Beck, V. Kreinovich, G.R. Keller, Towards automatic detection of erroneous measurement results in a gravity database, in *Proceedings of the 2001 IEEE Systems, Man, and Cybernetics Conference, Tucson, Arizona, October 7–10, 2001* (2001), pp. 2170–2175
14. H. Xie, N. Hicks, G.R. Keller, H. Huang, V. Kreinovich, An IDL/ENVI implementation of the FFT based algorithm for automatic image registration. Comput. Geosci. **29**(8), 1045–1055 (2003)
15. A.Q. Gates, V. Kreinovich, L. Longpré, P. Pinheiro da Silva, G.R. Keller, Towards secure cyberinfrastructure for sharing border information, in *Proceedings of the Lineae Terrarum: International Border Conference, El Paso, Las Cruces, and Cd. Juárez, March 27–30, 2006* (2006)

Chapter 20
In What Order to Present the Material Within Each Topic: Concrete-First Versus Abstract-First

Let us start our analysis of the order in which the material should be presented with the teaching of each individual topic. Each topic in math and sciences usually contains some new abstract notion(s). Several examples are usually given to motivate these notions and to illustrate their use. There are several possible ways of arranging the material of the topic:

- one possibility is to start with presenting concrete examples, and delay the introduction of abstract ideas and notions as much as possible;
- an alternative ideas is to introduce the abstract ideas from the very beginning, and to use examples to illustrate these ideas.

Intuitively, it seems that students would learn better if they are first presented with numerous concrete examples, and only learn abstract ideas later on. However, it turns out that empirically, the abstract-first approach for presenting the material often enhances learning; see, e.g., [1–4]. In this chapter, we provide a theoretical explanation for this seemingly counter-intuitive empirical phenomenon.

The results from this chapter first appeared in [5, 6].

Learning: a natural geometric representation. To facilitate reasoning about learning, let us start with a simple geometric representation of learning.

The process of learning means that we change the state of a student:

- from a state in which the student did not know the material (or does not have the required skill)
- to a state in which the student has (some) knowledge of the required material (or has the required skill).

Let s_0 denote the original state of a student, and let S denote the set of all the states corresponding to the required knowledge or skill.

- We start with a state which is not in the set S ($s_0 \notin S$), and
- we end up in a state s which is in the set S.

© Springer-Verlag GmbH Germany 2018
O. Kosheleva and K. Villaverde, *How Interval and Fuzzy Techniques Can Improve Teaching*, Studies in Computational Intelligence 750,
https://doi.org/10.1007/978-3-662-55993-2_20

On the set of all possible states, it is natural to define a metric $d(s, s')$ as the difficulty (time, effort, etc.) needed to go from state s to state s'. Our objective is to help the students learn in the easiest (fastest, etc.) way. In terms of the metric d, this means that we want to go from the original state $s_0 \notin S$ to the state $s \in S$ for which the effort $d(s_0, s)$ is the smallest possible.

In geometric terms, the smallest possible effort means the shortest possible distance. Thus, our objective is to find the state $s \in S$ which is the closest to s_0. Such closest state is called the *projection* of the original state s_0 on the set S.

Learning complex material: geometric interpretation. The above geometric description of learning as a transition from the original state s_0 to its projection on the desired set S describes learning *as a whole*. Our objective is to find out which order of presenting information is the best. Thus, our objective is to analyze the *process* of learning, i.e., learning as a multi-stage phenomenon. For this analysis, we must explicitly take into account that the material to be learned consists of several pieces.

Let S_i, $1 \le i \le n$, denote the set of states in which a student has learned the ith part of the material. Our ultimate objective is to make sure that the student learns all the parts of the material. In terms of states, learning the ith part of the material means belonging to the set S_i. Thus, in terms of states, our objective means that the student should end up in a state which belongs to all the sets $S_1, \ldots, S_n$ – i.e., in other words, in a state which belongs to the intersection

$$S \stackrel{\text{def}}{=} S_1 \cap \cdots \cap S_n \tag{20.1}$$

of the corresponding sets S_i.

In these terms, if we present the material in the order S_1, S_2, …, S_n, this means that:

- we first project the original state s_0 onto the set S_1, resulting is a state $s_1 \in S_1$ which is the closest to s_0;
- then, we project the state s_1 onto the set S_2, resulting is a state $s_2 \in S_2$ which is the closest to s_1;
- …
- at the last stage of the cycle, we project the state s_{n-1} onto the set S_n, resulting is a state $s_n \in S_n$ which is the closest to s_{n-1}.

In some cases, we end up learning all the material – i.e., in a state $s_n \in S_1 \cap \cdots \cap S_n$. However, often, by the time the students have learned S_n, they have somewhat forgotten the material that they learned in the beginning. So, it is necessary to repeat this material again (and again). Thus, starting from the state s_n, we again sequentially project onto the sets S_1, S_2, etc.

The above geometric interpretation makes computational sense. The above "sequential projections" algorithm is actually actively used in many applications; see, e.g., [7–9]. In the case when all the sets S_i are convex, the resulting Projections on Convex Sets (POCS) method actually guarantees (under certain reasonable conditions) that the corresponding projections converge to a point from the intersection $S_1 \cap \ldots \cap S_n$ – i.e., in our terms, that the students will eventually learn all parts of the necessary material.

In the more general non-convex case, the convergence is not always guaranteed – but the method is still efficiently used, and often converges.

The simplest case: two-part knowledge. Let us start with the simplest case when knowledge consists of two parts. In this simplest case, there are only two options:

- The first option is that:
  - we begin by studying S_1;
  - then, we study S_2,
  - then, if needed, we study S_1 again, etc.

- The second option is that:
  - we begin by studying S_2;
  - then, we study S_1,
  - then, if needed, we study S_2 again, etc.

We want to get from the original state s_0 to the state $\tilde{s} \in S_1 \cap S_2$ which is the closest to s_0. The effectiveness of learning is determined by how close we get to the desired set $S = S_2 \cap S_2$ in a given number of iterations.

In the case of two-part knowledge, it is natural to conclude that the amount of this knowledge is reasonably small – otherwise, we would have divided into a larger number of easier-to-learn pieces.

In geometric terms, this means that the original state s_0 is close to the desired intersection set $S_1 \cap S_2$, i.e., that the distance $d_0 \stackrel{\text{def}}{=} d(s_0, \tilde{s})$ is reasonably small.

Since all the states are close to each other, in the vicinity of the state $\tilde{s}$, we can therefore expand the formulas describing the borders of the sets S_i into Taylor series and keep only terms which are linear in the (coordinates of the) difference $s - \tilde{s}$. Thus, it is reasonable to assume that the border of each of the two sets S_i is described by a linear equation – and is hence a (hyper-)plane: a line in 2-D space, a plane in 3-D space, etc.

As a result, we arrive at the following configuration. Let 2α denote the angle between the borders of the sets S_1 and S_2, so that the angles between each of these borders and the midline is exactly α. Let β denote the angle between the direction from $\tilde{s}$ to s_0 and the midline. In this case, the angle between the border of S_1 and the midline is equal to $\alpha - \beta$. So, we arrive at the following configuration:

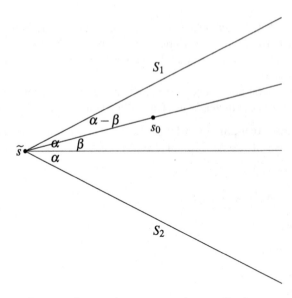

In the first option, we first project s_0 onto the set S_1. As a result, we get the following configuration:

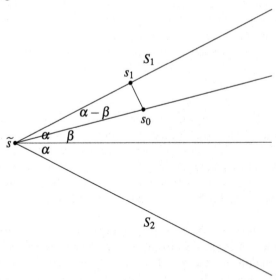

Here, the projection line $s_0 s_1$ is orthogonal to the border of S_1. From the right triangle $\triangle \widetilde{s} s_0 s_1$, we therefore conclude that the distance $d_1 \stackrel{\text{def}}{=} d(\widetilde{s}, s_1)$ from the projection point s_1 to the desired point $\widetilde{s}$ is equal to

$$d_1 = d_0 \cdot \cos(\alpha - \beta). \tag{20.2}$$

On the next step, we project the point s_1 from S_1 onto the line S_2 which is located at the angle 2α from S_1. Thus, for the projection result s_2, we will have

$$d_2 = d(s_2, \tilde{s}) = d_1 \cdot \cos(2\alpha) = d_0 \cdot \cos(\alpha - \beta) \cdot \cos(2\alpha). \qquad (20.3)$$

After this, we may again project onto S_2, then again project onto S_1, etc. For each of these projections, the angle is equal to 2α, so after each of them, the distance from the desired point $\tilde{s}$ is multiplied the same factor $\cos(2\alpha)$.

As a result, after k projection steps, we get a point s_k at a distance

$$d_k = d(s_k, \tilde{s}) = d_0 \cdot \cos(\alpha - \beta) \cdot \cos^{k-1}(2\alpha) \qquad (20.4)$$

from the desired state $\tilde{s}$.

In the second option, we start with teaching S_2, i.e., if we first project the state s_0 into the set S_2. In this option, we get the following configuration:

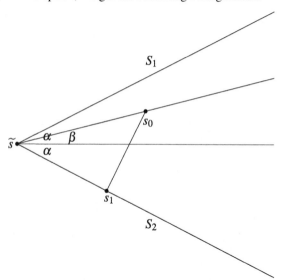

Here, we have

$$d_1 = d_0 \cdot \cos(\alpha + \beta), \qquad (20.5)$$

$$d_2 = d(s_2, \tilde{s}) = d_1 \cdot \cos(2\alpha) = d_0 \cdot \cos(\alpha + \beta) \cdot \cos(2\alpha), \qquad (20.6)$$

$$\cdots$$

$$d_k = d(s_k, \tilde{s}) = d_0 \cdot \cos(\alpha + \beta) \cdot \cos^{k-1}(2\alpha). \qquad (20.7)$$

Since, in general, $\cos(\alpha - \beta) \neq \cos(\alpha + \beta)$, we can see that a change in the presentation order can indeed drastically change the success of the learning procedure.

Conclusion: dependence explained. Thus, our simple geometric model explains why the effectiveness of learning depends on the order in which the material is presented.

Analysis. Let us extract more specific recommendations from our model. According to the above formulas, starting with S_1 leads to a more effective learning than starting with S_2 if and only if

$$d_0 \cdot \cos(\alpha - \beta) \cdot \cos^{k-1}(2\alpha) <$$

$$d_0 \cdot \cos(\alpha + \beta) \cdot \cos^{k-1}(2\alpha), \qquad (20.8)$$

i.e., equivalently, if and only if

$$\cos(\alpha - \beta) < \cos(\alpha + \beta). \qquad (20.9)$$

Since for the angles $x \in [0, \pi]$, the cosine $\cos(x)$ is a decreasing function, we conclude that projection of S_1 is better if and only if

$$\alpha - \beta > \alpha + \beta. \qquad (20.10)$$

Thus, we arrive at the following recommendation:

Recommendation. To make learning more efficient, we should start with studying the material which is further away from the current state of knowledge. In other words, we should start with a material that we know the least. This recommendation is also in a very good accordance: with the seemingly counter-intuitive conclusion from [1–4], that studying more difficult (abstract) ideas first enhances learning.

General advice. This recommendation also ties in nicely with a natural commonsense recommendation that to perfect oneself, one should concentrate on one's deficiencies.

Another conclusion: an explanation why early start can inhibit learning. The age at which we teach different topics change. If it turns out that students do not learn, say, reading by the time they should, a natural idea is to start teaching them earlier. Several decades ago, reading and writing started in the first grade, now they start at kindergarten and even earlier. At first glance, the earlier we start, the better the students will learn. Yes, they may play less, their childhood may be not as careless as it used to be, but at least they will learn better. In practice, however, this does not always work this way: early start often inhibits learning. For example, according to [10],

- human infants who started learning to turn their heads to specific sounds at the age of 31 days mastered this task, on average, at the age of 71 days, while
- infants who started learning this task at birth mastered this task, on average, at the age of 128 days.

This phenomenon is not limited to human infants: according to [11], an early start in training rhesus monkeys to discriminate objects decreased their peak performance level. Numerous examples when an early start inhibits learning are presented and discussed in [12–16].

In [12–15], an attempt is made to understand why early start can inhibit learning. However, the existing understanding is still mostly on the qualitative level, and even on this level, the proposed explanations are still not fully satisfactory; see, e.g., [16].

Our result provides a new explanation of this phenomenon: early start means a concentration on teaching, to human infants, skills that they can easily learn, skills which they can easily acquire and master. According to our geometric model, this is detrimental in the long run, since it is better to start with more challenging skills, skills that will not be acquired so easily—and will thus not lead to immediate spectacular results—but which will eventually lead to better success.

General case: analysis of the problem. What happens in the general case, when instead of only two knowledge components, we have a large number of different components? In the beginning, it still makes sense to project to the set S_{i_1} which is the farthest from the original state s_0.

After this original projection, in the general case, we still have a choice. We can project to any set S_{i_2}, $i_2 \neq i_1$, in which case the current distance d_1 to the desired state is multiplied by the cosine $\cos(\alpha_{i_1 i_2})$ of the angle between the corresponding sets S_{i_1} and S_{i_2}. After k steps, we get the original distance multiplied by the product of the corresponding cosines.

Our objective is to find the best order, i.e., the sequence $S_{i_1}, S_{i_2}, \ldots, S_{i_n}$ that covers all n sets $S_1, \ldots, S_n$ and for which the corresponding product

$$\cos(\alpha_{i_1 i_2}) \cdot \cos(\alpha_{i_1 i_2}) \cdot \cdots \cdot \cos(\alpha_{i_n i_1}) \tag{20.11}$$

attains the smallest possible value.

Usually, it is easier to deal with the sums than with the products. To transform the product into a sum, we can use the fact that:

- minimizing the product is equivalent to minimizing its logarithm, and
- the logarithm of the product is equal to the sum of the logarithms.

Thus, minimizing the product (20.11) is equivalent to minimizing the sum

$$D(i_1, i_2) + \cdots + D(i_n, i_1), \tag{20.12}$$

where

$$D(i, j) \stackrel{\text{def}}{=} \log(\cos(\alpha_{ij})). \tag{20.13}$$

In other words, we arrive at the following problem:

- we have n objects with known distances

$$D(i, j), \quad 1 \le i, j \le n;$$

- we must find a way to traverse all the objects and come back in such a way that the overall traveled distance is the smallest possible.

This is a well-known problem called a *traveling salesman* problem.
  It is known that

- in general, this problem is NP-hard (see, e.g., [17]), and
- in many cases, there exist reasonable algorithms for solving this problem; see, e.g., [18].

Recommendations and the need for expert (fuzzy) knowledge. Based on the above analysis, we can make the following recommendations:

- To find the optimal order of presenting the material, we must solve the corresponding instance of the traveling salesman problem, with the distances determined by the formula (20.13).
- Since in general, the traveling salesman problem is computationally difficult (NP-hard), to efficiently solve this problem, we must use expert knowledge – the knowledge for which fuzzy technique have been invented; see, e.g., [19, 20].

References

1. J.A. Kaminski, V.M. Sloutsky, A.F. Heckler, Do children need concrete instantiations to learn an abstract concept? in *Proceedings of the 27th Annual Conference of the Cognitive Science Society, Vancouver, BC, 26–29 July 2006*, ed. by R. Sun, N. Miyake (Lawrence Erlbaum, Mahwah, 2006), pp. 411–416
2. J.A. Kaminski, V.M. Sloutsky, A.F. Heckler, The advantage of abstract examples in learning math. Science **320**, 454–455 (2008)
3. L.M. Lesser, M.A. Tchoshanov, Selecting representations. Tex. Math. Teach. **53**(2), 20–26 (2006)
4. M.A. Tchoshanov, *Visual Mathematics* (ABAK Publisher, Kazan, 1997). (in Russian)
5. O. Kosheleva, Early start can inhibit learning: a geometric explanation, in *Proceedings of the 2009 World Congress of the International Fuzzy Systems Association IFSA'2009*, Lisbon, Portugal, 20–24 July (2009), pp. 438–442
6. O. Kosheleva, Early start can inhibit learning: a geometric explanation. Geombinatorics **19**(3), 108–118 (2010)
7. L.G. Gubin, B.T. Polyak, E.V. Raik, The method of projections for finding the common point of convex sets. USSR Comput. Math. Math. Phys. **7**, 1–24 (1967)
8. E.J. Kontoghiorghes, *Handbook of Parallel Computing and Statistics* (CRC Press, Boca Raton, 2006)
9. H. Stark, Y. Yang, *Vector Space Projections: A Numerical Approach to Signal and Image Processing* (Neural Nets and Optics, Wiley, 1988)
10. H. Papousek, M. Papousek, Mothering and the cognitive head-start: psychobiological considerations, in *Studies in Mother-Infant Interaction*, ed. by H.R. Schaffer (Academic Press, London, 1977)
11. H.F. Harlow, The development of learning in the rhesus monkey. Am. Sci. **47**, 459–479 (1959)
12. D.F. Bjorklund, B.L. Green, The adaptive nature of cognitive immaturity. Am. Psychol. **47**, 46–54 (1992)
13. D.F. Bjorklund, A. Pellegrini, *The Origins of Human Nature: Evolutionary Developmental Psychology* (American Psychological Association, Washington, 2002)

14. D.F. Bjorklund, *Why Youth is Not Wasted on the Young: Immaturity in Human Development* (Blackwell Publishing, Malden, 2007)

15. B.J. Ellis, D.F. Bjorklund (eds.), *Origins of the Social Mind: Evolutionary Psychology and Child Development* (Guilford Press, New York, 2005)

16. E. Remmel, The benefits of a long childhood. Am. Sci. **96**(3), 250–251 (2008)

17. C.H. Papadimitriou, *Computational Complexity* (Addison Wesley, Reading, California, 1994)

18. D.L. Applegate, R.M. Bixby, V. Chvátal, W.J. Cook, *The Traveling Salesman Problem* (Princeton University Press, Princeton, 2006)

19. G.J. Klir, B. Yuan, *Fuzzy Sets and Fuzzy Logic: Theory and Applications* (Prentice-Hall, Upper Saddle River, 1995)

20. H.T. Nguyen, E.A. Walker, *A First Course in Fuzzy Logic* (CRC Press, Boca Raton, 2006)

Part III
How to Select an Appropriate Way of Teaching Each Topic

Chapter 21
How to Select an Appropriate Way of Teaching Each Topic: An Overview of Part III

In the previous part of the book, we analyzed what is the best way to organize the topics and what is the best order of teaching different topics. In this part, we move to the analysis of what is the best way of teaching every time.

- We start, in Chap. 22, by taking into account the fact that the instructor's time is limited, and therefore, it is necessary to find an optimal way to allocate the instructor's efforts.
- The resulting analysis is supplemented, in Chap. 23, by analyzing a specific aspect of this problem: how much effort should the instructor allocate to perfecting his writing on the board.

The most effective way to teach is when teaching by an instructor is supplemented by students studying together and helping each other. This fact leads to a question of what is the optimal way to divide students into groups. This question is analyzed in Chaps. 24 and 25.

Finally, in Chap. 26, we consider a minor but important aspect of teaching large classes: when to let in late students?

© Springer-Verlag GmbH Germany 2018
O. Kosheleva and K. Villaverde, *How Interval and Fuzzy Techniques Can Improve Teaching*, Studies in Computational Intelligence 750,
https://doi.org/10.1007/978-3-662-55993-2_21

Chapter 22
What Is the Best Way to Distribute the Teacher's Efforts Among Students

In a typical class, we have students at different levels of knowledge, student with different ability to learn the material. In the ideal world, we should devote unlimited individual attention to all the students and make sure that everyone learns all the material. In real life, our resources are finite. Based on this finite amount of resources, what is the best way to distribute efforts between different students?

Even when we know the exact way each student learns, the answer depends on what is the objective of teaching the class. This can be illustrated on two extreme example: If the objective is to leave no student behind, then in the optimal resource arrangement all the effort goes to weak students who are behind, while more advanced students get bored. If the effort is to increase the school's rating by increasing the number of graduates who are accepted at top universities, then all the effort should go to the advanced students while weak students may fail.

An additional difficulty is that in reality, we do not have exact information about the cognitive ability of each student, there is a large amount of uncertainty. In this paper, we analyze the problem of optimal resource distribution under uncertainty. We hope that the resulting algorithms will be useful in designing teaching strategies.

The results from this chapter first appeared in [1–4].

Deciding which teaching method is better: formulation of the problem. Pedagogy is a fast developing field. New methods, new ideas and constantly being developed and tested. New methods and new idea may be different in many things:

- they may differ in the way material is presented,
- they may also differ in the way the teacher's effort is distributed among individual students: which of the students get more attention, etc.

Testing can also be different:

- Sometimes, the testing consists of comparing the new method with the method that is currently used.

© Springer-Verlag GmbH Germany 2018
O. Kosheleva and K. Villaverde, *How Interval and Fuzzy Techniques Can Improve Teaching*, Studies in Computational Intelligence 750,
https://doi.org/10.1007/978-3-662-55993-2_22

- Sometimes, the testing consists of comparing two (or more) different versions of the same pedagogical technique; such testing is needed to decide which version is better.

To perform a meaningful comparison, we need to agree on the criterion: how do we decide which method is better? Once we have selected a criterion, and we have performed enough experiments to get a good idea of how students will learn under different version of the method, a natural question is: what is the optimal way to teaching the students – optimal with respect to the selected criterion?

How this problem is usually solved now: a brief description. How can we gauge the efficiency of different teaching techniques? The success of each individual student i can be naturally gauged by this student's grade x_i. So, for two different techniques T and T', we know the corresponding grades $x_1, \ldots, x_n$ and $x'_1, \ldots, x'_{n'}$. Which method is better?

In some cases, the answer to this question is straightforward. For example, when $n' = n$ and when we can rearrange the grades in such a way that $x_i \leq x'_i$ for all i and $x_i < x'_i$ for some i, then clearly the method T' is better.

In practice, however, the comparison is rarely that straightforward. Often, some grades decrease while some other grades increase. In this case, how do we decide whether a new method is better or not?

In pedagogical experiments, the decision is usually made based on the comparison of the average grades

$$E \stackrel{\text{def}}{=} \frac{x_1 + \cdots + x_n}{n} \tag{22.1}$$

and

$$E' \stackrel{\text{def}}{=} \frac{x'_1 + \cdots + x'_{n'}}{n'}. \tag{22.2}$$

For example, we can use the t-test (see, e.g., [5]) and conclude that the method T' is better if the corresponding t-statistic

$$t \stackrel{\text{def}}{=} \frac{E' - E}{\sqrt{\dfrac{V}{n} + \dfrac{V'}{n'}}}, \tag{22.3}$$

where

$$V \stackrel{\text{def}}{=} \frac{1}{n-1} \cdot \sum_{i=1}^{n} (x_i - E)^2, \quad V' \stackrel{\text{def}}{=} \frac{1}{n'-1} \cdot \sum_{i=1}^{n'} (x'_i - E')^2, \tag{22.4}$$

exceeds the appropriate threshold t_α (depending on the level of confidence α with which we want to make this conclusion).

How This Problem is Usually Solved Now: Limitations. The average grade is not always the most adequate way to gauging the success of a pedagogical strategy. Whether the average grade is a good criterion or not depends on our objective.

Let us illustrate this dependence on a simplified example. Suppose that after using the original teaching method T, we get the grades $x_1 = 60$ and $x_2 = 90$. The average value of these grades is

$$E = \frac{60 + 90}{2} = 75. \tag{22.5}$$

Suppose that the new teaching method T' leads to the grades $x_1' = x_2' = 70$. The average of the new grades is $E' = 70$.

Since the average grade decreases, the traditional conclusion would be that the new teaching method T' is not as efficient as the original method T. However, one possible objective may be to decrease the failing rate. Usually, 70 is the lowest grade corresponding to C, and any grade below C is considered failing. In this case,

- in the original teaching method, one of the two students failed, while
- in the new teaching method, both students passed the class.

Thus, with respect to this objective, the new teaching method is better.

Towards Selecting the Optimal Teaching Strategy: Possible Objective Functions. Since the traditional approach – of using the average grade as a criterion – is not always adequate, let us formulate the general problem of optimal teaching.

General Description. To formulate this problem, we must know how the relative "quality" of a given teaching strategy can be determined from the grades $x_1, \ldots, x_n$. In this section, we will denote the corresponding dependence by $f(x_1, \ldots, x_n)$.

The Traditional Approach. In particular, the traditional approach corresponds to using the average

$$f(x_1, \ldots, x_n) = \frac{x_1 + \cdots + x_n}{n}. \tag{22.6}$$

Minimizing Failure Rate. The objective of minimizing the failure rate means that we minimize the number of students whose grade is below the passing threshold x_0:

$$f(x_1, \ldots, x_n) = \#\{i : x_i < x_0\}. \tag{22.7}$$

Comment. Since the general objective is to *maximize* the value of the objective function $f(x_1, \ldots, x_n)$, we can reformulate the criterion (22.7) as a maximization one: namely, minimizing (22.7) is equivalent to maximize the number of students whose grade is above (or equal to) the passing threshold x_0:

$$f(x_1, \ldots, x_n) = \#\{i : x_i \geq x_0\}. \tag{22.8}$$

No Child Left Behind. Other criteria are also possible. For example, the idea that no child should be left behind means, in effect, that we gauge the quality of a school by the performance of the worst student – i.e., of the student with the lowest grade $\min(x_1, \ldots, x_n)$. Thus, the corresponding objective is to maximize this lowest grade:

$$f(x_1, \ldots, x_n) = \min(x_1, \ldots, x_n). \tag{22.9}$$

Maximizing Success Rate. The quality of a high school is often gauged by the number of alumni who get into prestigious schools. In terms of the grades x_i, this means, crudely speaking, that we maximize the number of students whose grade exceeds the minimal entrance grade e_0 for prestigious schools:

$$f(x_1, \ldots, x_n) = \#\{i : x_i \geq e_0\}. \tag{22.10}$$

From the mathematical viewpoint, this criterion is equivalent to minimizing the number of students whose grade is below e_0 – and is, thus, equivalent to criterion (22.7), with $x_0 = e_0$,

Best School to Get In. There is a version of the above criterion which is not equivalent to (22.7), when the quality of a high school is gauged by the success of the best alumnus: e.g., "one of our alumni got into Harvard". In terms of the grades x_i, this means, crudely speaking, that we maximize the highest of the grades $\max(x_1, \ldots, x_n)$, i.e., that we take

$$f(x_1, \ldots, x_n) = \max(x_1, \ldots, x_n). \tag{22.11}$$

Case of Independence. An important practical case is when students are, in some reasonable sense, independent. This case has been actively analyzed in decision theory. In particular, it has been proven that the corresponding objective function can be represented as the sum of "marginal" objective functions representing different participants, i.e.,

$$f(x_1, \ldots, x_n) = f_1(x_1) + \cdots + f_n(x_n); \tag{22.12}$$

see, e.g., [6, 7].

In this case, increasing the grade of one of the students will make the situation better – so it is reasonable to assume that all the functions $f_i(x_i)$ are strictly increasing.

Criteria Combining Mean and Variance. Another possible approach comes from the fact that the traditional criterion– that only takes into account the average (mean) grade E is not always adequate. The reason for inadequacy is that the mean does not provide us any information about the "spread" of the grades, i.e., the information about how much the grades deviate from the mean. This information is provided by the standard deviation σ, or, equivalently, the sample variance $V = \sigma^2$. Thus, we arrive at criteria of the type $f(E, V)$.

When the mean is fixed, usually, we aim for the smallest possible variation – unless we gauge a school by its best students. Similarly, when the variance is fixed, we aim for the largest possible mean.

Thus, it is reasonable to require that the objective function $f(E, V)$ is an increasing function of E and a decreasing function of V.

Towards Selecting the Optimal Teaching Strategy: Formulation of the Problem.
Let $e_i(x_i)$ denote the amount of effort (time, etc.) that is need for i-th student to achieve the grade x_i. Clearly, the better grade we want to achieve, the more effort we need, so each function $e_i(x_i)$ is strictly increasing.

Let e denote the available amount of effort. In these terms, the problem of selecting the optimal teaching strategy means that we maximize the objective function under the constraint that the overall effort cannot exceed e:

$$\text{Maximize } f(x_1, \ldots, x_n) \tag{22.13}$$

under the constraint

$$e_1(x_1) + \cdots + e_n(x_n) \le e. \tag{22.14}$$

Explicit Solution: Case of Independent Students. For the case of independent students, when the objective function has the form (22.12), it is possible to derive an explicit solution to the corresponding constraint optimization problem (22.13) and (22.14).

First we note that, due to monotonicity, if the total effort is smaller than e, then we can spend more effort and get the better value of the objective function (22.12). In other words, the maximum is attained when all the effort is actually used, i.e., when we have the constraint

$$e_1(x_1) + \cdots + e_n(x_n) = e. \tag{22.15}$$

To maximize the objective function (22.12) under this constraint, we can use the Lagrange multiplier method. According to this method, the maximum of the function (22.12) under constraint (22.15) is attained when for some value λ, the auxiliary function

$$f_1(x_1) + \cdots + f_n(x_n) + \lambda \cdot (e_1(x_1) + \cdots + e_n(x_n)) \tag{22.16}$$

attains its (unconstrained) maximum. Differentiating this auxiliary function with respect to x_i and equating the derivatives to 0, we conclude that

$$f_i'(x_i) + \lambda \cdot e_i'(x_i) = 0, \tag{22.17}$$

where f_i' and e_i' denote the derivatives of the corresponding functions. From this formula, we can explicitly describe λ as

$$-\frac{f_i'(x_i)}{e_i'(x_i)} = \lambda. \tag{22.18}$$

So, once we know λ, we can find all the corresponding grades x_i – and the resulting efforts – by solving, for each i, a (non-linear) equation (22.18) with a single variable x_i.

The value λ can be found from the formula (22.15), i.e., from the condition that for the resulting values x_i, we get $\sum_{i=1}^{n} e_i(x_i) = e$.

Explicit Solution: "No Child Left Behind" Case. In the No Child Left Behind case, we maximize the lowest grade. For this objective function, there is also an explicit solution. Since our objective is to maximize the lowest grade, there is no sense to use the effort to get one of the student grades better than the lowest grade – because the lowest grade will not change. From the viewpoint of the objective function, it is more beneficial to use the same efforts to increase the grades of all the students at the same time – this will increase the lowest grade.

In this case, the common grade x_c that we can achieve can be determined from the condition (22.15), i.e., from the equation

$$e_1(x_c) + \cdots + e_n(x_c) = e. \tag{22.19}$$

Comment. A slightly more complex situation occurs when we start not at the beginning, but at the intermediate situation when some students already have some knowledge. Let us denote the starting grades by $x_i^{(0)}$. Without losing generality, let us assume that the students are sorted in the increasing order of their grades, i.e., that $x_1^{(0)} \leq \cdots \leq x_n^{(0)}$. In this case, the optimal effort distribution aimed at maximizing the lowest grade is as follows:

- first, all the efforts must go into increasing the original grade $x_1^{(0)}$ of the worst student to the next level $x_2^{(0)}$;
- if this attempt to increase consumes all available effort, then this is what we got;
- otherwise, if some effort is left, we raise the grades of the two lowest-graded students x_1 and x_2 to the yet next level $x_3^{(0)}$, etc.

In precise terms, the resulting optimal distribution of efforts can be described as follows. First, we find the largest value k for which all the grades $x_1, \ldots, x_k$ can be raised to the k-th original level $x_k^{(0)}$. In precise terms, this means the largest value k for which

$$e_1(x_k^{(0)}) + \cdots + e_k(x_k^{(0)}) \leq e. \tag{22.20}$$

This means that for the criterion $\min(x_1, \ldots, x_n)$, we can achieve the value $x_k^{(0)}$, but we cannot achieve the value $x_{k+1}^{(0)}$.

Then, we find the value $x \in [x_k^{(0)}, x_{k+1}^{(0)})$ for which

$$e_1(x) + \cdots + e_{k-1}(x) + e_k(x) = e. \tag{22.21}$$

This value x is the optimal value of the criterion $\min(x_1, \ldots, x_n)$.

Explicit Solution: "Best School to Get In" case. If the criterion is the Best School to Get In, i.e., in terms of grades, the largest possible grade x_i, then the optimal use of effort is, of course, to concentrate on a single individual and ignore the rest. Which individual to target depends on how much gain we will get. In other words,

- first, for each i, we find x_i for which $e_i(x_i) = e$, and then
- we choose the student with the largest value of x_i as a recipient of all the efforts.

Need to Take Uncertainty into Account: Reminder. In the above text, we assumed that:

- we know *exactly* the benefits $f(x_1, \ldots, x_n)$ of achieving the knowledge levels corresponding to the grades $x_1, \ldots, x_n$; for example, we know the exact expressions for the marginal functions $f_i(x_i)$;
- we know *exactly* how much effort $e_i(x_i)$ is needed to bring each student i to a given grade level x_i, and
- we know *exactly* the level of knowledge x_i of each student – it is exactly determined by the grade x_i.

In practice, we have *uncertainty*.

Average Benefit Function. First, we rarely know the exact marginal function $f_i(x_i)$ characterizing each individual student. At best, we know the *average* function $u(x)$ describing the average benefits of grade x to a student.

Average Effort Function. Second, we rarely know the exact effort function $e_i(x_i)$ characterizing each individual student. At best, we know the *average* function $e(x)$ describing the average effort needed to bring a student to the level of knowledge corresponding to the grade x.

Interval Uncertainty. Finally, the grade $\widetilde{x}_i$ is only an approximate indication of the student's level of knowledge. Once we know the grade $\widetilde{x}_i$, we cannot conclude that the level of knowledge x_i is exactly $\widetilde{x}_i$. At best, we know the accuracy ε_i of this representation. In this case, the actual (unknown) level of knowledge x_i can take any value from the interval $\mathbf{x}_i = [\underline{x}_i, \overline{x}_i] \overset{\text{def}}{=} [\widetilde{x}_i - \varepsilon_i, \widetilde{x}_i + \varepsilon_i]$.

Under interval uncertainty, instead of a single value of the objective function $f(x_1, \ldots, x_n)$, we get an *interval* of possible values

$$[\underline{f}, \overline{f}] = f(\mathbf{x}_1, \ldots, \mathbf{x}_n) \overset{\text{def}}{=} \{f(x_1, \ldots, x_n) \,|\, x_1 \in \mathbf{x}_1, \ldots, x_n \in \mathbf{x}_n\}. \qquad (22.22)$$

Fuzzy Uncertainty. In many practical situations, the estimates $\widetilde{x}_i$ come from experts. Experts often describe the inaccuracy of their estimates in terms of imprecise words from natural language, such as "approximately 0.1", etc. A natural way to formalize such words is to use special techniques developed for formalizing this type of estimates – specifically, the technique of fuzzy logic; see, e.g., [8, 9].

In this technique, for each possible value of $x_i \in [\underline{x}_i, \overline{x}_i]$, we describe the degree $\mu_i(x_i)$ to which this value is possible. For each degree of certainty α, we can determine the set of values of x_i that are possible with at least this degree of certainty – the α-cut

$x_i(\alpha) = \{x \mid \mu_i(x) \geq \alpha\}$ of the original fuzzy set. Vice versa, if we know α-cuts for every α, then, for each object x, we can determine the degree of possibility that x belongs to the original fuzzy set [8–12]. A fuzzy set can be thus viewed as a nested family of its (interval) α-cuts.

From the Computational Viewpoint, Fuzzy Uncertainty Can be Reduced to the Interval One. Once we know how to propagate interval uncertainty, then, to propagate the fuzzy uncertainty, we can consider, for each α, the fuzzy set y with the α-cuts

$$\mathbf{y}(\alpha) = f(\mathbf{x}_1(\alpha), \dots, \mathbf{x}_1(\alpha)); \tag{22.23}$$

see, e.g., [8–12]. So, from the computational viewpoint, the problem of propagating fuzzy uncertainty can be reduced to several interval propagation problems.

Because of this reduction, in the following text, we will mainly concentrate on algorithms for the interval case.

How to Take Uncertainty into Account. Let us analyze how we can take into account these different types of uncertainties.

Average Benefit Function: General Situation. Let us first consider the case when instead of the *individual* benefit functions $f_1(x_1), \dots, f_n(x_n)$, we only know the *average* benefit function $u(x)$. In this case, for a combination of grades $x_1, \dots, x_n$, the resulting value of the objective function is

$$f(x_1, \dots, x_n) = u(x_1) + \cdots + u(x_n). \tag{22.24}$$

Smooth Benefit Functions. Usually, the benefit function is reasonably smooth. In this case, if (hopefully) all grades are close, we can expand the function $u(x)$ in Taylor series around the average grade, and keep only quadratic terms in this expansion. The general form of this quadratic approximation is

$$u(x) = u_0 + u_1 \cdot x + u_2 \cdot x^2, \tag{22.25}$$

for some coefficients u_0, u_1, and u_2. For this function, the expression (22.24) for the objective function takes the form

$$f(x_1, \dots, x_n) = n \cdot u_0 + u_1 \cdot \sum_{i=1}^{n} x_i + u_2 \cdot \sum_{i=1}^{n} x_i^2, \tag{22.26}$$

i.e., the form

$$f(x_1, \dots, x_n) = f_0 + f_1 \cdot E + f_2 \cdot M, \tag{22.27}$$

where $f_0 \overset{\text{def}}{=} n \cdot u_0$, $f_1 \overset{\text{def}}{=} n \cdot u_1$, $f_2 \overset{\text{def}}{=} n \cdot u_2$, E is the average (22.1), and M is the second sample moment:

$$M \overset{\text{def}}{=} \frac{1}{n} \cdot \sum_{i=1}^{n} x_i^2. \tag{22.28}$$

Thus, for smooth benefit functions $u(x)$, to estimate the benefit of a given combination of grades $x_1, \ldots, x_n$, it is not necessary to know all these n grades, it is sufficient to know the average grade and the mean squared grade (or, equivalently, the standard deviation of the grades).

Comment. In general, the benefit function $u(x)$ is increasing with x_i. However, it is worth mentioning that this conclusion holds for every quadratic function $u(x)$, not necessarily a function which is increasing for all the values $x_1, \ldots, x_n$.

Case of Interval Uncertainty. Until now, we assumed that we know the exact values $x_1, \ldots, x_n$ of the students' knowledge levels. What will happen if instead, we only know intervals $[\underline{x}_i, \overline{x}_i]$ of possible values of x_i?

Since the benefit function $u(x)$ is increasing (the more knowledge the better),

- the largest possible value $\overline{f}$ of the objective function is attained when the values x_i are the largest possible $x_i = \overline{x}_i$, and
- the smallest possible value $\underline{f}$ of the objective function is attained when the values x_i are the smallest possible $x_i = \underline{x}_i$.

In other words, we get the following interval $[\underline{f}, \overline{f}]$ of possible values $f(x_1, \ldots, x_n)$ of the objective function:

$$[\underline{f}, \overline{f}] = \left[\sum_{i=1}^{n} u(\underline{x}_i), \sum_{i=1}^{n} u(\overline{x}_i) \right]. \tag{22.29}$$

Comment. We mentioned that for the case of smooth (quadratic) benefit function and exactly known x_i, we do not need to keep all n grades – it is sufficient to keep only the first and second sample moments of these grades. A natural question is: in the case of interval uncertainty, do we need to keep n intervals, or can we use a few numbers instead? At the end of the chapter, we show that under interval uncertainty, in the general case, all n values are needed.

Interval Uncertainty, Smooth Benefit Function: Analysis. In the main text, we mentioned that for the case of smooth (quadratic) benefit function $u(x)$ and exactly known x_i, we do not need to keep all n grades, it is sufficient to keep only the first and second sample moments of these grades. Let us show that for interval uncertainty, all n bounds are needed.

Specifically, we will prove the following.

Precise Formulation of the Result. Suppose that we have n intervals $[\widetilde{x}_i - \varepsilon_i, \widetilde{x}_i + \varepsilon_i]$. We will consider a *non-degenerate* case when all the grades $\widetilde{x}_i$ are different.

Let us assume that for every quadratic function $u(x)$, we know the range $[\underline{f}, \overline{f}]$ of the function $u(x_1) + \cdots + u(x_n)$ over the intervals $[\widetilde{x}_i - \varepsilon_i, \widetilde{x}_i + \varepsilon_i]$. Then, based on the ranges corresponding to different quadratic functions $u(x)$, we can uniquely reconstruct the original collection of intervals.

In other words, if two different non-degenerate collections of intervals lead to exact same ranges for every quadratic function, then these collections coincide – i.e., they differ only by permutations.

Comment. It is not known whether the same is true if we allow arbitrary – not necessarily non-degenerate – collections of intervals.

Proof For every quadratic function $u(x)$, the largest possible value $\overline{f}$ of the sum $\sum_{i=1}^{n} u(x_i)$ is attained when each of the terms $u(x_i)$ is the largest possible, and is equal to the sum of the corresponding n largest values:

$$\overline{f} = \overline{f}_1 + \cdots + \overline{f}_n. \tag{22.30}$$

For every real number a, the quadratic function $u(x) = (x - \alpha)^2$ attains its largest value on the interval $[\widetilde{x}_i - \varepsilon_i, \widetilde{x}_i + \varepsilon_i]$ at one of the endpoints $\widetilde{x}_i - \varepsilon_i$ or $\widetilde{x}_i + \varepsilon_i$. One can easily check that:

- when $a \leq \widetilde{x}_i$, then the largest possible value $\overline{f}_i$ of $u(x)$ on the interval

$$[\widetilde{x}_i - \varepsilon_i, \widetilde{x}_i + \varepsilon_i]$$

  is attained when $x_i = \overline{x}_i = \widetilde{x}_i + \varepsilon_i$ and is equal to $\overline{f}_i = (\overline{x}_i - a)^2$;
- when $a \geq \widetilde{x}_i$, then the largest possible value $\overline{f}_i$ of $u(x)$ on the interval

$$[\widetilde{x}_i - \varepsilon_i, \widetilde{x}_i + \varepsilon_i]$$

  is attained when $x_i = \underline{x}_i = \widetilde{x}_i - \varepsilon_i$ and is equal to $\overline{f}_i = (\underline{x}_i - a)^2$.

Let us use this fact to describe the dependence of $\overline{f}$ on the parameter a.

When $a \neq \widetilde{x}_i$, the value $\overline{f}$ is the sum of n smooth expressions.

At each point $a = \overline{x}_i$, all the terms $\overline{f}_j$ in the sum $\overline{f}$ are smooth except for the term $\overline{f}_i$ that turns from $(\overline{x}_i - a)^2$ to $(\underline{x}_i - a)^2$. The derivative of $\overline{f}_i$ with respect to a changes from $2 \cdot (a - \overline{x}_i)$ to $2 \cdot (a - \underline{x}_i)$, i.e., increases by

$$2 \cdot (a - \underline{x}_i) - 2 \cdot (a - \overline{x}_i) = 2 \cdot (\overline{x}_i - \underline{x}_i) = 4 \cdot \varepsilon_i. \tag{22.31}$$

Since all the other components $\overline{f}_j$ are smooth at $a = \widetilde{x}_i$, at $a = \widetilde{x}_i$, the derivative of the sum $\overline{f}(a)$ also increases by $4\varepsilon_i$.

Thus, once we know the value $\overline{f}$ for all a,

- we can find the values $\widetilde{x}_i$ as the values at which the derivative is discontinuous; and
- we can find each value ε_i as $1/4$ of the increase of the derivative at the corresponding point $\widetilde{x}_i$.

The statement is proven.

Estimating $f(E, V)$ Under Interval Uncertainty. Let us now consider the case when the objective function has the form $f(E, V)$, where $f(E, V)$ increases as a function of E and decreases as a function of V. How can we estimate the range $[\underline{f}, \overline{f}]$ of the values of this objective function under interval uncertainty $x_i \in [\underline{x}_i, \overline{x}_i]$?

In general, this range estimation problem is NP-hard already for the case $f(E, V) = -V$; see, e.g., [13]. This means, crudely speaking, that unless P = NP (and most computer scientists believe that P≠NP), no efficient (polynomial time) algorithm can always compute the exact range.

The maximum of the expression $f(E, V)$ can be found efficiently. For that, it is sufficient to consider all $2n + 2$ intervals $[\underline{r}, \overline{r}]$ into which the values $\underline{x}_i$ and $\overline{x}_i$ divide the real line, and for each of these intervals, and for each $r \in [\underline{r}, \overline{r}]$, take the values

- $x_i = \overline{x}_i$ when $\overline{x}_i \leq \underline{r}$;
- $x_i = r$ when $[\underline{r}, \overline{r}] \subseteq [\underline{x}_i, \overline{x}_i]$; and
- $x_i = \underline{x}_i$ when $\overline{r} \leq \underline{x}_i$.

(The proof is similar to the ones given in [13].)

For the minimum of $f(E, V)$, for reasonable cases, efficient algorithms are also possible. One such case is when none of the intervals $[\underline{x}_i, \overline{x}_i]$ is a proper subset of another one, i.e., to be more precise, when $\underline{x}_i, \overline{x}_i \not\subseteq (\underline{x}_j, \overline{x}_j)$.

In this case, a proof similar to the one from [13] shows that if we sort the intervals in lexicographic order

$$[\underline{x}_1, \overline{x}_1] \leq [\underline{x}_2, \overline{x}_2] \leq \cdots \leq [\underline{x}_n, \overline{x}_n], \tag{22.32}$$

where

$$[\underline{a}, \overline{b}] \leq [\underline{b}, \overline{b}] \leftrightarrow \underline{a} < \underline{b} \vee (\underline{a} = \underline{b} \,\&\, \overline{a} \leq \overline{b}), \tag{22.33}$$

then the minimum of f is attained at one of the combinations

$$(\underline{x}_1, \ldots, \underline{x}_{k-1}, x_k, \overline{x}_{k+1}, \ldots, \overline{x}_n) \tag{22.34}$$

for some $x_k \in [\underline{x}_k, \overline{x}_k]$. Thus, to find the minimum, it is sufficient to sort the values, and then find the smallest possible value of $f(E, V)$ for each of $n + 1$ such combinations.

References

1. O. Kosheleva, V. Kreinovich, Towards optimal effort distribution in process design under uncertainty, with application to education. Int. J. Reliab. Saf. 6(1–2), 148–166 (2012)
2. O. M. Kosheleva, V. Kreinovich, What is the best way to distribute efforts among students: towards quantitative approach to human cognition, in *Proceedings of the 28th North American Fuzzy Information Processing Society Annual Conference NAFIPS'09* Cincinnati, Ohio, 14–17 June, 2009
3. O. Kosheleva, V. Kreinovich, Towards Optimal Effort Distribution in Process Design under Uncertainty, with Application to Education, in *Proceedings of the 4th International Workshop on Reliable Engineering Computing REC'2010* Singapore, 3–5 March, 2010, pp. 509–525

4. O. Kosheleva, V. Kreinovich, What is the best way to distribute efforts among students: towards quantitative approach. Appl. Math. Sci. **4**(9), 417–429 (2010)
5. D.J. Sheskin, *Handbook of Parametric and Nonparametric Statistical Procedures* (Chapman and Hall/CRC, Boca Raton, Florida, 2007)
6. P.C. Fishburn, *Utility Theory for Decision Making* (Wiley, New York, 1969)
7. P.C. Fishburn, *Nonlinear Preference and Utility Theory* (The John Hopkins Press, Baltimore, MD, 1988)
8. G.J. Klir, B. Yuan, *Fuzzy Sets and Fuzzy Logic: Theory and Applications* (Prentice-Hall, New Jersey, 1995)
9. H.T. Nguyen, E.A. Walker, *A First Course in Fuzzy Logic* (CRC Press, Florida, 2006)
10. D. Dubois, H. Prade, Operations on fuzzy numbers. Int. J. Syst. Sci. **9**, 613–626 (1978)
11. R.E. Moore, W. Lodwick, Interval analysis and fuzzy set theory. Fuzzy Sets Syst. **135**(1), 5–9 (2003)
12. H.T. Nguyen, V. Kreinovich, Nested intervals and sets: concepts, relations to fuzzy sets, and applications, in *Applications of Interval Computations*, ed. by R.B. Kearfott, V. Kreinovich (Kluwer, Dordrecht, 1996), pp. 245–290
13. V. Kreinovich, G. Xiang, S.A. Starks, L. Longpré, M. Ceberio, R. Araiza, J. Beck, R. Kandathi, A. Nayak, R. Torres, J. Hajagos, Towards combining probabilistic and interval uncertainty in engineering calculations: algorithms for computing statistics under interval uncertainty, and their computational complexity. Reliab. Comput. **12**(6), 471–501 (2006)

Chapter 23
What is the Best Way to Allocate Teacher's Efforts: How Accurately Should We Write on the Board? When Marking Comments on Student Papers?

Writing on the board is an important part of a lecture. Lecturers' handwriting is not always perfect. Usually, a lecturer can write slower and more legibly, this will increase understandability but slow down the lecture. In this chapter, we analyze an optimal trade-off between speed and legibility.

The results from this chapter first appeared in [1].

People's handwriting is usually not perfect. Most people can write in a better (and more legible) handwriting if they make a special effort and write slower. Our handwritten notes to selves are usually less legible to others than notes to others. An extreme example is a difference between cursive – whose purpose is to make writing faster – against block letters which take longer to write but which make the text more legible.

Teachers and professors use a lot of handwriting:

- when writing on the board when lecturing,
- when answering individual student questions during office hours,
- when writing comments on the student papers, etc.

What is the appropriate degree of handwriting accuracy in each of these cases?

A natural objective for selecting the proper degree of accuracy is to minimize the total effort of the writer *and* of the readers, an effort weighted by the importance of the writer's and the readers' time for the society.

What effort do we need to hardwrite with a given accuracy: analysis of the problem. The inaccuracy of handwriting can be naturally described as noise added to the ideal writing result. In other words, at each moment of time, the actual position of the writing hand slightly differs from its ideal position. Let δt be the time quantum, i.e., the smallest period of time for which the the noise at each moment t and the noise at the next moment $t + \delta t$ are statistically independent.

Every small piece of a letter – that appears as a result of this writing – comes from averaging these positions over the time period that it takes to write this piece of the

© Springer-Verlag GmbH Germany 2018
O. Kosheleva and K. Villaverde, *How Interval and Fuzzy Techniques Can Improve Teaching*, Studies in Computational Intelligence 750,
https://doi.org/10.1007/978-3-662-55993-2_23

letter. During the time t that it takes to write a piece of the letter, we average $t/\delta t$ independent noise signals.

It is well known in statistics that if we take an average of n independent identically distributed random variables, then the standard deviation decreases by a factor of $\sqrt{n}$. Thus, the standard deviation (inaccuracy) σ of the resulting writing is equal to

$$\sigma = \frac{\sigma_0}{\sqrt{t/\delta t}},$$

where σ_0 is the standard deviation of the original noise.

It should be mentioned that the value σ_0 may be different for different people:

- some people write fast and still very accurately;
- for others, accurate handwriting is only possible when they write very slowly.

Let us find out how this accuracy depends on the speed with which we write. Let v_w be the speed of writing, i.e., the number of symbols per unit time that results from this writing. Let n be the average number of pieces contained in a symbol. Then, during a unit time, we write v_w symbols and thus, $n \cdot v_w$ pieces of symbols. So, the time t of writing one piece of a symbol is equal to $t = \dfrac{1}{n \cdot v_w}$. Substituting this expression into the above formula for σ, we conclude that

$$\sigma = \frac{\sigma_0}{\sqrt{(1/n \cdot v_w) \cdot \delta t}} = \sigma_0 \cdot \sqrt{\frac{n}{\delta t}} \cdot \sqrt{v_w}.$$

Thus, we arrive at the following conclusion:

What effort do we need to hardwrite with a given accuracy: result. If we write with a speed v_w, then the accuracy (standard deviation) of a person's handwriting is equal to $\sigma = c \cdot \sqrt{v_w}$, where the constant c depends on the individual writer.

What effort do we need to understand a handwriting: analysis of the problem. To understand the handwriting, we do not need to observe every single part of each symbol, it is sufficient to understand the symbol's general shape. For example, we usually, when we read a text, do not notice the details of the font – unless, of course, we are specifically paying attention to the font.

In other words, instead of reading every single pixel that forms a symbol, we only pay attention to a few pixels from this symbol. The speed with which we can understand the text depends on the number of pixels that we need to read in order to understand the symbol properly.

Within each part of the symbol (e.g., a linear part of a circular part), we read several pixels, and then take an average of their values to get the general impression about this part. The part is well recognized when the standard deviation of this average is smaller than (or equal to) some threshold value σ_t. If we read v_r symbols per unit time, this means that it takes time $1/v_r$ to read each symbol and thus, time $1/(v_r \cdot n)$

to read each part of the symbol. Let t_p be the time for reading one pixel. This means that for each part of the symbol, we read $n_p = 1/(v_r \cdot n \cdot t_p)$ pixels.

Each pixel is written with standard deviation σ, so when we average over n_p symbols, we get standard deviation

$$\sigma_r = \frac{\sigma}{\sqrt{n_p}} = \sigma \cdot \sqrt{v_r} \cdot \sqrt{n \cdot t_p}.$$

Substituting the above expression for σ in terms of the writing speed v_w, we conclude that

$$\sigma_r = c \cdot \sqrt{v_w} \cdot \sqrt{v_r} \cdot \sqrt{n \cdot t_p}.$$

The maximum reading speed is determine by the condition that this standard deviation is equal to the threshold value σ_t, i.e., that

$$\sigma_r = c \cdot \sqrt{v_w} \cdot \sqrt{v_r} \cdot \sqrt{n \cdot t_p},$$

hence the reading speed v_r can be found as

$$v_r = \frac{c^2 \cdot n \cdot t_p}{\sigma_t^2} \cdot \frac{1}{v_w}.$$

The coefficient at $\dfrac{1}{v_w}$ depends on the parameter c that describes the accuracy of the writer. Thus, this whole coefficient can be viewed as a parameter describing the writer's accuracy. Thus, we arrive at the following conclusion.

What effort do we need to understand a handwriting: result. The maximal speed v_r with which we can efficiently read a text written with speed v_w is equal to

$$v_r = \frac{C}{v_w},$$

where the parameter C describer the writer's accuracy.

How to describe efforts. Our objective is to find the writing speed v_w for which the overall effort E is the smallest. The overall efforts consists of the writer's effort and the reader's effort.

To write a text consisting of N symbols, the writer need time N/v_w. Let s_w be the societal value of one time unit of the writer – it can be gauged, e.g., by the writer's salary. Then, the total effort used by the writer is $s_w \cdot (N/v_w)$.

Let N_r be the number of intended readers of this text, and let s_r be the average societal value of the reader's time. Each reader needs time N/v_r to read all N symbols, so the total effort of all the readers is $N_r \cdot s_r \cdot (N/v_r)$. Substituting the above formula for v_r in terms of v_w, we conclude that the readers' effort is equal to $N_r \cdot s_r \cdot N \cdot (v_w/C)$. Thus, the overall effort E is equal to

$$E = s_w \cdot \frac{N}{v_w} + N_r \cdot s_r \cdot N \cdot \frac{v_w}{C}.$$

Formula describing the optimal writing speed. To minimize the total effort E with respect to v_w, we differentiate E relative to v_w and equate the resulting derivative to 0. As a result, we get

$$-s_w \cdot \frac{1}{v_w^2} + N_r \cdot s_r \cdot N \cdot \frac{1}{C} = 0,$$

hence

$$v_w = \sqrt{\frac{s_w \cdot C}{s_r \cdot N_r}}.$$

Here, s_w and s_r are the societal value of the writer and the reader, respectively, N_r is the number of intended readers, and C is a constant that describes the writer's handwriting ability, the constant that can be determined from the formula $v_r = \frac{C}{v_w}$ that describes the dependence of the reading speed v_r on the speed v_w with which the text was handwritten.

Conclusions. The slower the handwriting, the larger the effort. Thus, the formula leads to the following conclusions:

- The more important the writer and the less important the reader, the less he or she need to try to write more accurately. For example, a student handwriting a homework should use more effort in his or her handwriting than an instructor grading this homework.
- The more accurate the writer, the less effort he or she needs to spend writing.
- The larger the reading audience, the more effort the writer needs when writing. For example, the writing on a board when lecturing should be more accurate than a writing comments on an individual homework; when lecturing, the larger the class, the more accurate should be the handwriting.

Reference

1. M. Ceberio, O. Kosheleva, How accurately should we write on the board? when marking comments on student papers? J. Uncertain Syst. **6**(2), 89–91 (2012)

Chapter 24
How to Divide Students into Groups so as to Optimize Learning

To enhance learning, it is desirable to also let students learn from each other, e.g., by working in groups. It is known that such groupwork can improve learning, but the effect strongly depends on how we divide students into groups. In this chapter, based on a first approximation model of student interaction, we describe how to optimally divide students into groups so as to optimize the resulting learning. We hope that, by taking into account other aspects of student interaction, it will be possible to transform our solution into truly optimal practical recommendations.

The results from this chapter first appeared in [1].

Groupwork as a way to teach better. Traditionally, students mostly learn from their instructor. The instructor presents the new material, asks the students to solve some related problems, and then provides individual feedback to students – explaining, to each student, his or her possible misunderstandings. Such an individual feedback is extremely helpful to the student. However, providing such an individual feedback requires a lot of time – especially in a class of reasonable size. So, if this feedback only comes from the instructor (and a Teaching Assistant), the amount of such feedback is limited. Moreover, in such situations, a significant amount of time is needed to grade the assignments of the whole class, so there is a significant delay between the test time and the time when students get their feedback.

It is well known that we can increase the amount of feedback – and decrease the delay of producing this feedback – if we also ask students from the class to provide useful feedback to each other. Different students have somewhat different misconceptions, so when a small group of students starts solving a problem together, they can often see each other's mistakes and provide corrections – thus, teaching each other.

Groupwork is not a panacea. While in principle, groupwork is efficient, its efficiency depends on how we divide students into groups. If we simply allow students to group themselves together, often, strong students team together and weak students

© Springer-Verlag GmbH Germany 2018
O. Kosheleva and K. Villaverde, *How Interval and Fuzzy Techniques Can Improve Teaching*, Studies in Computational Intelligence 750,
https://doi.org/10.1007/978-3-662-55993-2_24

team together. Strong students already know the material, so they do not benefit from working together. Similarly, weak students are equally lost, so having them solve a problem together does not help; see, e.g., [2–11].

How to divide students into groups? Since the efficiency of groupwork depends on the subdivision into groups, to make groupwork as efficient as possible, it is desirable to find the optimal way to divide students into groups. This is the problem that we study in this chapter.

Need for an approximate description. A realistic description of student interaction requires that we take into account a multi-D *learning profile* of each student: how much the students knows of each part of the material, what is the student's learning style, etc. Such a description is difficult to formulate and even more difficult to optimize. Because of this difficulty, in this chapter, we consider a simplified description of student interaction. Already for this simplified description, the corresponding optimization problem is non-trivial – but we succeed in solving it under reasonable assumptions.

How to describe the current state of learning. In this chapter, we consider a simplified (first approximation) model of student interaction. Our main simplifying assumption is that for each student, the degree to which this student learned the material can be characterized by a *single* number – crudely speaking, this student's grade so far. In the following text, we will denote the number of students in the class by n, and we will denote the degree of knowledge of the ith student by d_i, $i = 1, \ldots, n$. So, we arrive at the following definition.

Definition 24.1 Let n be an integer; we will call this integer a number of students. By a state of knowledge, we mean a tuple $d = (d_1, \ldots, d_n)$ consisting of n non-negative numbers.

Subdivision into groups. The following describes the general subdivision into groups.

Definition 24.2 Let n be a number of students. By a subdivision into groups, we mean a subdivision of the set $\{1, \ldots, n\}$ into a finite number of non-intersecting subsets $G_1, \ldots, G_m$ for which $G_k \cap G_l = \emptyset$ for $k \neq l$ and $\bigcup_{k=1}^{m} G_k = \{1, \ldots, n\}$.

In this chapter, we will mostly consider subdivision into groups of equal size.

How groupwork helps students: a description. If two students with degrees $d_i < d_j$ work together, then the degree of knowledge of the ith student increases. As we have mentioned earlier, if two students are at the same level of knowledge, there is not much that they can learn from each other. The more the jth student knows that the ith student doesn't, the more the ith student will learn. So, it is reasonable to assume that the amount of material that the ith student learns is proportional to the difference $d_j - d_i$, with some known coefficient of proportionality α. Thus, after the groupwork, the new level of knowledge of the ith student is equal to $d_i' = d_i + \alpha \cdot (d_j - d_i)$.

If more than two students work together, then each student learns from all the students from the group who have a higher degree of knowledge. For example, if three students, with original degrees of knowledge $d_i < d_j < d_k$, work together, then after the groupwork, their new levels of knowledge are equal to

$$d_i' = d_i + \alpha \cdot (d_j - d_i) + (d_k - d_i), \quad d_j' = d_j + \alpha \cdot (d_k - d_j), \quad \text{and } d_k' = d_k.$$

In general, we arrive at the following definition.

Definition 24.3 Let n be the number of students, and let $\alpha > 0$ be a real number. For each state of knowledge $d = (d_1, \ldots, d_n)$ and for each subdivision into groups $G_1, \ldots, G_m$, the resulting state of knowledge $d' = (d_1', \ldots, d_n')$ is defined as follows: for every $k = 1, \ldots, m$ and for every $i \in G_k$, we have

$$d_i' = d_i + \alpha \cdot \sum_{j \in G_k, d_j > d_j} (d_j - d_i).$$

Dynamic groups. Subdivision into groups varies. Our objective is to find the subdivision which, at this moment of time, leads to the best gain. From this viewpoint, we need to find groups that work for a forthcoming short period of time, during which the change in grades – proportional to the coefficient α – is small. After this brief interaction, we can again gauge the student's knowledge and, of needed, change the subdivision into groups to reflect what students learned. From this viewpoint, it is sufficient to consider small positive values α.

Comment. Ideally, we should also take into account that there is a cost of group-changing: students spend some effort adjusting to their new groups.

Possible objective functions. Our goal is to find a subdivision into groups for which the overall degree of knowledge is optimal. This optimal subdivision depends on how we gauge the overall degree of knowledge. In this chapter, we will consider three possible criteria (see, e.g., [12]):

- first, we will consider the *average grade* $a \stackrel{\text{def}}{=} \dfrac{1}{n} \cdot \sum_{i=1}^{n} d_i$;
- another reasonable criterion is maximizing retention, i.e., minimizing the number of students who failed the course; in this case, most attention is paid to students who are at the largest risk of failing, i.e., to students with the smallest possible degree of knowledge; the better the knowledge of this worst performing student, the smaller the risk of failing; thus, from this viewpoint, we should maximize the *worst grade* $w \stackrel{\text{def}}{=} \min_{i=1,\ldots,n} d_i$;
- many high schools brag about the number of their graduates who get into Ivy League colleges; from this viewpoint, most attention is paid to the best students; from this viewpoint, we should maximize the *best grade* $b \stackrel{\text{def}}{=} \max_{i=1,\ldots,n} d_i$.

We will consider all three optimality criteria – and their combinations.

Straightforward results Let us consider the situation when we have $n = g \cdot m$ students, we know their degree of knowledge $d_1, \ldots, d_n$, and we want to subdivide these students into m subgroups of g students so as to maximize each of the three objective functions. Let us start with the simplest case $g = 2$, when we divide students into pairs.

Proposition 24.1 *To maximize the average grade a, we divide the students into pairs as follows:*

- *we sort the students by their knowledge, so that*

$$d_1 \leq d_2 \leq \cdots \leq d_n,$$

- *in each pair, we match one student from the lower half $L_0 \overset{\text{def}}{=} \{d_1, d_2, \ldots, d_{n/2}\}$ with one student from the upper half $L_1 \overset{\text{def}}{=} \{d_{(n/2)+1}, \ldots \leq d_n\}$.*

Comment. For reader's convenience, all the proofs are placed at the end.

Proposition 24.2 *To maximize the worst grade w, we divide the students into pairs as follows:*

- *we sort the students by their knowledge;*
- *we pair the worst-performing student (corresponding to d_1) with the best-performing student (corresponding to d_n), and,*
- *if there are other students with $d_i = d_1$, we match them with d_{n-1}, d_{n-2}, etc.*

Other students can be paired arbitrarily.

If we try to optimize the best grade, subdivision is useless: no matter how we subdivide the students, in this model, the best grade does not change; this is true for any group size $g \geq 2$:

Proposition 24.3 *No subdivision will improve the best grade b.*

For average-grade and worst-grade optimization, similar results hold for groups of general size $g \geq 2$:

Proposition 24.4 *For every $g \geq 2$, to maximize the average grade a, we divide the students into groups as follows:*

- *we sort the students by their knowledge;*
- *based on this sorting, divide the students into g sets:*

$$L_0 = \{d_1, d_2, \ldots, d_{n/g}\}; \ldots$$

$$L_k = \{d_{k \cdot (n/g)+1}, \ldots, d_{(k+1) \cdot (n/g)}\}, \ldots,$$

$$L_{g-1} = \{d_{(g-1) \cdot (n/g)+1}, \ldots, d_n\};$$

- *in each group, we pick one student from each of g sets $L_0, L_1, \ldots, L_{g-1}$.*

Editorial comment. For reader's convenience, all the proofs are placed at the end.

Comment. If we measure the students' performance accurately enough, then the degrees of knowledge of different students are different. In this case, we have the following result.

Proposition 24.5 *If all students' degree d_i are different, then, to maximize the worst grade w, we divide the students into groups as follows:*

- *we sort the students by their knowledge;*
- *we combine the worst-performing student (corresponding to d_1) with $g - 1$ best-performing students (corresponding to $d_n, d_{n-1}, \ldots, d_{n-2}$).*

Other students can be grouped arbitrarily.

Comment. The proposition also holds if several students have the same degree – as long as there is only one worst-performing student. If there are several equally low-performing students $d_1 = d_2 = \ldots = d_s$, then we need to divide $(g - 1) \cdot s$ top-performing students into s subgroups of $g - 1$ so as to maximize the minimum of the overall grade within each subgroups. Then, each of the low-performing students is matched with one of these subgroups.

More nuanced optimality criteria: discussion. When we use the above optimality criteria such as average grade, worst grade, or best grade, we end up with several different subdivisions which lead to the same optimal value of the selected criterion. We can use this non-uniqueness to optimize something else. In other words, instead of the original optimality criterion, we consider a *lexicographic* combination of these criteria. For example, we say that one subdivision is better than another one if either its average grade a is better ($a > a'$), or they have the same average grade $a = a'$ but the first subdivision has a better worst grade $w > w'$. Here are some related results.

Proposition 24.6 *Let us assume that we perform the following optimization:*

- *first, we optimize the average grade;*
- *if there are several subdivisions for which the average grade is optimal, then, among all the subdivisions, we select the one for which the worst grade is the largest;*
- *if we have several subdivisions with the largest worst grade, we select the one with the largest second worst grade,*
- *etc.*

Then, for $g = 2$, the following subdivision is optimal:

- *we sort the students by their knowledge, so that*

$$d_1 \leq d_2 \leq \cdots \leq d_n,$$

- *we then match d_1 with d_n, d_2 with d_{n-1}, and, in general, d_k with d_{n+1-k}.*

Comment. For group size $g = 2$, we get the exact same subdivision if we first maximize w, then a, and/or then maximize the second-worst grade, etc.

Proposition 24.7 *Let us assume that we perform the following optimization:*

- *first, we optimize the average grade;*
- *if there are several subdivisions for which the average grade is optimal, then, among all the subdivisions, we select the one for which the worst grade is the largest;*
- *if we have several subdivisions with the largest worst grade, we select the one with the largest second worst grade,*
- *etc.*

Then, for $g \geq 2$, if all the degrees are different, the following subdivision is optimal:

- *we sort the students by their knowledge,*
- *based on this sorting, divide the students into g sets:*

$$L_0 = \{d_1, d_2, \ldots, d_{n/g}\}; \ldots$$

$$L_k = \{d_{k \cdot (n/g)+1}, \ldots, d_{(k+1) \cdot (n/g)}\}, \ldots,$$

$$L_{g-1} = \{d_{(g-1) \cdot (n/g)+1}, \ldots, d_n\};$$

- *we match the smallest value $d_1 \in L_0$ with the largest values from $L_1, \ldots, L_{g-1}$,*
- *we match the second smallest value $d_2 \in L_0$ with the second largest values from $L_1, \ldots, L_{g-1}$,*
- *in general, we match $d_i \in L_0$ with the values $d_{(k+1) \cdot (n/g)+1-k} \in L_k$ for $k = 1, \ldots, g - 1$.*

A more nuanced model: main idea. In the above analysis, we used a simplified model in which only the weaker students, with $d_i < d_j$, benefit from the groupwork. In reality, stronger students, with $d_j > d_i$, benefit too: when they explain the material to the weaker students, they reinforce their knowledge, and they may see the gaps in their knowledge that they did not see earlier. The larger the difference $d_j - d_i$, the more the stronger student needs to explain and thus, the more this stronger student reinforces his or her knowledge. It is therefore reasonable to assume that the resulting increase in knowledge is proportional to the difference $d_j - d_i$, with a different coefficient $\beta > 0$. Thus, we arrive at the following definition:

Definition 24.4 Let n be the number of students, and let $\alpha > 0$ and $\beta > 0$ be real numbers. For each state of knowledge $d = (d_1, \ldots, d_n)$ and for each subdivision into groups $G_1, \ldots, G_m$, the resulting state of knowledge $d' = (d'_1, \ldots, d'_n)$ is defined as follows: for every $k = 1, \ldots, m$ and for every $i \in G_k$, we have

$$d'_i = d_i + \alpha \cdot \sum_{j \in G_k, d_j > d_j} (d_j - d_i) + \beta \cdot \sum_{j \in G_k, d_i > d_j} (d_i - d_j).$$

It turns out that if we maximize either the average grade or the worst grade, then the optimal subdivisions are exactly the same as for the previously used (less nuanced) model:

Proposition 24.8 *In the model described by Definition 24.4, to maximize the average grade a, we divide the students into pairs as follows:*

- *we sort the students by their knowledge;*
- *in each pair, we match one student from the lower half*

$$L_0 \stackrel{\text{def}}{=} \{d_1, d_2, \ldots, d_{n/2}\}$$

with one student from the upper half

$$L_1 \stackrel{\text{def}}{=} \{d_{(n/2)+1}, \ldots \leq d_n\}.$$

Proposition 24.9 *In the model described by Definition 24.4, for every $g \geq 2$, to maximize the average grade a, we divide the students into groups as follows:*

- *we sort the students by their knowledge;*
- *based on this sorting, divide the students into g sets:*

$$L_0 = \{d_1, d_2, \ldots, d_{n/g}\}; \ldots$$

$$L_k = \{d_{k \cdot (n/g)+1}, \ldots, d_{(k+1) \cdot (n/g)}\}, \ldots,$$

$$L_{g-1} = \{d_{(g-1) \cdot (n/g)+1}, \ldots, d_n\};$$

- *in each group, we pick one student from each of g sets $L_0, L_1, \ldots, L_{g-1}$.*

Proposition 24.10 *In the model described by Definition 24.4, to maximize the worst grade w, we divide the students into pairs as follows:*

- *we sort the students by their knowledge;*
- *we pair the worst-performing student (corresponding to d_1) with the best-performing student (corresponding to d_n), and,*
- *if there are other students with $d_i = d_1$, we match them with d_{n-1}, d_{n-2}, etc.*

Other students can be paired arbitrarily.

Proposition 24.11 *In the model described by Definition 24.4, if all students' degree d_i are different, then, to maximize the worst grade w, we divide the students into groups as follows:*

- *we sort the students by their knowledge, so that*

$$d_1 \leq d_2 \leq \cdots \leq d_n,$$

- *we combine the worst-performing student (corresponding to d_1) with $g - 1$ best-performing students (corresponding to $d_n, d_{n-1}, \ldots, d_{n-2}$).*

Other students can be grouped arbitrarily.

Proposition 24.12 *Let us assume that we perform the following optimization:*

- *first, we optimize the average grade;*
- *if there are several subdivisions for which the average grade is optimal, then, among all the subdivisions, we select the one for which the worst grade is the largest;*
- *if we have several subdivisions with the largest worst grade, we select the one with the largest second worst grade,*
- *etc.*

Then, in the model described by Definition 24.4, for $g = 2$, the following subdivision is optimal:

- *we sort the students by their knowledge, so that*

$$d_1 \leq d_2 \leq \cdots \leq d_n,$$

- *we then match d_1 with d_n, d_2 with d_{n-1}, and, in general, d_k with d_{n+1-k}.*

Proposition 24.13 *Let us assume that we perform the following optimization:*

- *first, we optimize the average grade;*
- *if there are several subdivisions for which the average grade is optimal, then, among all the subdivisions, we select the one for which the worst grade is the largest;*
- *if we have several subdivisions with the largest worst grade, we select the one with the largest second worst grade,*
- *etc.*

Then, in the model described by Definition 24.4, for $g \geq 2$, if all the degrees are different, the following subdivision is optimal:

- *we sort the students by their knowledge, so that*

$$d_1 < d_2 < \cdots < d_n,$$

- *based on this sorting, divide the students into g sets:*

$$L_0 = \{d_1, d_2, \ldots, d_{n/g}\}; \ldots$$

$$L_k = \{d_{k \cdot (n/g)+1}, \ldots, d_{(k+1) \cdot (n/g)}\}, \ldots,$$

$$L_{g-1} = \{d_{(g-1) \cdot (n/g)+1}, \ldots, d_n\};$$

- *we match the smallest value $d_1 \in L_0$ with the largest values from $L_1, \ldots, L_{g-1}$,*
- *we match the second smallest value $d_2 \in L_0$ with the second largest values from $L_1, \ldots, L_{g-1}$,*
- *in general, we match $d_i \in L_0$ with the values $d_{(k+1)\cdot(n/g)+1-k} \in L_k$ for $k = 1, \ldots, g - 1$.*

Interestingly, now we can optimize the best grade b. For $g = 2$, the result is the same as for optimizing worst grades, but for $g > 2$, the result is different:

Proposition 24.14 *In the model described by Definition 24.4, to maximize the best grade w, we divide the students into pairs as follows:*

- *we sort the students by their knowledge;*
- *we pair the best-performing student (corresponding to d_n) with the worst-performing student (corresponding to d_1), and,*
- *if there are other students with $d_i = d_n$, we match them with d_2, d_3, etc.*

Other students can be paired arbitrarily.

Proposition 24.15 *In the model described by Definition 24.4, if all students' degree d_i are different, then, to maximize the worst grade w, we divide the students into groups as follows:*

- *we sort the students by their knowledge;*
- *we group the best-performing student (corresponding to d_n) with $g - 1$ worst-performing students (corresponding to $d_1, d_2, \ldots, d_{g-1}$).*

Other students can be grouped arbitrarily.

Case of uncertainty. In practice, we rarely know the exact values of d_i, we only know approximately values $\widetilde{d}_i$ – and, e.g., we know the accuracy Δ of these estimates, i.e., we know that $d_i \in [\widetilde{d}_i - \Delta, \widetilde{x}_i + \Delta]$. In this case, we do not know the exact gain, so it is reasonable to select a "maximin" subdivision, i.e., a subdivision for which the guaranteed (= worst-case) gain is the largest. One can prove that the subdivisions obtained by applying the above algorithms to the approximate value $\widetilde{d}_i$ are optimal in this minimax sense as well.

Proof of Proposition 24.1.

$1°$. First, we note that maximizing the average grade is equivalent to maximizing the sum $n \cdot a = \sum\limits_{i=1}^{n} g'_i$ of the new grades, which is, in turn, equivalent to maximizing the overall gain $\sum\limits_{i=1}^{n} g'_i - \sum\limits_{i=1}^{n} g_i = \sum\limits_{i=1}^{n} (g'_i - g_i)$.

$2°$. Let us take the optimal subdivision, and show that it has the form described in the formulation of Proposition 24.1.

Indeed, in each pair, with degrees $d_i \leq d_j$, we have a weaker student i and a stronger student j. Let us prove that the optimal subdivision into groups, each stronger student is stronger (or of the same strength) than each weaker student. In other words, if we have two pairs $d_i \leq d_j$ and $d_{i'} \leq d_{j'}$, then $d_i \leq d_{j'}$. We will prove this by contradiction. Let us assume that $d_i > d_{j'}$. Let us then swap the ith and the j'th students, i.e., instead of the original pairs (i, j) and (i', j'), let us consider two new pairs (i, j') and (i', j). The corresponding two terms in the overall gain are changed from $\alpha \cdot (d_j + d_{j'} - d_i - d_{i'})$ to $\alpha \cdot (d_j - d_{j'} + d_i - d_{i'})$. The difference between the two expressions is equal to $2\alpha \cdot (d_i - d_{j'})$. Since we assumed that $d_i > d_{j'}$, this difference is positive, which means that the above swap increases the overall gain. The possibility of such an increase contradicts to the fact that we have selected the subdivision for which the overall gain is already the largest possible. This contradiction shows that our assumption $d_i > d_{j'}$ is wrong, and thus, $d_i \leq d_{j'}$.

Since every weaker-of-pair student is weaker than every stronger-of-pair student, all weaker-of-pair students form the bottom of the ordering of the degrees d_i, while all the stronger-of-pair students form the top of this ordering – exactly as the formulation of Proposition 24.1 suggests.

$3°$. To complete the proof, we need to prove that every subdivision satisfying the condition of Proposition 24.1 leads to the optimal average grade. Indeed, we know that one optimal subdivision satisfies this condition. One can check that for each such subdivision, the overall gain is equal to $\sum_{i \in L_1} d_i - \sum_{j \in L_0} d_j$, where L_1 is the set of all the indices i from the upper half, and L_0 is the set of all the indices from the lower half. Thus, the overall gain for all such subdivisions is the same – and it is therefore exactly equal to the gain corresponding to the optimal subdivision. So, all subdivisions satisfying the condition of Proposition 24.1 indeed lead to the optimal average grade.

The proposition is proven.

Proof of Proposition 24.2 The worst grade w is the grade of the worst-performing student. For each of these students i, the larger the difference $d_j - d_i$, the more their grade will increase. So, if there is only one such student, we pair him or her with the best-performing one. If there are other students with the same worst grade, we pair them with the best-performing among the not-yet-paired students, etc.

Proof of Proposition 24.3 One can easily see that in the above model, in each group, the largest grade does not change. So, the best grade b, which is the largest of the group-wide best grades – also does not change.

Proof of Proposition 24.4

$1°$. Let us first prove that an optimal group subdivision satisfies the property described in the formulation of Proposition 24.4. Indeed, let us start with an optimal subdivision. Within each group, we can sort its g students in the increasing order of their grades; thus, every student gets assigned a rank in the corresponding group. Now, we can prove that for every two ranks $r < r'$, a grade of a student of rank r is always less than or equal to the grade of a student of rank r' – even when they are from different groups. Similarly to the proof of Proposition 24.1, this can be proven by contradiction: if a grade d_i of a student of rank r is larger than the grade d_j of a student of rank r', then we can swap these two students and improve the overall gain.

2°. To complete the proof, we must show that for all subdivisions that satisfy the condition from the formulation of Proposition 24.4, the gain is the same. Indeed, the overall gain is equal to the sum of gains obtained in each group. Let us therefore calculate the gain in each group.

For $g = 3$, once we have $d_i \leq d_j \leq d_k$, the gain is equal to α times the sum

$$(d_k - d_j) + (d_k - d_i) + (d_j - d_i) = 2d_k - 2d_i.$$

Thus, the overall gain is equal to $2 \sum\limits_{i \in L_2} d_i - 2 \sum\limits_{i \in L_0} d_i$, and hence, it indeed does not depend on the subdivision – as long as the subdivision satisfies the condition from the formulation of Proposition 24.4.

For $g = 4$, once we have $d_i \leq d_j \leq d_k \leq d_l$, the gain is equal to

$$(d_l - d_k) + (d_l - d_j) + (d_l - d_i) + (d_k - d_j) + (d_k - d_i) + (d_j - d_i) =$$

$$3d_l + d_k - d_j - 3d_i.$$

Thus, the overall gain is equal to

$$3 \sum_{i \in L_3} d_i + \sum_{i \in L_2} d_i - \sum_{i \in L_1} d_i - 3 \sum_{i \in L_0} d_i.$$

For a general group size g, one can prove, by induction, that once we have $d_{i_1} \leq \cdots \leq d_{i_g}$, then the gain of this group is equal to

$$(g - 1) \cdot d_{i_g} + (g - 3) \cdot d_{i_{g-1}} + \ldots + (2k - g - 1) \cdot d_{i_k} + \ldots$$

$$-(g - 3) \cdot d_{i_2} - (g - 1) \cdot d_{i_1}.$$

Thus, the overall gain is equal to

$$(g - 1) \cdot \sum_{i \in L_{g-1}} d_i + (g - 3) \cdot \sum_{i \in L_{g-2}} d_i + \ldots + (2k - g + 1) \cdot \sum_{i \in L_k} d_i + \ldots$$

$$-(g - 3) \cdot \sum_{i \in L_1} d_i - (g - 1) \cdot \sum_{i \in L_0} d_i.$$

So, the sum does not depend on the subdivision – as long as the subdivision satisfies the condition from the formulation of Proposition 24.4.

The statement is proven, and so is the proposition.

Proof of Proposition 24.5 (and of the Comment after Proposition 24.5) is similar to the proof of Proposition 24.2.

Proof of Proposition 24.6 (and of the Comment after Proposition 24.6). According to Proposition 24.1, we select between subdivisions in which in each pair, one element is taken from the lower half L_0 and the other element is taken from the upper half L_1. Similarly to the proof of Proposition 24.2, among such subdivisions, we select a one for which the student with the smallest grade d_1 is matched with the student with the largest grade d_n. To maximize the smallest of the remaining grades, we need to match the smallest of the remaining grades d_2 with the largest of the remaining grades d_{n-1}, etc. The proposition is proven.

Proof of Proposition 24.7 According to Proposition 24.4, we select between subdivisions in which in each group, one element is taken from the lower set L_0 and one element is taken from upper-level sets $L_1, \ldots, L_{g-1}$. Similarly to the proof of Proposition 24.2, among such subdivisions, we select a one for which the student with the smallest grade d_1 is matched with the students with the largest grades from the sets $L_1, \ldots, L_{g-1}$. To maximize the smallest of the remaining grades, we need to match the smallest of the remaining grades d_2 with the largest of the remaining grades, etc. The proposition is proven.

Proof of Propositions 24.8–24.13 We have already mentioned, in the proof of Proposition 24.1, that optimizing the average grade is equivalent to optimizing the overall gain.

In model described by Definition 24.4, the gain coming from interaction between the ith and the jth students with $d_i < d_j$ is equal to

$$\alpha \cdot (d_j - d_i) + \beta \cdot (d_j - d_i) = \alpha' \cdot (d_j - d_i),$$

where $\alpha' \overset{\text{def}}{=} \alpha + \beta$. Thus, in the new model, the overall gain is described by the same formula as in the old model, but with a new coefficient α' instead of the original coefficient α. Since the formula for the overall gain is the same, the optimal subdivisions are also the same.

With respect to optimizing worst grades, the relation with the original model is even easier: the new formula does not change the grades of the worst-performing students. The propositions are thus proven.

Proof of Propositions 24.14–24.15 is similar to the proof of Proposition 24.2.

References

1. O. Kosheleva, V. Kreinovich, How to divide students into groups so as to optimize learning: towards a solution to a pedagogy–related optimization problem, in *Proceedings of the IEEE International Conference on Systems, Man, and Cybernetics IEEE SMC'2012, Seoul, Korea, October 14–17, 2012* (2012), pp. 1948–1953

2. D. Bhattacharya, Inferring optimal peer assignment from experimental data. J. Am. Stat. Assoc. **104**, 486–500 (2009)
3. S.E. Carrell, R.L. Fullerton, J.E. West, Does your cohort matter? Estimating peer effects in college achievement. J. Labor Econ. **27**(3), 439–464 (2009)
4. S.E. Carrell, F.V. Malmstrom, J.E. West, Peer effects in academic cheating. J. Hum. Resour. **43**(1), 173–207 (2008)
5. S.E. Carrell, B.I. Sacerdote, J.E. West, From Natural Variation to Optimal Policy? An Unsuccessful Experiment in Using Peer Effects Estimates to Improve Student Outcomes, UC Davis Working Paper (2012), http://www.econ.ucdavis.edu/faculty/scarrell/sortexp.pdf
6. G. Foster, It's not your peers, and it's not your friends: some progress toward understanding the educational peer effect mechanism. J. Public Econ. **90**(8–9), 1455–1475 (2006)
7. B.S. Graham, G.W. Imbens, G. Ridder, Complementarity and Aggregate Implications of Assortative Matching: A Nonparametric Analysis, Working Paper 14860, National Bureau of Economic Research, 2009
8. D.S. Lyle, Estimating and interpreting peer and role model effects from randomly assigned social groups at West Point. Rev. Econ. Stat. **89**(2), 289–299 (2007)
9. B.I. Sacerdote, Peer effects with random assignment: results for Dartmouth Roommates. Q. J. Econ. **116**(2), 681–704 (2001)
10. R. Stinebrickner, T.R. Stinebrickner, What can be learned about peer effects using college roomates? Evidence from new survey data and students from disadvantaged backgrounds. J. Public Econ. **90**(8–9), 1435–1454 (2006)
11. D.J. Zimmerman, Peer effects in academic outcomes: evidence from a natural experiment. Rev. Econ. Stat. **85**(1), 9–23 (2003)
12. O. Kosheleva, V. Kreinovich, Towards optimal effort distribution in process design under uncertainty, with application to education. Int. J. Reliab. Safety **6**(1–3), 148–166 (2012)

Chapter 25
How to Divide Students into Groups: Importance of Diversity and Need for Intelligent Techniques to Further Enhance the Advantage of Groups with Diversity in Problem Solving

In practice, there are many examples when the diversity in a group enhances the group's ability to solve problems – and thus, leads to more efficient groups, firms, schools, etc. Several papers, starting with the pioneering research by Scott E. Page from the University of Michigan at Ann Arbor, provide a theoretical justification for this known empirical phenomenon. However, when the general advise of increasing diversity is transformed into simple-to-follow algorithmic rules (like quotas), the result is not always successful. In this chapter, we prove that the problem of designing the most efficient group is computationally difficult (NP-hard). Thus, in general, it is not possible to come up with simple algorithmic rules for designing such groups: to design optimal groups, we need to combine standard optimization techniques with intelligent techniques that use expert knowledge.

The results from this chapter first appeared in [1].

Introduction to the problem. In real life, there are many examples that diversity in a group enhances the group's ability to solve problems – and thus, leads to more efficient groups, firms, schools, etc. Several papers, starting with the pioneering research by Scott E. Page from the University of Michigan at Ann Arbor, provide a theoretical justification for this known empirical phenomenon; see, e.g., [2–5] and references therein. Specifically, these papers have shown that groups of diverse problem solvers *can* outperform groups of high-ability problem solvers.

The word *can* is here (and in the title of the paper [3]) for a good reason: when the general advise of increasing diversity is transformed into simple-to-follow algorithmic rules (like quotas), the result is not always successful.

In this chapter, we consider the problem of designing the most efficient group as a precise optimization problem. We show that this optimization problem is computationally difficult (NP-hard). Thus, in general, it is not possible to come up with simple algorithmic rules for designing such groups: to design optimal groups, we need to combine standard optimization techniques with intelligent techniques that use expert knowledge.

© Springer-Verlag GmbH Germany 2018
O. Kosheleva and K. Villaverde, *How Interval and Fuzzy Techniques Can Improve Teaching*, Studies in Computational Intelligence 750,
https://doi.org/10.1007/978-3-662-55993-2_25

Comment. Similar results are known: e.g., the problem of maximizing diversity and the problem of finding a group which is most representative of the population are both NP-hard [6, 7]. In this chapter, we extent these results further – from gauging and maintaining the *degree* of diversity to gauging and maintaining the *positive effects* of diversity – such as the increased ability of a group to solve problems.

Towards the formulation of the problem in exact terms. Let us assume that we have a population consisting of n individuals. From this population $\{1, \ldots, n\}$, we need to select a group $G \subseteq \{1, \ldots, n\}$ which is the most efficient in solving a given problem.

In mathematical terms, to describe a group G, we must describe, for each individual i ($i = 1, \ldots, n$), whether this individual is selected for this group or not. In computational terms, for each $i = 1, \ldots, n$, we thus need to select a Boolean ("true"-"false") value x_i for which:

- $x_i =$ "true" means that we select the ith individual into the group, and
- $x_i =$ "false" means that we do not select the ith individual into the group.

Inside the computer, "true" is usually represented as 1, while "false" is usually represented as 0. Thus, we can describe each group by selecting, for each $i = 1, \ldots, n$, a value $x_i \in \{0, 1\}$ for which

- $x_i = 1$ means that we select the ith individual into the group, and
- $x_i = 0$ means that we do not select the ith individual into the group.

In order to select the most efficient group, we must describe how the group's efficiency p depends on the selections x_i.

For simple mechanical tasks like digging trenches or doing simple menial work, people perform these tasks individually. For such tasks, the efficiency p of a group is simply the sum of the productivity values p_i of all the individuals who form this group G: $p = \sum_{i \in G} p_i$. In terms of the variables x_i, this formula means that we add p_i if $x_i = 1$ and that we do not add p_i if $x_i = 0$. In other words, this formula can be described as $p = \sum_{i=1}^{n} p_i \cdot x_i$. In this simple model, the more people work on a project, the larger the productivity.

Most practical problems are not that simple. In solving these problems, interaction between the individuals can enhance their productivity. In mathematical terms, this means that

- in addition to the above terms $p_i \cdot x_i$ which are linear in x_i,
- we also have terms $p_{ij} \cdot x_i \cdot x_j$, $i \neq j$, which are quadratic in x_i; these terms describe pair-wise interaction between the individuals;
- we may also have cubic terms $p_{ijk} \cdot x_i \cdot x_j \cdot x_k$ which describe triple interactions,
- and we can also have higher order terms, which describe the effect of larger subgroups.

In other words, in general, the formula describing the productivity of a group takes a more complex form $p = \sum_{i=1}^{n} p_i \cdot x_i + \sum_{i \neq j} p_{ij} \cdot x_i \cdot x_j + \ldots$.

For example, for a group consisting of two individuals, i and j, the productivity is equal to $p = p_i + p_j + p_{ij} + \cdots$

It should be mentioned that interaction is not always helpful. For example, if we are interested in solving a complex problem, and we bring together two individuals with similar ways of thinking and with similar skills, then there is not much that these individuals can learn from each other.

In some cases, they may speed up the process by dividing the testing of possible approaches between themselves. In such cases, they can solve the problem twice faster, the productivity increases twice – so there is, in effect, no interaction terms p_{ij}.

In other cases, when the problem is not easy to subdivide, the fact that we have two similar solvers solving the same problem does not help at all – the overall time is the same as for each individual solver. In this case, $p \approx p_i \approx p_j$ and thus $p < p_i + p_j$, i.e., $p_{ij} < 0$.

On the hand, in a diverse group, individuals complement each other, learn from each other, and as a result, their productivity increases above what would have happened if they worked on their own: $p > p_i + p_j$, so $p_{ij} > 0$.

Such "negative" and "positive" interactions (i.e., $p_{ij} < 0$ and $p_{ij} > 0$) are not just a negative possibility – this is exactly the reason why, as we have mentioned, groups of diverse problem solvers can outperform groups of high-ability problem solvers.

Because of the interaction, the problem of selecting the optimal group becomes non-trivial. In this chapter, we show that the problem of selecting an optimal group is computationally difficult (NP-hard). Moreover, we will show that this problem is NP-hard already if we take into account the simplest possible non-linear terms – i.e., quadratic terms. In other words, the problem becomes NP-hard already for the following productivity expression: $p = \sum_{i=1}^{n} p_i \cdot x_i + \sum_{i \neq j}^{n} p_{ij} \cdot x_i \cdot x_j$. So, we arrive at the following problem.

Definition 25.1 By a problem of selecting the most efficient group, we mean the following problem. We are given:

- an integer $n > 0$;
- rational numbers $p_1, \ldots, p_n$, and
- rational numbers r_{ij}, $1 \leq i, j \leq n, i \neq j$.

We must find the combination of n values $x_1 \in \{0, 1\}, \ldots, x_n \in \{0, 1\}$ for which the expression $p = \sum_{i=1}^{n} p_i \cdot x_i + \sum_{i \neq j} p_{ij} \cdot x_i \cdot x_j$ is the largest possible.

Instead of trying to find the most efficient group, we can also formulate a less ambitious problem of finding a group with a given efficiency.

Definition 25.2 By a problem of selecting a group with a given efficiency, we mean the following problem. We are given:

- an integer $n > 0$;
- rational numbers $p_1, \ldots, p_n$,

- rational numbers r_{ij}, $1 \leq i, j \leq n, i \neq j$, and
- a rational value p_0.

We must find the combination of n values $x_1 \in \{0, 1\}, \ldots, x_n \in \{0, 1\}$ for which

$$p \stackrel{\text{def}}{=} \sum_{i=1}^{n} p_i \cdot x_i + \sum_{i \neq j} p_{ij} \cdot x_i \cdot x_j \geq p_0.$$

Proposition 25.1 *The problem of selecting the most efficient group is NP-hard.*

Proposition 25.2 *The problem of selecting a group with a given efficiency is NP-hard.*

Comment. The proof of this result is placed in the special Proofs section. For readers who are not very familiar with the notion of NP-hardness and of Np-hardness proofs, we precede the Proof section with a special section describing the NP-hardness notions and ideas. Readers who are familiar with these notions and ideas can skip this preceding section and go directly to the Proofs section.

Examples For simplicity, let us assume that we have two groups of people. People from each groups are equally productive, and their productivity increases when they team together with people from the same group. On the other hand, due to cultural differences, interaction between people from different groups is not as productive. Examples of such groups are easy to find: e.g., in big projects where, in principle, problems can be solved both by an appropriate hardware and by an appropriate software, hardware and software folks often experience difficulties communicating with other, and these difficulties can drag down the collective performance.

Let us assume that a company has m folks from one group; (we will assign them indices $i = 1, \ldots, m$), and the same company has the same number m of folks from the second group (with indices $i = m + 1, \ldots, 2m$). Let us assume that the productivity p_i of every person from both groups is the same, i.e., that $p_1 = \cdots = p_m = p_{m+1} = \cdots = p_{2m} = 1$. Let us also assume that each collaboration between people from the same groups adds one extra unit to the overall productivity, e.g., $p_{ij} = p_{m+i,m+j} = 1$ for all $i, j = 1, \ldots, m$. Let a denote the loss of productivity caused by a tension between the two representatives of opposite groups, i.e., that $p_{i,m+j} = p_{m+i,j} = -a$ for all $i, j = 1, \ldots, m$. Our objective is to select a group consisting of some (maybe none) representatives of the first group and some (maybe none) representatives of the second group so that the overall productivity of the resulting group is the largest.

In this simplified situation, in terms of productivity, all the persons from the first groups are equivalent – and all the persons from the second group are also equivalent to each other. Thus, the overall productivity does not depend on which exactly persons from the first group we select, and it does not depend on who exactly we select from the second group. The overall productivity only depends on the number of people m_1 chosen from the first group and on the number of people m_2 chosen from the second group.

If we only select m_1 people from the first group (i.e., if we choose $m_2 = 0$), then the productivity becomes equal to

$$p = \sum_{i=1}^{n} p_i \cdot x_i + \sum_{i \neq j}^{n} p_{ij} \cdot x_i \cdot x_j = m_1 + m_1 \cdot (m_1 - 1) = m_1^2.$$

Similarly, if we only select m_2 people from the second group (i.e., if we choose $m_1 = 0$), then the overall productivity is equal to m_2^2. In the general case, we have

$$p = \sum_{i=1}^{n} p_i \cdot x_i + \sum_{i \neq j}^{n} p_{ij} \cdot x_i \cdot x_j = m_1^2 + m_2^2 - a \cdot m_1 \cdot m_2.$$

For example, if we select one person from each group, i.e., if we take $m_1 = m_2 = 1$, then the overall productivity is $p = 1^2 + 1^2 - a \cdot 1 \cdot 1 = 2 - a$. If we select two people from each group, then the overall productivity is $p = 2^2 + 2^2 - a \cdot 2 \cdot 2 = 8 - 4a$. The larger $2 + 2$ group is more productive than the smaller $1 + 1$ group if $8 - 4a > 2 - a$, i.e., if $6 > 3a$ and $a < 2$. When $a = 2$, these two groups are equally productive, and when $a > 2$, the smaller group is more productive.

In other words, if the coefficient a that describes the tension between the groups is not too high (smaller than the threshold $a = 2$), then it is more beneficial to use a larger group (in spite of this tension). On the other hand, if the tension is too high (exceeds the threshold), then larger groups are no longer more productive.

A similar threshold $a = 1$ appears when we compare the "maximally diverse" group $m_1 = m_2 = m$ and the "minimally diverse" groups (either $m_1 = m$ and $m_2 = 0$ or $m_1 = m$ and $m_2 = 0$). Indeed, in the first case, we have $p = m^2 + m^2 - a \cdot m^2 = (2 - a) \cdot m^2$, while in the second case, we have $p = m^2$. The first value is larger than the second one when $2 - a > 1$, i.e., when $a < 1$.

It turns out that in our simple example, these groups are actually optimal; namely:

- when $a < 1$, the most productive group is the one that includes all the persons from both groups: $m_1 = m_2 = m$;
- on the other hand, when $a > 1$, then the most productive group is a one that includes all the persons from one group and no one from the other group: either $m_1 = m$ and $m_2 = 0$ or $m_1 = m$ and $m_2 = 0$.

For $a = 1$, both the "maximally diverse" group $m_1 = m_2 = m$ and the "minimally diverse groups" (either $m_1 = m$ and $m_2 = 0$ or $m_1 = m$ and $m_2 = 0$) are equally productive.

Indeed, we will show that the corresponding maximum is attained even when we formally allow real (nor necessarily integer) values of $m_1 \in [0, m]$ and $m_2 \in [0, m]$. The maximum of a differentiable function p over each of the variable m_i taking values over an interval is attained either at one of the endpoints $m_i = 0$ or $m_i = m$ of this interval, or when the partial derivative relative to m_i is equal to 0, i.e., when $\dfrac{\partial p}{\partial m_i} = 0$. Here, $\dfrac{\partial p}{\partial m_1} = 2m_1 - a \cdot m_2$, so, for m_1, we have three possible values where the maximum can be attained: when $m_1 = 0$, when $m_1 = m$, and when $2m_1 - a \cdot m_2$ (i.e., when $m_1 = (a/2) \cdot m_2$). Similar three cases are possible for m_2, so, to find the maximum, we have to consider $3 \times 3 = 9$ combinations of these cases. Due to symmetry, we only need to only consider 6 cases:

- $m_1 = m_2 = 0$; in this case, $p = 0$ – smaller than for the minimally diverse group; so this cannot be the maximum;
- $m_1 = 0$ and $m_2 = (a/2) \cdot m_1$; here too, $m_1 = m_2 = 0$ and $p = 0$;
- $m_1 = 0$ and $m_2 = m$, where $p = m^2$;
- $m_1 = (a/2) \cdot m_2$ and $m_2 = (a/2) \cdot m_1$; here, $m_1 = (a/2)^2 \cdot m_1$, so either $m_1 = m_2 = 0$ or (for $a = 2$) $m_1 = m_2$, in which case the maximum is attained when $m_1 = m_2 = 0$ or $m_1 = m_2 = m$;
- $m_1 = (a/2) \cdot m_2$ and $m_2 = m$; in this case, $p = (a/2)^2 \cdot m^2 + m^2 - a \cdot (a/2) \cdot m^2 = (1 - a/4) \cdot m^2 < m^2$, so this cannot be the maximum either;
- $m_1 = m_2 = m$, in which case $p = (2 - a) \cdot m^2$.

Thus, the maximum is either when $m_1 = m_2 = m$ or when $m_1 = 0$ and $m_2 = m$ (or vice versa).

Comment. In this example, we have explicitly found the optimal group. Our main result says that in general, finding the most productive group is a computationally difficult problem.

What is NP-hardness: a brief informal reminder. As we have mentioned earlier, the intent of this section is that the readers who are not very familiar with NP-hardness and related notions will be able to understand our proofs. Readers who are already well familiar with NP-hardness and related notions can skip this section and go directly to the next (Proofs) section.

Informally, a problem $\mathscr{P}_0$ is called *NP-hard* if it is at least as hard as all other problems from a certain reasonable class. Let us describe this notion in more detail.

When is an algorithm feasible? The notion of NP-hardness is related to the known fact that some algorithms are feasible and some are not. Whether an algorithm is feasible or not depends on how many computational steps it needs.

For example, if for some input x of length $\text{len}(x) = n$, an algorithm requires 2^n computational steps, then for an input of a reasonable length $n \approx 300$, we would need 2^{300} computational steps. Even if we use a hypothetical computer for which each step takes the smallest physically possible time (the time during which light passes through the smallest known elementary particle), we would still need more computational steps than can be performed during the (approximately 20 billion years) lifetime of our Universe.

A similar estimate can be obtained for an arbitrary algorithm whose running time $t(n)$ on inputs of length n grows at least as an exponential function, i.e., for which, for some $c > 0$, $t(n) \geq \exp(c \cdot n)$ for all n. As a result, such algorithms (called *exponential-time*) are usually considered *not feasible*.

Comment. The fact that an algorithm is not feasible, does not mean that it can never be applied: it simply means that there are cases when its running time will be too large for this algorithm to be practical; for other inputs, this algorithm can be quite useful.

On the other hand, if the running time grows only as a polynomial of n (i.e., if an algorithm is *polynomial-time*), then the algorithm is usually quite feasible.

As a result of the above two examples, researchers have arrived at the following idea: An algorithm $\mathcal{U}$ is called *feasible* if and only if it is *polynomial-time*, i.e., if and only if there exists a polynomial $P(n)$ such that for every input x of length $\mathrm{len}(x)$, the computational time $t_{\mathcal{U}}(x)$ of the algorithm $\mathcal{U}$ on the input x is bounded by $P(\mathrm{len}(x))$: $t_{\mathcal{U}}(x) \leq P(\mathrm{len}(x))$.

In most practical cases, this idea *adequately* describes our intuitive notion of feasibility: *polynomial-time* algorithms are usually *feasible*, and *non-polynomial-time* algorithms are usually *not feasible*. However, the reader should be warned that in some (rare) cases, it does not work:

- Some algorithms are polynomial-time but not feasible: e.g., if the running time of an algorithm is $10^{300} \cdot n$, this algorithm is polynomial-time, but, clearly, not feasible.
- Vice versa, there exist algorithms whose computation time grows, say, as exp $(0.000 \ldots 01 \cdot \mathrm{len}(x))$. Legally speaking, such algorithms are exponential time and thus, not feasible, but for all practical purposes, they are quite feasible.

It is therefore desirable to look for a *better* formalization of feasibility, but as of now, "polynomial-time" is the best known description of feasibility.

Definition 25.3 An algorithm U is called feasible if there exists a polynomial $P(n)$ such that for every input x, the running time $t_U(x)$ of this algorithm does not exceed $P(\mathrm{len}(x))$, where by $\mathrm{len}(x)$, we denoted the length of the input x (i.e., the number of bits that form this input).

When is a problem tractable? At first glance, now, that we have a definition of a feasible algorithm, we can describe which problems are tractable and which problems are intractable: If there exists a polynomial-time algorithm that solves all instances of a problem, this problem is tractable, otherwise, it is intractable.

In some cases, this ideal solution is possible, and we either have an explicit polynomial-time algorithm, or we have a proof that no polynomial-time algorithm is possible.

Unfortunately, in many cases, we do not know whether a polynomial-time algorithm exists or not. This does not mean, however, that the situation is hopeless: instead of the missing *ideal* information about intractability, we have another information that is almost as good.

Namely, for some cases, we do not know whether the problem can be solved in polynomial time or not, but we do know that this problem is as hard as practical problems can get: if we can solve *this* problem easily, then we would have an algorithm that solves *all* problems easily, and the existence of such universal solves-everything-fast algorithm is very doubtful. We can, therefore, call such "hard" problems *intractable*.

In order to formulate this notion in precise terms, we must describe what we mean by a problem, and what we mean by the ability to *reduce* other problems to this one.

What is a practical problem? When we say that there is a practical problem, we usually mean that:

- we have some information (we will denote its computer representation by x), and
- we know the relationship $R(x, y)$ between the known information x and the desired object y.

In the computer, everything is represented by a binary sequence (i.e., sequence of 0's and 1's), so we will assume that x and y are binary sequences.

In this section, we will trace all the ideas on two examples, one taken from mathematics and one taken from physics.

- (Example from *mathematics*) We are given a mathematical statement x. The desired object y is either a proof of x, or a "disproof" of x (i.e., a proof of "not x"). Here, $R(x, y)$ means that y is a proof either of x, or of "not x".
- (Example from *physics*) x is the results of the experiments, and the desired y is the formula that fits all these data. Imagine that we have a series of measurements of voltage and current: e.g., x consists of the following pairs $(x_1^{(k)}, x_2^{(k)})$, $1 \le k \le 10$: $(1.0, 2.0)$, $(2.0, 4.0)$, …, $(10.0, 20.0)$; we want to find a formula that is consistent with these experiments (e.g., y is the formula $x_2 = 2 \cdot x_1$).

For a problem to be practically meaningful, we must have a way to check whether the proposed solution is correct. In other words, we must assume that there exists a feasible algorithm that checks $R(x, y)$ (given x and y). If no such feasible algorithm exists, then there is no criterion to decide whether we achieved a solution or not.

Another requirement for a real-life problem is that in such problems, we usually know an *upper bound* for the length $\text{len}(y)$ of the description of y. In the above examples:

- In the *mathematical* problem, a proof must be not too huge, else it is impossible to check whether it is a proof or not.
- In the *physical* problem, it makes no sense to have a formula $x_2 = f(x_1, C_1, \dots, C_{40})$ with, say, 40 parameters to describe the results $(x_1^{(1)}, x_2^{(1)}), \dots, (x_1^{(10)}, x_2^{(10)})$ of 10 experiments, for two reasons:

  - First, one of the goals of physics is to discover the laws of nature. If the number of parameters exceeds the number of experimental data, then no matter what dependency $f(x_1, C_1, \dots)$ we choose, in order to determine C_i, we have, say, 10 equations with 40 unknowns. Such under-determined system usually has a solution, so the fact that, say, a linear formula with many parameters fits all the experimental data does not mean that the dependency is proven to be linear: a quadratic or cubic formula with as many parameters will fit the same data as well.
  - Second, another goal of physics (definitely related to the first one) is to find a way to *compress* the data, so that we will not need to store all billions of experimental results in order to make predictions. A dependency y that requires more storage space than the original data x is clearly not satisfying this goal.

In all cases, it is necessary for a user to be able to read the desired solution symbol-after-symbol, and the time required for that reading must be feasible. In the previous section, we have formalized "feasible time" as a time that is bounded by some polynomial of len(x). The reading time is proportional to the length len(y) of the answer y. Therefore, the fact the reading time is bounded by a polynomial of len(x) means that the length of the output y is also bounded by some polynomial of len(x), i.e., that len(y) $\leq P_L(\text{len}(x))$ for some polynomial P_L.

So, we arrive at the following formulation of a problem:

Definition 25.4 By a general practical problem (or simply a *problem*, for short), we mean a pair $\langle R, P_L \rangle$, where $R(x, y)$ is a feasible algorithm that transforms two binary sequences into a Boolean value ("true" or "false"), and P_L is a polynomial.

Definition 25.5 By an instance of a (general) problem $\langle R, P_L \rangle$, we mean the following problem:

GIVEN: a binary sequence x.
GENERATE

- either y such that $R(x, y)$ is true and len(y) $\leq P_L(\text{len}(x))$,
- or, if such a y does not exist, a message saying that there are no solutions.

For example, for the general mathematical problem described above, an instance would be: given a statement, find its proof or disproof.

Comments. What we called "general practical problems" is usually described as "problems from the class NP" (to separate them from more complicated problems in which the solution may not be easily verifiable). Problems for which there is a feasible algorithm that solves all instances are called *tractable, easily solvable,* or "problems from the class P" (P from *Polynomial*). It is widely believed that not all (general practical) problems are easily solvable (i.e., that NP $\neq$ P), but it has never been proved.

One way to solve an NP problem is to check $R(x, y)$ for all binary sequences y with len(y) $\leq P_L(\text{len}(x))$. This algorithm (called *British Museum* algorithm) requires $2^{P_L(\text{len}(x))}$ checks. This algorithm takes exponential time and is therefore, not feasible.

Reducing a problem to another one. Let us start with an example. Suppose that we can have an algorithm that checks whether a given system of linear inequalities

$$a_{i1} \cdot x_1 + \cdots + a_{im} \cdot x_m \geq b_i, \quad 1 \leq i \leq n,$$

with known a_{ij} and b_i, has a solution. A problem of checking whether a given system of inequalities *and equalities* $c_{k1} \cdot x_1 + \cdots + c_{km} \cdot x_m = d_k$ is consistent can be *reduced* to the problem of checking inequalities if we replace each equality by two inequalities: $c_{k1} \cdot x_1 + \cdots + c_{km} \cdot x_m \geq d_k$ and $(-c_{k1}) \cdot x_1 + \cdots + (-c_{km}) \cdot x_m \geq -d_j$ (the latter being equivalent to $c_{k1} \cdot x_1 + \cdots + c_{km} \cdot x_m \leq d_k$).

In general, we can say that a problem $\mathscr{P} = \langle R, P_L \rangle$ can be *reduced* to a problem $\mathscr{P}' = \langle R', P'_L \rangle$ if there exist three feasible algorithms U_1, U_2, and U_3 with the following properties:

- The (feasible) algorithm U_1 transforms each input x of the first problem into an input of the second problem.
- The (feasible) algorithm U_2 transforms each solution y of the first problem into the solution of the corresponding case of the second problem: i.e., if $R(x, y)$ is true, then $R'(U_1(x), U_2(y))$ is also true.
- The (feasible) algorithm U_3 transforms each solution y' of the corresponding instance of the second problem into the solution of the first problem: i.e., if $R'(U_1(x), y')$ is true, then $R(x, U_3(y'))$ is also true.

(In the above example, U_1 transforms each equality into two inequalities, and U_2 and U_3 simply do not change the values x_i at all.)

If there exists a reduction, then an instance x of the first problem is solvable if and only if the corresponding instance $U_1(x)$ of the second problem is solvable. Moreover, if we can actually solve the second instance (and find a solution y'), we will then be able to find a solution to the original instance x of the first problem (as $U_3(y')$). Thus, if we have a *feasible* algorithm for solving the second problem, we would thus design a *feasible* algorithm for solving the first problem as well.

Comment. We only described the simplest way of reducing one problem to another one: when a *single* instance of the first problem is reduced to a *single* instance of the second problem. In some cases, we cannot reduce to a *single* case, but we can reduce to *several* cases, solving which helps us solve the original instance of the first problem.

Definition 25.6

- A problem (not necessarily from the class NP) is called *NP-hard* if every problem from the class NP can be reduced to it.
- If a problem from the class NP is NP-hard, it is called *NP-complete*.

If a problem $\mathscr{P}$ is NP-hard, then every feasible algorithm for solving *this* problem $\mathscr{P}$ would lead to feasible algorithms for solving *all* problems from the class NP, and this is generally believed to be hardly possible.

- For example, mathematicians believe that not only there is *no algorithm* for checking whether a given statement is provable or not (the famous Gödel's theorem has proven that), but also they believe that there is *no feasible way* to find a proof of a given statement even if we restrict the lengths of possible proofs. (In other words, mathematicians believe that computers cannot completely replace them.)
- Similarly, physicists believe that what they are doing cannot be completely replaced by computers.

In view of this belief, NP-hard problems are also called *intractable*.

Comment. It should be noted that although most scientists *believe* that intractable problems are not feasible, we still *cannot prove* (or disprove) this fact. If a NP-hard problem *can* be solved by a feasible algorithm, then (by definition of NP-hardness) *all* problems from the class NP will be solvable by feasible algorithms and thus, P = NP. Vice versa, if P = NP, then all problems from the class NP (including all NP-complete problems) can be solved by polynomial-time (feasible) algorithms.

So, if P $\neq$ NP (which is a common belief), then the fact that the problem is NP-hard means that *no matter what algorithm we use, there will always be some cases for which the running time grows faster than any polynomial*. Therefore, for these cases, the problem is truly intractable.

Examples of NP-hard problems. Historically the NP-complete problem proved to be NP-complete was the so-called *propositional satisfiability* (*3-SAT*) problem for 3-CNF formulas.

This problem consists of the following: Suppose that an integer v is fixed, and a formula F of the type $F_1 \& F_2 \& \ldots \& F_k$ is given, where each of the expressions F_j has the form $a \lor b$ or $a \lor b \lor c$, and a, b, c are either the variables $z_1, \ldots, z_v$, or their negations $\neg z_1, \ldots, \neg z_v$ (these a, b, c, $\ldots$ are called *literals*)

For *example*, we can take a formula $(z_1 \lor \neg z_2) \& (\neg z_1 \lor z_2 \lor \neg z_3)$.

If we assign arbitrary Boolean values ("true" or "false") to v variables $z_1, \ldots, z_v$, then, applying the standard logical rules, we get the truth value of F. We say that a formula F is *satisfiable* if there exist truth values $z_1, \ldots, z_v$ for which the truth value of the expression F is "true". The problem is: given F, check whether it is satisfiable.

In the *subset sum* problem, given n integers $s_1, \ldots, s_n$, we must check whether there exist values $x_1, \ldots, x_n \in \{-1, 1\}$ for which $s_1 \cdot x_1 + \cdots + s_n \cdot x_n = 0$.

How NP-hardness is usually proved. The original proof of NP-hardness of certain problems $\mathscr{P}_0$ is rather complex, because it is based on explicitly proving that *every* problem from the class NP can be reduced to the problem $\mathscr{P}_0$. However, once we have proven NP-hardness of a problem $\mathscr{P}_0$, the proof of NP-hardness of other problems $\mathscr{P}_1$ is much easier.

Indeed, from the above description of a reduction, one can easily see that reduction is a transitive relation: if a problem $\mathscr{P}$ can be reduced to a problem $\mathscr{P}_0$, and the problem $\mathscr{P}_0$ can be reduced to a problem $\mathscr{P}_1$, then, by combining these two reductions, we can prove that $\mathscr{P}$ can be reduced to $\mathscr{P}_1$.

Thus, to prove that a new problem $\mathscr{P}_1$ is NP-hard, it is sufficient to prove that one of the known NP-hard problems $\mathscr{P}_0$ can be reduced to this problem $\mathscr{P}_1$. Indeed, since $\mathscr{P}_0$ is NP-hard, every other problem $\mathscr{P}$ from the class NP can be reduced to this problem $\mathscr{P}_0$. Since $\mathscr{P}_0$ can be reduced to $\mathscr{P}_1$, we can now conclude, by transitivity, that every problem $\mathscr{P}$ from the class NP can be reduced to this problem $\mathscr{P}_1$ – i.e., that the problem $\mathscr{P}_1$ is indeed NP-hard.

Comment. As a consequence of the definition of NP-hardness, we can conclude that if a problem $\mathscr{P}_0$ is NP-hard, then every more general problem $\mathscr{P}_1$ is also NP-hard.

Indeed, the fact that $\mathscr{P}_0$ is NP-hard means that every instance p of every problem $\mathscr{P}$ can be reduced to some instance p_0 of the problem $\mathscr{P}_0$. Since the problem $\mathscr{P}_1$ is more general than the problem $\mathscr{P}_0$, every instance p_0 of the problem $\mathscr{P}_0$ is also an instance of the more general problem $\mathscr{P}_1$.

Thus, every instance p of every problem $\mathscr{P}$ can be reduced to some instance p_0 of the problem $\mathscr{P}_1$ – i.e., that the more general problem $\mathscr{P}_1$ is indeed NP-hard.

Reduction in our proof: to subset sum, a known NP-hard problem. We prove NP-hardness of our problem by reducing a known NP-hard problem to it: namely, a *subset sum* problem, in which we are given n positive integers $s_1, \ldots, s_n$, and we must find the signs $\varepsilon_i \in \{-1, 1\}$ for which $\sum_{i=1}^{n} \varepsilon_i \cdot s_i = 0$; see, e.g., [8, 9].

A reduction means that to every instance $s_1, \ldots, s_n$ of the subset sum problem, we must assign (in a feasible, i.e., polynomial-time way) an instance of our problem in such a way that the solution to the new instance will lead to the solution of the original instance.

Reduction: idea. In our reduction, we would like to transform each variable ε_i from the subset sum problem into a variable x_i from our problem, so that our problem (formulated in terms of x_i) is optimal if and only if the original problem has a solution.

For that, we need to transform each variable x_i which takes the values 0 and 1 into a variable ε_i that takes values -1 and 1 (and vice versa). The simplest way to perform this reduction is to take a linear function $\varepsilon_i = a \cdot x_i + b$, where the coefficients a and b are selected in such as way that $a \cdot 0 + b = -1$ and $a \cdot 1 + b = 1$. In other words, we have $b = -1$ and $a + b = 1$. Substituting $b = -1$ into the equation $a + b = 1$, we conclude that $a = 2$, i.e., that $\varepsilon_i = 2 \cdot x_i - 1$.

Let us select an integer $p_0 > 0$ and consider the formula

$$p = p_0 - \left(\sum_{i=1}^{n} \varepsilon_i \cdot s_i \right)^2 .$$

This expression is always $\leq p_0$, and it attains the value p_0 if and only if $\sum_{i=1}^{n} \varepsilon_i \cdot s_i = 0$. In terms of x_i, we have

$$p = p_0 - \left(\sum_{i=1}^{n} (2 \cdot x_i - 1) \cdot s_i \right)^2 ,$$

i.e.,

$$p = p_0 - \left(\sum_{i=1}^{n} x_i \cdot (2 \cdot s_i) - s_0 \right)^2 ,$$

where we denoted $s_0 \stackrel{\text{def}}{=} \sum_{i=1}^{n} s_i$. By using the formula for the square of the difference, we conclude that

$$p = p_0 - \left(\sum_{i=1}^{n} x_i \cdot (2 \cdot s_i)\right)^2 + 2 \cdot s_0 \cdot \sum_{i=1}^{n} x_i \cdot (2 \cdot s_i) - s_0^2,$$

i.e.,

$$p = p_0 - \left(\sum_{i=1}^{n} x_i \cdot (2 \cdot s_i)\right)^2 + \sum_{i=1}^{n} x_i \cdot (4 \cdot s_0 \cdot s_i) - s_0^2.$$

The square of the sum takes the form

$$\left(\sum_{i=1}^{n} x_i \cdot (2 \cdot s_i)\right)^2 = \sum_{i=1}^{n} x_i^2 \cdot (4 \cdot s_i^2) + \sum_{i \neq j} (4 \cdot s_i \cdot s_j) \cdot x_i \cdot x_j.$$

Since $x_i = 0$ or $x_i = 1$, we always have $x_i^2 = x_i$ and thus,

$$\left(\sum_{i=1}^{n} x_i \cdot (2 \cdot s_i)\right)^2 = \sum_{i=1}^{n} x_i \cdot (4 \cdot s_i^2) + \sum_{i \neq j} (4 \cdot s_i \cdot s_j) \cdot x_i \cdot x_j.$$

Substituting this expression into the above formula for p, we get

$$p = p_0 - \sum_{i=1}^{n} x_i \cdot (4 \cdot s_i^2) - \sum_{i \neq j} (4 \cdot s_i \cdot s_j) \cdot x_i \cdot x_j + \sum_{i=1}^{n} x_i \cdot (4 \cdot s_0 \cdot s_i) - s_0^2.$$

By grouping together terms independent on x_i and terms proportional to p_i, we get

$$p = (p_0 - s_0^2) + \sum_{i=1}^{n} x_i \cdot (4 \cdot s_0 \cdot s_i - 4 \cdot s_i^2) + \sum_{i \neq j} (-4 \cdot s_i \cdot s_j) \cdot x_i \cdot x_j.$$

Thus, if we choose $p_0 = s_0^2$, then the above expression takes the desired form

$$p = \sum_{i=1}^{n} p_i \cdot x_i + \sum_{i \neq j} p_{ij} \cdot x_i \cdot x_j,$$

with $p_i = 4 \cdot s_0 \cdot s_i - 4 \cdot s_i^2$ and $p_{ij} = -4 \cdot s_i \cdot s_j$.

Resulting reduction. To each particular case of the subset sum problem, described by parameters $s_1, \ldots, s_n$, we assign the following particular case of our problem. First, we compute $p_0 = s_0 = \sum_{i=1}^{n} s_i$; then, we compute

$$p_i = 4 \cdot s_0 \cdot s_i - 4 \cdot s_i^2; \quad p_{ij} = -4 \cdot s_i \cdot s_j.$$

For these values, the quadratic function $p = \sum\limits_{i=1}^{n} p_i \cdot x_i + \sum\limits_{i \neq j} p_{ij} \cdot x_i \cdot x_j$ has the form

$$p = p_0 - \left(\sum_{i=1}^{n} \varepsilon_i \cdot s_i \right)^2, \text{ where } \varepsilon_i = 2 \cdot x_i - 1 \in \{-1, 1\}.$$

The above argument shows that for this selection, the quadratic function p attains the value $p_0 = s_0$ if and only if the original instance of the subset sum problem $\sum\limits_{i=1}^{n} \varepsilon_i \cdot s_i = 0$ has a solution with $\varepsilon_i \in \{-1, 1\}$, and thus, with $x_i = \dfrac{\varepsilon_i + 1}{2} \in \{0, 1\}$.

The reduction is proven, so our problem is indeed NP-hard.

Comment. Strictly speaking, we have proved NP-hardness of a specific choice of the quadratic function $p(x_1, \ldots, x_n)$. However, we have already mentioned earlier that if a problem $\mathscr{P}_0$ is NP-hard, then a more general problem $\mathscr{P}_1$ is NP-hard as well. Thus, we have indeed proved that the (more general) problem is also NP-hard.

Conclusions. One of the applications of fuzzy techniques is to formalize the meaning of words from natural language such as "efficient", "diverse", etc. The main idea behind fuzzy techniques is that they formalize expert knowledge expressed by words from natural language; see, e.g., [10, 11].

In this chapter, we have shown that if we do not use this knowledge, i.e., if we only use the data, then selecting the most efficient group (or even selecting a group with a given efficiency) becomes a computationally difficult (NP-hard) problem. Thus, the need to select such groups in reasonable time justifies the use of fuzzy (intelligent) techniques – and, moreover, the need to combine intelligent techniques with more traditional optimization techniques.

References

1. O. Castillo, P. Melin, E. Gamez, V. Kreinovich, O. Kosheleva, Intelligence techniques are needed to further enhance the advantage of groups with diversity in problem solving, in *Proceedings of the 2009 IEEE Workshop on Hybrid Intelligent Models and Applications HIMA'2009, Nashville, Tennessee, 30 March–2 April 2009* (2009), pp. 48–55
2. L. Hong, S.E. Page, Problem solving by heterogeneous agents. J. Econ. Theory **97**(1), 123–163 (2001)
3. L. Hong, S.E. Page, Groups of diverse problem solvers can outperform groups of high-ability problem solvers. Proc. Natl. Acad. Sci. **101**(46), 16385–16389 (2004)
4. J.H. Miller, S.E. Page, *Complex Adaptive Social Systems: The Interest in Between* (Princeton University Press, Princeton, 2006)
5. S.E. Page, *The Difference: How the Power of Diversity Creates Better Groups, Firms, Schools, and Societies* (Princeton University Press, Princeton, 2007)
6. J.E. Gamez, F. Modave, O. Kosheleva, Selecting the most representative sample is NP-hard: need for expert (fuzzy) knowledge, in *Proceedings of the IEEE World Congress on Computational Intelligence WCCI'2008, Hong Kong, China, 1–6 June 2008* (2008), pp. 1069–1074
7. C.C. Kuo, F. Glover, K.S. Dhir, Analyzing and modeling the maximum diversity problem by zero-one programming. Decision Sci. **24**(6), 1171–1185 (1993)

8. V. Kreinovich, E. Johnson-Holubec, L.K. Reznik, M. Koshelev, Cooperative learning is better: explanation using dynamical systems, fuzzy logic, and geometric symmetries, in *Proceedings of the Vietnam–Japan Bilateral Symposium on Fuzzy Systems and Applications VJFUZZY'98, HaLong Bay, Vietnam, 30th September–2nd October, 1998*, ed. by H.P. Nguyen, A. Ohsato (1998), pp. 154–160
9. C.H. Papadimitriou, *Computational Complexity* (Addison Wesley, Reading, 1994)
10. G.J. Klir, B. Yuan, *Fuzzy Sets and Fuzzy Logic: Theory and Applications* (Prentice-Hall, Upper Saddle River, 1995)
11. H.T. Nguyen, E.A. Walker, *A First Course in Fuzzy Logic* (CRC Press, Boca Raton, 2006)

Chapter 26
A Minor but Important Aspect of Teaching Large Classes: When to Let in Late Students?

Some students are late for classes. If we let in these late, this disrupts the class and decreases the amount of effective lecture time for the students who arrived on time. On the other hand, if many students are late and we do not let them in, these students will miss the whole lecture period. It is therefore reasonable to sometimes let students in, but restrict the times when late students can enter the class. In this chapter, we show how, depending on the number of late students – and depending on how late they are – we can find the optimal schedule of letting in late students.

The results from this chapter first appeared in [1].

Letting in late students is disruptive. Some students are late for class. Letting them walk in all the time disrupts others. As a result, some teachers in schools and even some professors at the universities do not let late students in at all.

Comment. This is not only about classes. The famous Russian theater reformer Stanislasvky started his reform by not letting late spectators in – and thus, minimizing disruptions for others. This tradition is held in many theaters now.

Not letting in late students is probably too harsh. On the other hand, such a no-late policy may be too harsh, especially if we take into account that lateness is often caused by things beyond a student's control – e.g., on a commuter campus like ours, an accident on a freeway that caused traffic delays make students arrive late to their first class of the day.

Resulting problem: when to let in late students? Based on the above discussion, we conclude that:

- in principle, it is desirable to let late students in, but
- we cannot let them in all the time.

So, we should select specific times when the students will be allowed to enter.

© Springer-Verlag GmbH Germany 2018
O. Kosheleva and K. Villaverde, *How Interval and Fuzzy Techniques Can Improve Teaching*, Studies in Computational Intelligence 750,
https://doi.org/10.1007/978-3-662-55993-2_26

How this problem is solved now? Sometimes, these times are determined by the event. For example, in a symphony concert, late patrons have to wait for the end of the first musical piece to enter. What shall we do in a lecture where there are no such easily determined least-disruption times?

There are many heuristic ways of dealing with such situations. For example, a recent recollection volume by students from the Mathematical Department of St. Petersburg University, Russia, mentions that some professors teaching big Calculus classes would ask late students to wait until it is exactly 10 min after the beginning of the lecture, and let in all accumulated late students [2].

Need for an optimal solution. Instead of relying on such heuristic rules, it is desirable to come up with a precise solution to the problem – a solution obtained by optimizing an appropriately chosen objective function.

This is what we do in this chapter.

Towards formalization of the problem. Let T be the duration of the lecture, and let Δ denote the disruption time cause by letting late students in. For every real number $t \in [0, T]$, Let $f(t)$ denote the total proportion of students who arrive between the beginning of the lecture and time t after the lecture started.

Our objective is to minimize the total disruption, i.e., minimize the disruption time per student.

When we do not let in late students. If we do not let any late students in, then the only disruption comes from late students missing the class. Each of the late students missed time T, and the proportion of late students is $f(T)$. Thus, the resulting disruption per student is

$$D_0 = T \cdot f(T).$$

When we let in late students at one single moment of time. If we let in late students at a single moment of time t_1, then there are three sources of disruption:

- there is a disruption $\Delta \cdot (1 - f(T))$ (caused by letting students in) for all students who arrived on time,
- there is a disruption $t_1 \cdot f(t_1)$ caused by the fact that students who arrive between times 0 and t_1 miss time t_1;
- finally, there us a disruption $T \cdot (f(T) - f(t_1))$ caused by the fact that students who arrive after moment t_1 miss the whole lecture.

The resulting overall disruption is equal to

$$d_1(t_1) = \Delta \cdot (1 - f(T)) + t_1 \cdot f(t_1) + T \cdot (f(T) - f(t_1)).$$

The time t_1 should be selected from the condition that the resulting overall disruption is the smallest possible. For such an optimal value t_1, the resulting disruption is equal to

$$D_1 = \min\{\Delta \cdot (1 - f(T)) + t_1 \cdot f(t_1) + T \cdot (f(T) - f(t_1)) : 0 \le t_1 \le T\}.$$

The optimal value t_1 can be determined by the condition that the derivative of the minimized function is equal to 0, i.e., that

$$f(t_1) + t_1 \cdot f'(t_1) - T \cdot f'(t_1) = 0,$$

where $f'(t)$ denoted the derivative of the function $f(t)$. This condition can be equivalently reformulated as

$$f(t_1) = (T - t_1) \cdot f'(t_1).$$

Shall we let in students or shall we not? Whether we should let in students at all or not depends on whether $D_1 = d_1(t_1) < D_0$. The corresponding inequality has the form

$$\Delta \cdot (1 - f(T)) + t_1 \cdot f(t_1) + T \cdot (f(T) - f(t_1)) < T \cdot f(T),$$

i.e., equivalently,

$$\Delta \cdot (1 - f(T)) + t_1 \cdot f(t_1) < T \cdot f(t_1),$$

which is, in turn, equivalent to:

$$\Delta \cdot (1 - f(T)) < (T - t_1) \cdot f(t_1).$$

General case. Let us now consider the general case, in which we let students in at several ($k \geq 0$) moments of time $0 < t_1 < t_2 < \cdots < t_k < T$. To simplify the description of this inequality, it makes sense to set $t_0 = 0$ and $t_{k+1} = T$, then this inequality has the form

$$0 = t_0 < t_1 < t_2 \cdots < t_k < t_{k+1} = T.$$

For every i from 0 to k, students who arrive between times t_i and t_{i+1} lose time $t_{i+1} -$ the next time late students are let in. The proportion of such students is $f(t_{i+1}) - f(t_i)$, so the disruption for all these students is equal to $t_{i+1} \cdot (f(t_{i+1}) - f(t_i))$. To those students who have already been sitting in class by the time of the i-th disruption – their proportion is $1 - f(T) + f(t_{i-1})$ – the disruption is equal to

$$\Delta \cdot (1 - f(T) + f(t_{i-1})).$$

Thus, the overall disruption is equal to

$$d_k(t_1, \ldots, t_k) = \sum_{i=1}^{k} \Delta \cdot (1 - f(T) + f(t_{i-1})) + \sum_{i=0}^{k} t_{i+1} \cdot (f(t_{i+1}) - f(t_i)).$$

The times $t_1, \ldots, t_k$ should be selected from the condition that the resulting disruption is the smallest possible. For such optimal values $t_1, \ldots, t_k$, the resulting disruption is equal to

$$D_k = \min \left\{ \sum_{i=1}^{k} \Delta \cdot (1 - f(T) + f(t_{i-1}))+ \right.$$

$$\left. \sum_{i=0}^{k} t_{i+1} \cdot (f(t_{i+1}) - f(t_i)) : 0 < t_1 < t_2 < \ldots < T \right\}.$$

The optimal values $t_1, \ldots, t_i, \ldots, t_k$ can be determined by the condition that the partial derivative of the minimized function with respect to each variable t_i, $1 \le i \le k$, is equal to 0. For $i < k$, we get

$$\Delta \cdot f'(t_i) + (f(t_i) - f(t_{i-1}) + t_i \cdot f'(t_i) - t_{i+1} \cdot f'(t_i) = 0.$$

This condition can be equivalently reformulated as

$$f(t_i) - f(t_{i-1}) = (t_{i+1} - t_i - \Delta) \cdot f'(t_i).$$

For $i = k$, we similarly get

$$f(t_k) - f(t_{k-1}) = (T - t_k) \cdot f'(t_k).$$

Whether we should let students in at k different moments of time depends on whether $D_k \le D_\ell$ for all $\ell \ne k$.

As a result, we arrive at the following solution.

General solution. For each integer k, we find values

$$0 = t_0 < t_1 < t_2 < \ldots < t_k < t_{k+1} = T$$

for which the expression

$$\sum_{i=1}^{k} \Delta \cdot (1 - f(T) + f(t_{i-1})) + \sum_{i=0}^{k} t_{i+1} \cdot (f(t_{i+1}) - f(t_i))$$

is the smallest possible. This can be done, e.g., by solving the following system of equations:

$$f(t_i) - f(t_{i-1}) = (t_{i+1} - t_i - \Delta) \cdot f'(t_i), \quad i < k;$$

$$f(t_k) - f(t_{k-1}) = (T - t_k) \cdot f'(t_k).$$

Let us denote the corresponding smallest value of the minimized expression by D_k.

Then, we select k for which the value D_k is the smallest possible, and for this k, take the corresponding minimizing values $t_1, \ldots, t_k$. These are the times at which we let late students in.

Comment. When k increases, the second term in the optimization function tends to the Stiltjes integral $\int t \cdot df(t)$ describing the overall disruption in the case when every late student is let in right away.

Example Let us illustrate the above idea on the example when the students arrive uniformly, i.e., when $f(t) = f_0 \cdot t$ for some f_0. In this case, the above equation for determining t_i for i_k takes the form

$$f_0 \cdot (t_i - t_{i-1}) = (t_{i+1} - t_i - \Delta) \cdot f_0,$$

i.e., equivalently, that $\Delta t_i \stackrel{\text{def}}{=} t_{i+1} - t_i$ satisfies the condition $\Delta t_i = \Delta t_{i-1} + \Delta$.

In other word, in this case, if there are several moments of time when we let students in, the waiting time before each letting-in increases by Δ from the previous waiting time.

For $i = k$, we similarly conclude that $\Delta t_k = \Delta t_{k-1}$.

In this case, $\Delta t_i = \Delta t_0 + i \cdot \Delta$ for $i < k$, in particular, $\Delta t_{k-1} = \Delta t_0 + (k-1) \cdot \Delta$. Thus, $\Delta t_k = \Delta t_{k-1}$ implies that $\Delta t_k = \Delta t_0 + (k-1) \cdot \Delta$.

We can now express the optimal values t_i in terms of the difference Δt_i. Indeed, since $t_0 = 0$, we have

$$t_i = t_i - t_0 = (t_i - t_{i-1}) + (t_{i-1} - t_{i-2}) + \cdots + (t_1 - t_0) = \Delta t_{i-1} + \Delta t_{i-2} + \cdots + \Delta t_0 =$$

$$\Delta t_0 + (i-1) \cdot \Delta + \Delta t_0 + (i-2) \cdot \Delta + \cdots + \Delta t_0 =$$

$$i \cdot \Delta t_0 + \Delta \cdot (1 + 2 + \cdots + (i-1)),$$

hence

$$t_i = i \cdot \Delta t_0 + \Delta \cdot \frac{(i-1) \cdot i}{2}.$$

To complete our description of the optimal schedule corresponding to the given number k of letting students in, we need to determine the value Δt_0. This value can be determined from the fact that

$$\Delta t_k = T - t_k = \Delta_0 + (k-1) \cdot \Delta.$$

From the above formula, we know that

$$t_k = k \cdot \Delta t_0 + \Delta \cdot \frac{(k-1) \cdot k}{2}.$$

Thus, we conclude that

$$T = t_k + \Delta t_k = (k+1) \cdot \Delta_0 + \left(\frac{(k-1) \cdot k}{2} + (k-1) \right) \cdot \Delta =$$

$$(k+1) \cdot \Delta_0 + \frac{(k-1) \cdot (k+2)}{2} \cdot \Delta.$$

So, we have

$$\Delta t_0 = \frac{T - \dfrac{(k-1) \cdot (k+2)}{2} \cdot \Delta}{k+1}.$$

Substituting the resulting optimal values t_i into the corresponding expression for $d_k(t_1, \ldots, t_k)$, we can find the value D_k for each k and thus, find the optimal number of disruptions k.

References

1. O. Kosheleva, When to let in late students? J. Uncertain Syst. **6**(2), 114–117 (2012)
2. D. Epstein, Y. Shapiro, S. Ivanov (eds.), Mathematics Department of St, Petersburg University, St. Petersburg, Russia, 2011 (in Russian)

Part IV
How to Assess Students, Teachers, and Teaching Techniques

Chapter 27
How to Assess Students, Teachers, and Teaching Techniques: An Overview of Part IV

It is important to assess the results of teaching. The main objective of teaching is that students learn the material. So, the most important assessment task is to assess the students. Different aspects of this assessment are analyzed in Chaps. 28–33.

- In Chap. 28, we use the fact that education can be viewed as a particular case of control to explain why effective assessment procedures assess not only the resulting student's knowledge, but also the amount of their effort in pursuing this knowledge.
- In Chap. 29, we use a more detailed version of the general optimal control model of education to come up with a recommendation to assess frequently.
- In Chap. 30, we use specific features of young student's psychology to argue that assessment should have an element of surprise.

Chapters 28–30 deal with assessment of student's individual work. In modern education, a large amount of effort goes into group projects. The question of how to assess individual contributions to a group project is analyzed in Chap. 31.

- For some classes, there are no follow-ups, so all we need is for the students to gain the corresponding knowledge – and the grade for this class should reflect the student's knowledge (and efforts in achieving this knowledge).
- However, in many other cases, the class is simply a pre-requisite for the following class. In this case, a more adequate assessment of the student's success is the degree to which a student is ready for the following class. In Chap. 32, we analyze how to assess this readiness.

In Chap. 32, we encounter situations in which, to get a more adequate assessment, it is necessary to go beyond the usual weighted average grade and use special non-linear formulas for combining grades of individual assignments into a single grade for the class. In Chap. 33, we consider all possible non-linear combination formulas, and we show that fuzzy-motivated combination formulas are (in some reasonable sense) the best.

© Springer-Verlag GmbH Germany 2018
O. Kosheleva and K. Villaverde, *How Interval and Fuzzy Techniques Can Improve Teaching*, Studies in Computational Intelligence 750,
https://doi.org/10.1007/978-3-662-55993-2_27

- Once we have grades for each of the students, it is desirable to assess the performance of the class as a whole. This task is analyzed in Chap. 34.
- Student's assessments also provide the data based on which we can assess:

  – teachers (Chap. 35),
  – teaching techniques (Chap. 36), and
  – universities as a whole (Chap. 37).

Chapter 28
How to Assess Students: Rewarding Results or Rewarding Efforts?

The main objective of teaching is to make sure that the students learn the required material. From this viewpoint, it seems reasonable to reward students when they learn this material. In other words, it seems reasonable to assign the student's rewards (such as grades) based on their level of knowledge. However, experiments show that, contrary to this seemingly intuitive conclusion, we achieve much better results if we base rewards not only on the student's results, but also on the student's efforts. In this chapter, we use the known fact about optimal control to explain this seemingly counterintuitive phenomenon.

The results from this chapter first appeared in [1].

Normally, grades reflect results. The main objective of teaching is to make sure that the students learn the material. From this viewpoint, it seems reasonable to assign grades based on the results, based on how well the students learned – and this is how we normally grade.

This does not mean that effort is not taken into account: in the borderline cases, when a grade is between A ("excellent") and B ("good"), or between B ("good") and C ("satisfactory"), we often give extra points for extra effort. However, these extra points form a small portion of the overall grade.

Seemingly counter-intuitive empirical observation. At first glance, it seems reasonable to reward results not efforts. However, contrary to this seemingly reasonable conclusion, experiments have shown that if we give more weight to *efforts* when deciding on the reward, students achieve better *results*; see, e.g., [2–4] and references therein.

What we do in this chapter. In this chapter, we provide an explanation for this seemingly counter-intuitive phenomenon.

Main idea of this chapter: teaching can be viewed as a particular case of the general engineering control problem. The main objective of teaching is to optimally change the student's state of the mind. From the general engineering viewpoint, the problem of (optimally) changing the system's state is called the *control* problem.

© Springer-Verlag GmbH Germany 2018
O. Kosheleva and K. Villaverde, *How Interval and Fuzzy Techniques Can Improve Teaching*, Studies in Computational Intelligence 750,
https://doi.org/10.1007/978-3-662-55993-2_28

Thus, the problem of optimal teaching can be viewed as a particular case of a general problem of optimal control.

What is known about optimal control. Different control problems have different objectives. In the simplest case, the objective is to keep the value of a certain quantity at its desired level. For example, it is desirable to control the electric grid so that the voltage remains the same irrespective of the load. When we control a chemical plant, we want to make sure that we maintain the temperature at which the corresponding chemical reactions are the most efficient. An autopilot tries to preserve the same direction and speed of the plane, etc.

In such cases, what we want is to maintain the desired value x_0 of the corresponding quantity x. At first glance, it seems like the best way to do it is to react to deviations from x, i.e., to set up a system that engages the control force every time the actual value x deviates from the desired value x_0 – and the larger the deviation $q \stackrel{\text{def}}{=} x - x_0$, the larger the corresponding control force.

However, both mathematical analysis of the problem and control practice show that much better results are achieved when we use a proportional-integral-derivative (PID) controller, i.e., a controller whose control value depends not only on the deviation q, but also on the time derivative $\dot{q}$ and on the integral $\int^t q(s)\, ds$.

This explain the need to reward efforts. When we apply this general control conclusion to the teaching situation, we conclude that the best teaching is achieved when the reward (i.e., control) depends not only on the results q, but also on the first derivative $\dot{q}$ of q – i.e., on efforts. This is exactly what is empirically observed.

What we did. Thus, we explained the seemingly counter-intuitive empirical fact as a natural consequence of optimal control theory.

How this is useful. The fact that general control theory ideas lead to recommendations that are consistent with the educational experience makes us believe that a general use of control theory results can lead to better education.

How to interpret the integral part of the optimal control? In the above text, we used the proportional (P) and the derivative (D) components of the optimal control. To make the control really optimal, we also need to take into account the integral (I) part.

This part actually has a very good education interpretation: namely, it corresponds to the difference between the traditional Russian grading system and the American system. In Russia, the grade for the class used to be determined exclusively by the grade on the final exam, i.e., by the amount of knowledge gained by the student. In contrast, in the US, the grade on the final exam constitutes only a part of the final grade, other points come from averaging the grades obtained at all the previous class assignments, quizzes, and tests. Each such previous grade reflects the student's previous state of knowledge $q(s)$, $s < t$; thus, averaging over all such grades means, in effect, that we integrate $q(s)$ over such past time moments.

The fact that an integral term is important for optimal control indicates that the accumulating-grade American system is potentially better than the traditional Russian system – in which the grade depends only on the final knowledge.

References

1. O. Kosheleva, Rewarding results or rewarding efforts. Appl. Math. Sci. **6**(15), 707–709 (2012)
2. P.O. Bronson, A. Merryman, *NurtureShock: New Thinking About Children* (Twelve Publishing, New York, 2009)
3. E. Galinsky, *Mind in the Making: The Seven Essential Life Skills Every Child Needs* (Harper-Collins, New York, 2010)
4. K. Ginsburg, S. FitzGerald, *Letting Go with Love and Confidence: Raising Responsible, Resilient, Self-Sufficient Teens in the 21st Century* (Avery Trade, New York, 2011)

Chapter 29
How to Assess Students: Assess Frequently

Students do not always spend enough time studying. How can we encourage them to study more? In this chapter, we show that a lot depends on the grading policy. At first glance, the problem of grading may seem straightforward: since our objective is that the students gain the largest amount of knowledge and skills at the end of the class, the grade should describe this amount. We show, however, that it is exactly this seemingly straightforward grading policy that often leads to an unfortunate learning behavior. To improve the students' learning, it is therefore necessary to use a grading policy which goes beyond the straightforward approach. In this chapter, we use fuzzy-motivated intuition to formulate selection of a grading policy as a precise optimization problem, and, in the first approximation, provide a solution to this optimization problem. This solution is in line with what experienced instructors are actually doing when grading the class.

The results from this chapter first appeared in [1].

Empirical fact. It is known that not all students spend as much time studying as they should. As a result, their knowledge (as judged by their grades) is not as good as it could be if they studied more.

How to improve the situation. It is desirable to make sure that the students spend more time studying. To be able to do that, we first need to understand the students' reasoning, to understand why they do not always study enough, and then use this understanding to propose ways to encourage them study more.

What we do in this chapter. In this chapter, we show that the amount of time that the students spend studying depends on the grading policy. It is therefore desirable to come up with a grading policy under which the students will, at the end of the class, gain the largest possible amount of knowledge.

© Springer-Verlag GmbH Germany 2018
O. Kosheleva and K. Villaverde, *How Interval and Fuzzy Techniques Can Improve Teaching*, Studies in Computational Intelligence 750,
https://doi.org/10.1007/978-3-662-55993-2_29

Qualitative analysis of the problem. Before we formulate quantitative models, let us start with a qualitative analysis of the problem.

Students usually want to succeed, to gain the *maximum* amount of knowledge in the class – at least in the classes from their major, classes that are of interest to them. On the other hand, students usually do not want to spend all their time studying, they also want to "have a life"; from this viewpoint, the students want to *minimize* their study efforts.

In decision making terms, this means that the students solve a multi-objective optimization problem.

Need for a fuzzy approach. Strictly speaking, the two objectives of a student, while reasonable, are inconsistent. From the purely mathematical viewpoint, if a student really wants to gain the maximum amount of knowledge, this student should spend all free time studying.

Vice versa, if a student really wants to minimize the study efforts, the student should not study at all.

What these weird conclusions mean is that the above "maximization" and "minimization" should not be taken literally, in the crisp sense, they should be taken informally, in fuzzy sense.

Fuzzy logic provides a natural description of such "fuzzy optimization"; see, e.g., [2, 3].

Comment. It is worth mentioning that fuzzy-related optimization techniques have been successfully used in education: e.g., in selecting the optimal way of presenting the material; see, e.g., [4]. In this chapter, we apply these techniques to the problem of designing the optimal grading policy.

Simplifying assumptions. To describe this formalization, let us agree to gauge the amount V knowledge that a student gains in the class as the portion of what this student can potentially gain. In this description, having learned nothing means $V = 0$ and having learned everything that was taught in this class means $V = 1$.

For simplicity, we will also assume:

- that the grade is proportional to the amount of knowledge, and
- that the effort needed to achieve this grade is proportional to this amount as well.

The second of these assumptions means that the students study in a correct way (and study the correct things), so that their study efforts are not wasted. It also means that we gauge the amount of knowledge not simple the number of learned ideas, skills, and results, but that we also take into account the relative difficulty of these learned ideas, skills, and results – so that a more complex idea is worth more than a simpler one.

Case study: US grading system. In this chapter, we will illustrate our ideas on the example of the typical US grading system, but, of course, the same ideas are applicable to any grading system.

In the US, usually, the instructor first computes the numerical grade from 0 to 100, 0 meaning no knowledge at all and 100 means perfect knowledge. Based on this numerical grade, the instructor then computes a letter grade that goes into the student's transcript:

- A numerical grade of 90 is above usually means the "excellent" letter grade A.
- A numerical grade between 80 and 90 usually means the "good" letter grade B.
- A numerical grade between 70 and 80 usually means the "satisfactory" letter grade C.
- A numerical grade between 60 and 70 usually means the "poor" letter grade D.
- A numerical grade below 60 usually means the "unsatisfactory" letter grade F.

(Some universities have more subtle schemes, with pluses and minuses.)

Membership function for knowledge. For many classes, the passing grade is D, which means, crudely speaking, that the minimal amount of knowledge that enables a student to pass the class corresponds to the value $V = 0.6$.
In this case:

- For values a little bit smaller than 0.6, the student did not pass the class, so his or her degree of satisfaction with the knowledge acquired in this class is 0: $\mu_k(0.6 - \varepsilon) = 0$. By continuity, when ε tends to 0, we conclude that $\mu_k(0.6) = 0$ – and that $\mu_k(V) = 0$ for all $V \leq 0.6$.
- When a student achieved perfect knowledge $V = 1$, the degree of satisfaction with the knowledge acquired in the class is 1.0: $\mu_k(1.0) = 1$.

We now need to interpolate this membership function to the whole interval $[0.6, 1.0]$. Since we are performing a qualitative analysis anyway, it is reasonable to take the simplest possible interpolation, i.e., a linear function that takes the value 0 for $V = 0.6$ and the value 1 for $V = 1.0$; this approach works very well in many practical applications of fuzzy, e.g., in fuzzy control, where the resulting trapezoid and triangular membership functions lead to good results. In our cases, this interpolation leads to the membership function

$$\mu_k(V) = \frac{V - 0.6}{1.0 - 0.6} = \frac{V - 0.6}{0.4}.$$

Membership function for effort. In a similar way, we can propose a membership function $\mu_e(V)$ for minimizing effort.

- When a student does not study at all, i.e., when $V = 0$, the goal of minimizing effort is absolutely fulfilled, i.e., $\mu_e(0) = 1$.
- On the other hand, when a student spends all the time studying, i.e., when $V = 1$, then the goal of minimizing the effort is not satisfied at all, i.e., $\mu_e(1) = 0$.

We can now use linear interpolation to find the values of this membership function $\mu_e(V)$ for all intermediate values $V \in (0, 1)$. As a result, we get the membership

function
$$\mu_e(V) = 1 - V.$$

Combining the two objectives into a single objective function. A student selects a value V that satisfies both objectives: maximizes knowledge *and* minimizes effort.

- The degree to which the value V satisfies the first objective, i.e., to which this value maximizes knowledge, is described by the value $\mu_k(V)$.
- The degree to which the value V satisfies the second objective, i.e., to which this value saves the efforts, is described by the value $\mu_e(V)$.

Thus, in accordance with the general fuzzy techniques, the degree to which the value V satisfies *both* objectives can be described by the value $f_\&(\mu_k(V), \mu_e(V))$, where $f_\&(a, b)$ is a t-norm ("and"-operation).

Similarly to the our choice of the simplest membership function, it is reasonable to choose the simplest t-norm, i.e., the t-norm $f_\&(a, b) = \min(a, b)$. For this t-norm, the degree to which a given value V satisfies both student's objectives is equal to $\mu(V) = \min(\mu_k(V), \mu_e(V))$. The student then selects a value V for which this degree is the largest possible.

Resulting student's behavior. For values $V \le 0.7$, we have $\mu_k(V) = 0$ and thus, $\mu(V) = \min(\mu_k(V), \mu_e(V)) = 0$. One can easily check that:

- The function $\mu(V)$ first coincides with the first membership function $\mu_k(V)$ (and thus, increases) while
$$\mu_k(V) \le \mu_e(V).$$

- Then, the function $\mu(V)$ coincides with the second membership function $\mu_e(V)$ (and thus, decreases) while
$$\mu_e(V) \le \mu_k(V).$$

So, this function $\mu(V) = \min(\mu_k(V), \mu_e(V))$ attains its maximum at a value V_m for which $\mu_k(V_m) = \mu_e(V_m)$, i.e., for which

$$\frac{V_m - 0.6}{0.4} = 1 - V_m.$$

Multiplying both sides by 0.4, we get $V_m - 0.6 = 0.4 - 0.4 \cdot V_m$, hence $1.4 \cdot V_m = 0.6 + 0.4 = 1$ and

$$V_m = \frac{1}{1.4} \approx 0.7.$$

So, the average grade is close to C, which is exactly what we observe in many classes.

What if only C is a passing grade. In some classes, a student needs to get at least a grade of C to pass. For example, in the undergraduate Computer Science program of the University of Texas at El Paso, each class which is a pre-requisite for another required class has to be passed with at least a C grade. As we have mentioned, the C grade corresponds to 0.7. Thus, in this case, $\mu_k(0.7 - \varepsilon) = 0$ and $\mu_k(1.0) = 1.0$, so linear interpolation leads to

$$\mu_k(V) = \frac{V - 0.7}{1.0 - 0.7} = \frac{1.0 - 0.7}{0.3}.$$

Thus, the function $\mu(V) = \min(\mu_k(V), \mu_m(V))$ attains its maximum at a value V_m for which $\mu_k(V_m) = \mu_e(V_m)$, i.e., for which

$$\frac{V_m - 0.7}{0.3} = 1 - V_m.$$

Multiplying both sides by 0.3, we get $V_m - 0.7 = 0.3 - 0.3 \cdot V_m$, hence $1.3 \cdot V_m = 0.7 + 0.3 = 1$ and

$$V_m = \frac{1}{1.3} \approx 0.78.$$

So, in such classes, the average grade is close to B, which is also exactly what we observe in many such classes.

Towards a quantitative description. Let $v(t)$ denote a rate with which a student studies at moment t, i.e., the amount of new knowledge per unit time acquired in the vicinity of this moment.

In this notation, from moment t to moment $t + \Delta t$, a student acquires the amount of knowledge $v(t) \cdot \Delta t$. Thus, the total amount of knowledge acquired by a moment t_0 can be described by adding all the amounts acquired at the previous moments of time, i.e., as $\sum v(t) \cdot \Delta t$. From the mathematical viewpoint, this is an integral sum, and in the limit $\Delta t \to 0$, it tends to the integral

$$\int_0^{t_0} v(t)\,dt.$$

Comment. In this simplified description, we assume that once a student learned some subject, the student will retain this knowledge, i.e., that there is no forgetting.

Describing effort. We have assumed that the effort used is proportional to the amount learned. Let us denote the corresponding proportionality coefficient by K.

Of course, the displeasure caused by studying should be smaller than the pleasure caused by acquiring knowledge, otherwise the student will not study at all. So, we must have $K < 1$.

First idea: straightforward evaluation. Our objective is that the student learns as much as possible at the end of the class, i.e., during the time T that a class takes. In other words, our objective is to maximize the amount that the student learned during the class, i.e., the value $\int_0^T v(t) \, dt$. Therefore, it seems reasonable to make the final grade for this class proportional to this value.

Comment. This idea is used, e.g., in Russian universities, where all the midterm exams and quizzes only determine whether a person is allowed to take the final exam, and the grade for the class is determined exclusively by this comprehensive final exam.

Towards analysis of the straightforward approach. In the straightforward approach, a student gets, at time T, the grade proportional to the integral $\int_0^T v(t) \, dt$.

Discounting of future utilities: general description. In decision theory, it is usually assumed that expected future events that will occur at moment f in the future are "discounted" with the factor $e^{-\alpha \cdot f}$, where the parameter α depends on the individual; see, e.g., [5].

Discounting: financial explanation. This discounting is easy to explain for financial gains. Indeed, if we are promised an amount A at time f in the future, then we can gain the same amount at this future time if we get now an amount a and invest it so that it will bring some interest (e.g., place it into the saving account). After f years, the original amount a will turn into $(1 + r)^f$, where r is the yearly interest rate. Thus, to get the amount A in the future, we need now to select the amount a for which $a \cdot (1 + r)^f = A$, i.e., the amount $a = (1 + r)^{-f} \cdot A$. This is exactly the above discounting, with $\alpha = \ln(1 + r)$.

Comment. At the end of this chapter, we show that the same formula for discounting can be derived from first principles, without using specifics of financial situations.

Analysis of the straightforward approach (cont-d). Because of the discounting, the student's expected gain at the moment the class starts is equal to

$$ e^{-\alpha \cdot T} \cdot \int_0^T v(t) \, dt. $$

At each moment of time t, the rate $v(t)$ with which a student studies leads to this student's rate of effort spending equal to $K \cdot v(t)$. Because of the discounting, the effect of all these efforts at the beginning of the class is equal to

$$ K \cdot \int_0^T e^{-\alpha \cdot t} \cdot v(t) \, dt. $$

At the beginning of the semester, the overall utility of a study plan $v(t)$ to a student is equal to the sum of the positive utility caused by knowledge and the negative utility caused by efforts, i.e., to

$$U = e^{-\alpha \cdot T} \cdot \int_0^T v(t)\, dt - K \cdot \int_0^T e^{-\alpha \cdot t} \cdot v(t)\, dt.$$

In general, at a moment t_0, when the student makes a decision about the learning rate $v(t_0)$ to select at this moment of time, the part of the utility that depends on the remaining selections $v(t)$ ($t \geq t_0$) takes the form

$$U(t_0) = e^{-\alpha \cdot (T - t_0)} \cdot \int_{t_0}^T v(t)\, dt - K \cdot \int_{t_0}^T e^{-\alpha \cdot (t - t_0)} \cdot v(t)\, dt.$$

This expression is a linear function of all the values $v(t)$:

$$U = \int_{t_0}^T c(t) \cdot v(t)\, dt,$$

where

$$c(t) = e^{-\alpha \cdot (T - t_0)} - K \cdot e^{-\alpha \cdot (t - t_0)}.$$

Thus, maximizing this expression over all possible non-negative values $v(t_0)$ means that we select $v(t_0) = 0$ when $c(t_0) = e^{-\alpha \cdot (T - t_0)} < K \cdot e^{-\alpha \cdot (t_0 - t_0)}$ or, equivalently,

$$e^{-\alpha \cdot (T - t_0)} < K.$$

Specifics of the young people decision making: brief reminder. Many students are young people, and for young people, the discount coefficient α is often high.

This is shown by the fact that they often do not consider possible negative future consequences of their actions and vice versa, they do not value possible future positive effects, they live more in the today and they do not care that much about their future.

Unexpected negative consequence of the straightforward approach. As a result of the high value α typical for young people, we get $e^{-\alpha \cdot T} \ll K$ (and thus, $\alpha \cdot T \gg |\ln(K)|$). Hence, for $t_0 = 0$, we get $c(t_0) < 0$ and thus, $v(t_0) = 0$. This means that at the beginning, students do not study – because they have no incentive to study.

Moreover, the students do not start studying until the moment t_s at which the coefficient $c(t_s)$ turns from negative to positive, i.e., at which $e^{-\alpha \cdot (T - t_s)} = K$ and thus, $\alpha \cdot (T - t_s) = |\ln(K)|$. Since $\alpha \cdot T \gg |\ln(K)|$, this means that $\alpha \cdot T \gg \alpha \cdot (T - t_s)$, i.e., $T \gg T - t_s$. Thus, the time period $T - t_s$ during which the students actually study is much smaller than the total period T during which they could potentially study.

Since a human ability to learn the material in a given time is limited, this means that the students end up not learning as much as they potentially could.

Possible solution to the problem. The above problem was caused by the way we assign the grade. So, to resolve the problem, it is therefore desirable to modify the way we assign the grade for the class: assigning the grade simply based on the knowledge at the end of the class often does not lead to good results.

Main idea. The main problem with the straightforward grading scheme is that the final grade is only determined by the final knowledge. This grade becomes known only a long time after the class starts, and this long delays drastically decreases the effect of this grading on the student behavior.

A way to solve this problem is to decrease the time delay between the student learning the material and the student being tested on this material. To achieve this objective, we must introduce earlier (intermediate) tests.

Comment. In the US, such tests are often known – somewhat confusingly – as *midterm exams*. This term originated from the setting in which there is one such exam, given exactly in the middle of the semester, but it is also used to describe situations in which there are several intermediate exams given at different times during the semester.

Sequence of tests. Instead of a single exam at the end of the class, let us assume that we have m exams which are equally spaced in time, at moments

$$T_1, \quad T_2 = 2T_1, \quad \ldots, \quad T_m = m \cdot T_1 = T.$$

We consider a simplified model of learning in which there is no forgetting, i.e., in which, once a student learned some subject, the student will retain this knowledge. In this simplified model, once we have checked that a student knows some material, there is no need to check this knowledge again at a later moment of time. It therefore makes sense to arrange the tests in such a way that the i-th test tests only the knowledge that would normally acquired between the previous test and this one, i.e., between the moments T_{i-1} and T_i. For the 1-st test, there is no previous test, so we set $T_0 = 0$.

How to determine the grade for each test. The grade for each test should be determined based on the material that a student should have learned between moments T_{i-1} and T_i: the ratio of the material that a student has actually learned to the amount that the student could have learned.

Ideal learning schedule. In the ideal situation, students study at a uniform speed $v(t) = v_{\text{ideal}}$. By time T, they should be able to learn all the material, i.e., teach the value

$$V = \int_0^T v_{\text{idea}}(t)\, dt = \int_0^T v_{\text{idea}}\, dt = v_{\text{ideal}} \cdot T = 1.$$

Thus, the ideal learning rate is $v_{\text{ideal}} = \dfrac{1}{T}$. With this ideal rate, by the time T_{i-1}, a student would learn the amount

$$v_{\text{ideal}} \cdot T_{i-1} = \frac{1}{T} \cdot T_{i-1}.$$

In this ideal case, between the times T_{i-1} and T_i, the student should have learned the amount $v_{\text{ideal}} \cdot (T_i - T_{i-1})$.

Actual learning. The actual amount that a student learns by the time T_i is equal to $\int_0^{T_i} v(t)\,dt$. Thus, if we only count the material that a student should have learned after time T_{i-1}, we get the amount

$$\max \left(\int_0^{T_i} v(t)\,dt - v_{\text{ideal}} \cdot T_{i-1}, 0 \right).$$

(Maximum with 0 is introduced since if a student is behind, i.e., the student have not even started learning the material that he or she was supposed to learn between moments T_{i-1} and T_i, the difference will be negative, but the student's grade cannot be negative, so it 0.)

Grade for the i-th test. As we have mentioned, the grade g_i for the i-th test is equal to the ratio of what the student actually learned to what he or she should have learned, i.e., to the ratio

$$g_i = \frac{\max \left(\int_0^{T_i} v(t)\,dt - v_{\text{ideal}} \cdot T_{i-1}, 0 \right)}{v_{\text{ideal}} \cdot (T_i - T_{i-1})}.$$

The overall grade. The overall grade g is usually computed as a weighted average of grades g_i of different tests, i.e., as a value $g = \sum_{i=1}^{m} w_i \cdot g_i$, where $w_1, \ldots, w_m$ be the corresponding weights, $w_i \geq 0$, $\sum_{i=1}^{m} w_i = 1$.

Utility caused by the overall grade. A student learns his or her grade for the i-th test practically right after the test, i.e., at the moment T_i. Thus, the student's utility of test i is discounted with a factor $e^{-\alpha \cdot T_i}$.

Resulting utility. As a result, the student's total utility for the learning schedule $v(t)$ is equal to

$$U = \sum_{i=1}^{m} w_i \cdot e^{-\alpha \cdot T_i} \cdot \frac{\max \left(\int_0^{T_i} v(t)\,dt - v_{\text{ideal}} \cdot T_{i-1}, 0 \right)}{v_{\text{ideal}} \cdot (T_i - T_{i-1})} -$$

$$K \cdot \int_0^{T} e^{-\alpha \cdot t} \cdot v(t)\,dt.$$

Remaining problem: selecting weights. The remaining problem is to select appropriate weights w_i. We want to select the weights in such a way that a student has an incentive to study all the time.

At each moment t_0 between the tests $j - 1$ and j, the student's utility is similarly equal to

$$U = \sum_{i=j}^{m} w_i \cdot e^{-\alpha \cdot (T_i - t_0)} \frac{\max\left(\int_0^{T_i} v(t)\, dt - v_{\text{ideal}} \cdot T_{i-1}, 0\right)}{v_{\text{ideal}} \cdot (T_i - T_{i-1})} -$$

$$K \cdot \int_{t_0}^{T} e^{-\alpha \cdot (t - T_0)} \cdot v(t)\, dt.$$

This is a linear function of the values $v(t)$. A student is encouraged to study at this moment of time t_0 if the coefficient at $v(t_0)$ in the above expression is non-negative, i.e., when

$$\frac{w_j}{v_{\text{ideal}} \cdot (T_j - T_{j-1})} \cdot e^{-\alpha \cdot (T_j - t_0)} \geq K.$$

We want this inequality to be satisfied for all the values $t_0 \in [T_{j-1}, T_j]$. The right-hand side of the desired inequality is a constant not depending on t_0 at all; thus, for this inequality to hold for all $t_0 \in [T_{j-1}, T_j]$, it is sufficient to check that it is true when the left-hand side attains its smallest possible value.

The left-hand side of the desired inequality is an increasing function of t_0. Thus, this left-hand side attains its smallest value when the variable t_i attains its smallest possible value $t_0 = T_{j-1}$. For this value, the above inequality takes the form

$$\frac{w_j}{v_{\text{ideal}} \cdot (T_j - T_{j-1})} \cdot e^{-\alpha \cdot (T_j - T_{j-1})} \geq K,$$

i.e., equivalently,

$$w_j \geq v_{\text{ideal}} \cdot (T_j - T_{j-1}) \cdot e^{\alpha \cdot (T_j - T_{j-1})} \cdot K.$$

Since the tests are equally spaced, the right-hand side is the same for all j. Let us denote this common value by w_0; under this notation, the above inequality takes the form $w_j \geq w_0$ for all j, i.e., equivalently, $\min_{j=1,\ldots,m} w_j \geq w_0$.

Towards optimal selection of the weights corresponding to different tests. Our objective is to make sure that the testing schedule encourages as many students as possible to study – of course, within the time limitations caused by the fact that a student is taking several classes at the same time.

Different students have different discount parameters α. The value w_0 depends on the discount parameter α: the larger α, the larger w_0. We can thus say that different students are characterized by different values of the parameter w_0. A grading scheme works for a student with the value w_0 if and only if $w_0 \leq \min_{j=1,\dots,m} w_j$. Thus, to make sure that this scheme works for as many students as possible, we must make sure that the threshold $\min_{j=1,\dots,m} w_j$ is as large as possible.

Since $\sum_{j=1}^{m} w_j = 1$, we cannot have $\min_{j=1,\dots,m} w_j > \dfrac{1}{m}$, since then we would have

$$\sum_{j=1}^{m} w_j \geq m \cdot \min_{j=1,\dots,m} w_j > m \cdot \frac{1}{m} = 1.$$

Thus, we should have $\min w_j \leq \dfrac{1}{m}$. When all the weights are equal, we have $w_j = \dfrac{1}{m}$ and thus, $\min_{j=1,\dots,m} w_j = \dfrac{1}{m}$. Thus, the threshold is the largest when all the weights are equal.

Let us show that equal weights are the only case when the threshold is equal to its largest value $\dfrac{1}{m}$. Indeed, if the weights are not equal, one of these weights must be smaller than $\dfrac{1}{m}$: otherwise, if all are larger than or equal to $\dfrac{1}{m}$ and some are different, the sum will be larger than 1. Thus, we arrive at the following conclusion.

Conclusion. The grading scheme is optimal if and only if we assign equal weights to all the midterm exams.

Comment. This conclusion is in good accordance with how the class grade is usually calculated in the US universities.

Discussion. One of the results of our chapter is that we proved mathematically what has been known in education for some time – the need for frequent student activities. An instructor has to keep giving quizzes, exams, assignments, etc., to keep the students engaged on the material; otherwise, the students will just wait for the final or the midterm and not do much of anything else to learn the material. That is why it is so important to have learning activities for students.

The need to have learning activities is well understood in both grading systems described in this chapter:

• in the US system, where the overall grade for the course is a combination of grades for intermediate assignments, and
• in the Russian system, where the overall grade for the course is the grade for the final exam – provided that a student passed all intermediate assignments with a passing grade.

The difference between these systems is how much weight is assigned to the intermediate assignments (such as midterm exams):

• in the US system, we assign non-zero weights to all the exams, while

- in the Russian system, midterm exams are assigned 0 weight and the final exam the full weight.

Within the US system, different instructors assign different weights to different midterm exams:

- some instructors assign equal weight to all midterms,
- other instructors assign different weights to different midterm exams – e.g., assign more weight to later midterm exams, so that students who did not do well on the first midterm exam still have a chance to learn the material by the second exam and get both a good knowledge and a good grade.

In this chapter, we show that the optimal learning occurs when all midterm exams are given non-zero weight – moreover, when different midterm exams are assigned the same weight.

Possible future work. In our approach, we only discusses exams. Many courses use several different kinds of graded student work besides exams, such as quizzes, homeworks, lab assignments, etc. It is desirable to take into account the difference between these types of work.

It is also desirable to not only describe the optimal grading policy for a given schedule of graded work, but also to optimize the schedule itself. For example, it is desirable to find out what is the optimal number of exams for a semester, what is the optimal number of quizzes (and are these quizzes beneficial at all), what emphasize should be placed on the final comprehensive exam – and should we have such an exam at all, etc.

Appendix. Discounting: general explanation. Let us show that the formula for discounting can be derived from first principles, without using specifics of financial situations. Indeed, let $d(t)$ be a discounting after time t, meaning that one unit of utility at a future time t is equivalent to $d(t) < 1$ units at present.

The longer time delay, the less effect the future utility has on the present decisions; thus, the discounting function $d(t)$ should be decreasing with time.

Let $s > 0$ and $s' > 0$ be two real numbers. Then, we can describe one unit of utility at time $s + s'$ in two different ways:

- First, we can directly use the definition of the discounting function $d(t)$ and conclude that this unit is equivalent to $d(s + s')$ units at present.
- Alternatively:
  - We can first conclude that 1 unit at time $s + s'$ is equivalent to $d(s')$ units s moments earlier, at the moment s.
  - Then, since one unit at moment s is equivalent to $d(s)$ units at present, we conclude that $d(s')$ units at moment s are equivalent to $d(s) \cdot d(s')$ units at present.
  - Thus, one unit of utility at moment $s + s'$ is equivalent to $d(s) \cdot d(s')$ units at present.

From these two conclusions, we deduce that

$$d(s + s') = d(s) \cdot d(s').$$

It is known that every monotonically decreasing solution to this functional equation has the desired form

$$d(t) = \exp(-\alpha \cdot t)$$

for some real value $\alpha > 0$; see, e.g., [6].

Thus, we get the desired justification of the standard discounting formula.

References

1. O. Kosheleva, K. Villaverde, How to make sure that students study well: fuzzy-motivated optimization approach, in *Proceedings of the 2012 IEEE World Congress on Computational Intelligence WCCI'2012*, Brisbane, Australia, June 10–15, 2012, pp. 641–647
2. G.J. Klir, B. Yuan, *Fuzzy Sets and Fuzzy Logic: Theory and Applications* (Prentice-Hall, Upper Saddle River, 1995)
3. H.T. Nguyen, E.A. Walker, *A First Course in Fuzzy Logic* (CRC Press, Boca Raton, 2006)
4. D. Pritchard, M.G. Negoita, A fuzzy-GA hybrid technique for optimization of teaching sequences presented in ITSs, in *Computational Intelligence, Theory and Applications, Proceedings of the 8th Fuzzy Days International Conference, Dortmund, Germany, September 29 – October 1, 2004*, vol. 33, Advances in Soft Computing, ed. by B. Reusch (Springer, Berlin, 2004), pp. 311–316
5. H. Raiffa, *Decision Analysis: Introductory Lectures on Choices under Uncertainty* (McGraw Hill, New York, 1997)
6. J. Aczél, *Lectures on Functional Equations and their Applications* (Dover, New York, 2006)

Chapter 30
How to Assess Students: Surprise Them

Most education efforts are aimed at educating young people. So, to make education as effective as possible, it is desirable to take into account psychological features of young people. One of the typical features of their psychology – as distinct from the psychology of more mature population – is that they are much more risk-prone. Based on the fact that young people prefer risky situations, in which the results depend on a random selection, a good strategy is to introduce as much randomness into teaching as possible, e.g., use surprise quizzes – in addition to normally scheduled tests and quizzes. To appropriately take into account student's attitude to risk, we need first to explain it within a quantitative model. Such a basic explanation is provided in this chapter.

The results from this chapter first appeared in [1, 2].

Young people are risk-prone. This is a lot of anecdotal evidence that young people are risk-prone. This was unexpectedly confirmed at one of our universities during a research project sponsored by the Texas Department of Transportation [3].

In traffic planning, it is important to take into account the driver's acceptance of risk. For example, in El Paso, there are two ways to get to the university from faraway places on the Westside:

- by the Interstate I-10 that passes through the city and
- by Mesa Street that also passes through this part of the city.

On average, I-10 is faster, but sometimes during rush hours, there are traffic jams or accidents there. If you are stuck on a freeway between two exists, it can take a while to get to the nearest exit.

On the other hand, on Mesa, even when there is an accident, there are usually side street that let you reach UTEP faster.

As a result, while on average, I-10 is faster, there is a low-probability risk that taking I-10 will drastically delay the travel time. So, whether a driver will take I-10 depends on his or her tolerance to risk.

© Springer-Verlag GmbH Germany 2018
O. Kosheleva and K. Villaverde, *How Interval and Fuzzy Techniques Can Improve Teaching*, Studies in Computational Intelligence 750,
https://doi.org/10.1007/978-3-662-55993-2_30

Risk tolerance is known to be somewhat different at different geographical locations. So, to decide on the best traffic planning, the researchers decided to quantify to what extent the local population accepts risk.

At the university, we have more than 20,000 students – enthusiastic responders to different surveys, so the researchers decided to first ask the students. The results showed a drastic difference with risk-tolerance of drivers in different geographic locations: students showed full tolerance to risk. According to their responses, all they care about is the average travel time, and the possibility of long delays did not negatively affect their decision at all. Actually, the research got the result which is opposite to what one would expect based from a rational decision-maker: that the more risk, the more preferable the alternative. The researchers got the same result when they extended the survey to other young people, beyond the university students.

However, once the researchers repeated their survey with a general population, the results became fully in line with what was observed in other cities: that many people are risk-averse. The only non-risk-averse part of the population were young people.

They are so much non-risk-averse that they often seem to be risk-prone: they prefer a much risker route.

Our objective: use this feature in education. As university faculty, we are very much interested in effectiveness of the university education – and of course, when we prepare teachers for schools, we are also interested in making these future teachers as effective as possible.

The majority of students are young people. From this viewpoint, the more we know about the young people psychology, the better we can adjust our lesson so that they are most efficient. Being risk-prone is such a big part of the young people culture that it is definitely imperative to take this feature into account when developing teaching strategies.

Why are students risk-prone? To take this feature into account, it is desirable to understand it better. So, we arrive at a natural question: why are young people risk-prone?

To answer this question, let us analyze how people make decisions, and what is so different about young people.

Decision making: brief reminder. Let us recall what decision theory tells us re how decisions are made in the first place; see, e.g., [4, 4–8].

Intuitively, a decision maker selects an alternative which is the best. So, to describe decision making in quantitative terms, we need to have a numerical description of how good different alternatives are to a given decision maker.

Decision theory is based on the following natural scale. We select two alternatives:

- a very bad alternative A_- and
- a very good alternative A_+,

so that every other alternative is in between A_- and A_+.

For every real number $p \in [0, 1]$, we can form a lottery $L(p)$ in which we get A_+ with probability p and A_- with the remaining probability $1 - p$. The larger the probability p of getting a good alternative, the more preferable the lottery. So, if $p < p'$, then $L(p) < L(p')$, where $A < A'$ means that to the decision maker, the alternative A' is better than the alternative A.

Let us now describe the quality of a given alternative A. When $p = 0$, the lottery $L(p)$ coincides with the very bad alternative A_- for which $A_- < A$, i.e., $L(0) < A$. When $p = 1$, the lottery $L(p)$ coincides with the very good alternative A_+ for which $A < A_+$, i.e., $A < L(1)$. Let $u(A)$ denote the supremum (least upper bound) of the set of all the values p for which $L(p) < A$. Then, one can show that:

- for all values $p < u(A)$, we have $L(p) < A$; and
- for all values $p > u(A)$, we have $A < L(p)$.

Indeed, let us assume that $p < u(A)$. In this case, the midpoint $\dfrac{p + u(A)}{2}$ is between p and $u(A)$:

$$ p < \frac{p + u(A)}{2} < u(A). $$

Since $u(A)$ is the least upper bound of the set

$$ \{q : L(q) < A\}, $$

the smaller value $\dfrac{p + u(A)}{2}$ is not an upper bound for this set, i.e., there exists a value q for which $L(q) < A$ and $q \nleq \dfrac{p + u(A)}{2}$, i.e., $\dfrac{p + u(A)}{2} < q$. From

$$ p < \frac{p + u(A)}{2} < q, $$

we conclude that $p < q$ and thus, $L(p) < L(q)$. Hence $L(q) < A$ implies $L(p) < A$.

Similarly, when $p > u(A)$, the midpoint $\dfrac{p + u(A)}{2}$ is between $u(A)$ and p: $u(A) < \dfrac{p + u(A)}{2} < p$. Since $u(A)$ is the least upper bound for the set $\{q : L(q) < A\}$, and $\dfrac{p + u(A)}{2} \nleq u(A)$, this means that $L\left(\dfrac{p + u(A)}{2}\right) \nleq A$, i.e., $A \leq L\left(\dfrac{p + u(A)}{2}\right)$. From $\dfrac{p + u(A)}{2} < p$, we can now conclude that $L\left(\dfrac{p + u(A)}{2}\right) < L(p)$, so from $A \leq L\left(\dfrac{p + u(A)}{2}\right)$, we conclude that $A < L(p)$.

For each real number $\varepsilon > 0$, the alternative A is better than the lottery $L(u(A) - \varepsilon)$ and worse than the lottery $L(u(A) + \varepsilon)$. This is true for values which are as small as we want. It is thus reasonable to say that A is *equivalent* to $L(u(A))$; we will denote this equivalence by $A \sim L(u(A))$. The corresponding value $u(A)$ is called the *utility* of the alternative A.

Suppose now that we have a lottery L in which we get alternative A_1 with probability p_1, alternative A_2 with the probability p_2, ..., and the alternative A_n with probability p_n. For every i from 1 to n, let $u(A_i)$ be the utility of the alternative A_i. What is the utility of the lottery L?

By definition of the utility, each alternative A_i is equivalent to the lottery $L(u(A_i))$ in which we get A_+ with probability $u(A_i)$ and A_- with the remaining probability $1 - u(A_i)$. Thus, the lottery L is equivalent to a two-stage lottery, in which:

- we first select i from 1 to n with probability p_i, and
- then, depending on the selection of i, select A_+ with probability $u(A_i)$ and A_- with the remaining probability.

As a result of this two-stage lottery, we get either A_+ or A_-. So, the utility of the lottery L is equal to the probability $u(L)$ of getting A_+ in this two-stage lottery. This probability can be estimated as the sum of the probabilities to get A_+ under the condition that different values $i = 1, \ldots, n$ were selected at the first stage:

$$u(L) = p(1 \text{ selected at 1st stage \& } A_+ \text{ selected at 2nd stage}) + \cdots +$$

$$p(n \text{ selected at 1st stage \& } A_+ \text{ selected at 2nd stage}).$$

Each of the corresponding probabilities can be described in terms of conditional probabilities:

$$p(i \text{ selected at 1st stage \& } A_+ \text{ selected at 2nd stage}) =$$

$$p(i \text{ selected at 1st stage}) \times p(A_+ \text{ selected at 2nd stage} \,|\, i \text{ selected at 1st stage}),$$

i.e.,

$$p(i \text{ selected at 1st stage \& } A_+ \text{selected at 2nd stage}) = p_i \cdot u(A_i).$$

Thus, the overall probability $u(L)$ to get A_+ is equal to

$$u(L) = p_1 \cdot u(A_1) + \ldots + p_n \cdot u(A_n).$$

In mathematical terms, the right-hand side of this formula is the expected value of the utility $u(A_i)$. Thus, the utility of the lottery is equal to the expected value of the utilities of different alternatives.

Utility is not uniquely defined. The numerical value of the utility depends on the selection of two alternatives A_- and A_+. What happens if we select different alternatives? For example, what if we select alternatives $A'_- < A_-$ and $A'_+ > A_+$, and select lotteries $L'(p)$ and define utilities u' based on these new alternatives?

In this case, both original selections A_- and A_+ are equivalent to lotteries in terms of A'_- and A'_+, i.e., $A_- \sim L'(p_-)$ and $A_+ \sim L'(p_+)$ for some p_- and p_+.

By definition, a utility u of an event is the probability for which the event is equivalent to a lottery $L(p)$, a lottery in which A_+ appears with probability u and A_- appears with probability $1 - u$. Since:

- the original alternative A_- is, in its turn, equivalent to the lottery $L'(p_-)$ in which A'_+ appears with probability p_-, and
- the original alternative A_+ is, in its turn, equivalent to the lottery $L'(p_+)$ in which A'_+ appears with probability p_+,

the original event is equivalent to the lottery $L'(u')$, in which the new alternative A'_+ appears with probability

$$u' = u \cdot p_+ + (1 - u) \cdot p_i.$$

Thus, when we change a scale, the new utility u' is a linear function of the old one:

$$u' = a \cdot u + b$$

for some $a > 0$ and b.

Strictly speaking, we only proved this for the case when $A'_- < A_-$ and $A'_+ > A_+$, but if we have two other scales, we can always compare each of them with a new scale in which $A''_- < A_-$, A'_- and $A''_+ > A_+$, A'_+. In this case,

- transition from u to u'' is linear,
- transition from u'' to u is also linear, and thus,
- the transition from u to u' is linear as well – as a composition of two linear functions.

Vice versa, for every $a > 0$ and b, we can find the new alternatives A'_- and A'_+ for which $u' = a \cdot u + b$. Thus, the utility is defined modulo a linear transformation.

Comment. This non-uniqueness is similar to non-uniqueness in describing the numerical values of such quantities as time or temperature. Indeed, to describe different values of these quantities by numbers, we need to select:

- a starting point and
- a measuring unit.

Once we change the starting point and/or the measuring unit, we get different numerical values that are related to the original ones by a similar linear transformation $x' = a \cdot x + b$.

For example, the transformation from the temperature t_C in the Celsius scale and the temperature t_F in the Fahrenheit scale is described by the known formula $t_F = 1.8 \cdot t_C + 32$.

Decision making: summary. Each action has several possible consequences. Let n denote the number of such possible consequences, and let u_i denote the utility of the ith consequence. Then, if p_i is the probability of the ith consequence, we select an action for which the expected value $\sum_{i=1}^{n} p_i \cdot u_i$ is the largest.

What we need to know to make a decision. According to the above description, to make a decision, for each possible consequence, we need to know two things:

- its utility u_i describing how beneficial this consequence is for the decision maker, and
- its probability p_i.

The utility u_i describes to what extent the outcome is beneficial for the decision maker, something that any decision maker, experienced or not, can judge for him- or herself. On the other hand, the probability of different events is something about which we can have more knowledge or less knowledge, and how well we know these probabilities depends on our experience.

How can we determine the probabilities of different consequences? A usual way to find the probability p_i of an event is to make several (N) observations and to estimate p_i as the frequency with which this event occurs, i.e., as the ratio

$$p_i \approx \widetilde{p}_i \stackrel{\text{def}}{=} \frac{N_i}{N},$$

where N_i is the number of cases when the ith event occurred.

It is known (see, e.g., [9]) that the expected value of this estimate is $p_i = E[\widetilde{p}_i]$, and the standard deviation is equal to

$$\sigma_i = \sqrt{E[(\Delta p_i)^2]} = \sqrt{\frac{p_i \cdot (1 - p_i)}{N}},$$

where we denoted $\Delta p_i \stackrel{\text{def}}{=} \widetilde{p}_i - p_i$. The values Δp_i corresponding to different events are, for large n, practically independent.

How the uncertainty of these estimates affects the decision making. Let us describe how the uncertainty of these events affect decision-making. In the ideal world, we should take into account the actual probabilities p_i, and base our decisions based on the expected utility $u = \sum_{i=1}^{n} p_i \cdot u_i$.

In reality, we only know the approximate values $\widetilde{p}_i$ of the probabilities, i.e., we know that $p_i = \widetilde{p}_i - \Delta p_i$ for some random (unknown) differences Δp_i. Substituting this expression into the above formula for utility, we conclude that

$$u = \sum_{i=1}^{n} p_i \cdot u_i = \sum_{i=1}^{n} \widetilde{p}_i \cdot u_i - \sum_{i=1}^{n} \Delta p_i \cdot u_i,$$

i.e., $u = \widetilde{u} - \Delta u$, where $\widetilde{u} \stackrel{\text{def}}{=} \sum_{i=1}^{n} \widetilde{p}_i \cdot u_i$ and $\Delta u \stackrel{\text{def}}{=} \sum_{i=1}^{n} \Delta p_i \cdot u_i$.

For many practical problems, the number of alternatives n is large. For such problems, the value Δu is a linear combination of a large number of independent random variables Δp_i. According to the Central Limit Theorem (see, [9]) this implies that the distribution of Δu is close to normal. Since the mean value of each Δp_i is 0, the mean value of the linear combination is also 0, and its variance is equal to

$$E[(\Delta u)^2] = \sum_{i=1}^{n} E[(\Delta p_i)^2] \cdot u_i^2,$$

i.e., using the known formulas for $E[(\Delta p_i)^2]$,

$$E[(\Delta u)^2] = \sum_{i=1}^{n} \frac{p_i \cdot (1 - p_i)}{N} \cdot u_i^2.$$

For most of the events, the probabilities p_i are small, so in the first approximation, $p_i \cdot (1 - p_i) \approx p_i$, and

$$E[(\Delta u)^2] = \frac{1}{N} \cdot \sum_{i=1}^{n} p_i \cdot u_i^2.$$

So, the standard deviation $\sigma = \sqrt{E[(\Delta u)^2]}$ is equal to

$$\sigma \approx \frac{1}{\sqrt{N}} \cdot \sqrt{\sum_{i=1}^{n} p_i \cdot u_i^2}.$$

Thus, the only information that we have about the actual (unknown) value of the expected utility u of an action is that u is (approximately) normally distributed with a known mean $\tilde{u}$ and a known standard deviation σ.

It is known that for a normal distribution with mean a and standard deviation σ:

- with probability 90%, the actual value of the random variable is in the interval

$$[a - 2\sigma, a + 2\sigma];$$

- with probability 99.9%, the actual value of the random variable is in the interval

$$[a - 3\sigma, a + 3\sigma];$$

- with probability $1 - 10^{-8}$, the actual value of the random variable is in the interval

$$[a - 6\sigma, a + 6\sigma];$$

- etc.

Thus, based on N observations, we can conclude, with certain confidence, that the actual (unknown) value of the expected utility u belongs to the interval

$$[\widetilde{u} - k_0 \cdot \sigma, \widetilde{u} + k_0 \cdot \sigma],$$

where

- for $k_0 = 2$, we get confidence 90%;
- for $k_0 = 3$, we get confidence 99.9%;
- for $k_0 = 6$, we get confidence $1-10^{-8}$, etc.

So, instead of the *exact* values of the utility, we now have an *interval* of possible values of the utility. How can we make decisions based on such intervals?

Decision making under interval uncertainty: brief reminder. In the previous section, we encountered a situation in which we do not know the exact value u of the utility, we only know that this value belongs to the interval $[\underline{u}, \overline{u}]$. The problem of decision making under such interval uncertainty was first handled by the future Nobelist L. Hurwicz in [10].

As we have mentioned earlier, the preference of each situation can be described by a utility value. Thus, to describe decisions under interval uncertainty, we must assign, to each such interval $[\underline{u}, \overline{u}]$, a utility value $u(\underline{u}, \overline{u})$.

No matter what value we get from this interval, this value will be larger than or equal to $\underline{u}$ and smaller than or equal to $\overline{u}$. Thus, the equivalent utility value $u(\underline{u}, \overline{u})$ must satisfy the same inequalities: $\underline{u} \le u(\underline{u}, \overline{u}) \le \overline{u}$. In particular, for $\underline{u} = 0$ and $\overline{u} = 1$, we get $0 \le \alpha \le 1$, where we denoted $\alpha \stackrel{\text{def}}{=} u(0, 1)$.

We have mentioned that the utility is determined modulo a linear transformation $u' = a \cdot u + b$. It is therefore reasonable to require that the equivalent utility does not depend on what scale we use, i.e., that for every $a > 0$ and b, we have

$$u(a \cdot \underline{a} + b, a \cdot \overline{u} + b) = a \cdot u(\underline{u}, \overline{u}) + b.$$

In particular, for $\underline{u} = 0$ and $\overline{u} = 1$, we get

$$u(b, a + b) = a \cdot u(0, 1) + b = a \cdot \alpha + b.$$

So, for every $\underline{u}$ and $\overline{u}$, we can take $b = \underline{u}$, $a = \overline{u} - \underline{u}$, and get

$$u(\underline{u}, \overline{u}) = \underline{u} + \alpha \cdot (\overline{u} - \underline{u}) = \alpha \cdot \overline{u} + (1 - \alpha) \cdot \underline{u}.$$

This expression is called *Hurwicz optimism-pessimism criterion*, because:

- when $\alpha = 1$, we make a decision based on the most optimistic possible values $u = \overline{u}$;
- when $\alpha = 0$, we make a decision based on the most pessimistic possible values $u = \underline{u}$;
- for intermediate values $\alpha \in (0, 1)$, we take a weighted average of the optimistic and pessimistic values.

It is worth mentioning that most people are more optimists than pessimists in the sense that the weight α of the optimistic case is usually larger than the weight $1 - \alpha$ of the pessimistic case: $\alpha < 1 - \alpha$, i.e., equivalently, $2 \cdot \alpha > 1$ and $\alpha > 0.5$.

Let us apply Hurwicz criterion to our problem. In our case, $\underline{u} = \widetilde{u} - k_0 \cdot \sigma$ and $\underline{u} = \widetilde{u} + k_0 \cdot \sigma$, where

$$\widetilde{u} = \sum_{i=1}^{n} \widetilde{p}_i \cdot u_i.$$

Hence, the equivalent utility is equal to

$$\alpha \cdot \overline{u} + (1 - \alpha) \cdot \underline{u} = \alpha \cdot (\widetilde{u} - k_0 \cdot \sigma) + (1 - \alpha) \cdot (\widetilde{u} + k_0 \cdot \sigma) = \widetilde{u} + (2\alpha - 1) \cdot k_0 \cdot \sigma,$$

where $k_0 = 2$, 3, or 6 (or some other similar number, depending on the desired confidence level), and

$$\sigma \approx \frac{1}{\sqrt{N}} \cdot \sqrt{\sum_{i=1}^{n} p_i \cdot u_i^2}.$$

This formula enables us to produce the desired explanation of why young people are risk-prone.

In these terms, what distinguishes young people from others. The main difference between young people and the general population is that the young people are less experienced, i.e., in our terms, they have encountered few situations N.

Consequences of the above formula for the general population. For people with experience, the value N is large, thus σ is small, and the resulting effective utility $u \approx \widetilde{u}$ is determined by the usual expected utility formula $\sum_{i=1}^{n} \widetilde{p}_i \cdot u_i$.

Consequences for young people. For young people, the value N is small, so we can no longer ignore the σ terms in comparison with the expected utility term $\widetilde{u}$.

In the extreme case, when the σ-term is dominant, we select the alternative for which this term is the largest, i.e., equivalently, for which the expected value

$$\sum_{i=1}^{n} p_i \cdot u_i^2$$

of the *squared* utility is the largest.

And here is where risk-proneness comes into picture. Let us assume, e.g., that we are talking about monetary outcomes, and that the utility value is proportional to the monetary amount. Let us assume that we have 10 different alternatives the probability of each of which is 0.1. Let us consider the following two situations.

In the first situation, in each alternative, the person get the amount 0.5. In this situation, there is no risk, we get 0.5 no mater what. In this case, the expected utility is 0.5, and

$$\sum_{i=1}^{n} p_i \cdot u_i^2 = 0.25.$$

In the second situation, in the first five alternatives, the person gets the amount 0, and in the second five alternatives, she gets the utility 1. In this situation, there is a high risk: with probability 0.5, we get nothing. The expected utility is still the same

$$5 \cdot 0.1 \cdot 0 + 5 \cdot 0.1 \cdot 1 = 0.5,$$

but now we have

$$\sum_{i=1}^{n} p_i \cdot u_i^2 = 5 \cdot 0.1 \cdot 0 + 5 \cdot 0.1 \cdot 1 = 0.5.$$

From the viewpoint of the sum $\sum_{i=1}^{n} p_i \cdot u_i^2$, the second (risk-prone situation) is clearly preferable to the previous one, so a young person will prefer it. Similarly, if we take into account both terms, the risk-prone strategy is clearly preferable. This explains why young people are risk-prone.

In more general mathematical terms, when we maximize the expected value of some function $y = f(x)$ of the monetary value x, risk-averse means that a person prefers to receive the average $\sum_{i=1}^{n} p_i \cdot x_i$ with probability 1 rather than participate in the lottery in which she gets x_i with probability p_i:

$$f\left(\sum_{i=1}^{n} p_i \cdot x_i\right) \geq \sum_{i=1}^{n} p_i \cdot f(x_i).$$

Functions with this property are called *concave*. In contrast, for the function $f(x) = x^2$, the opposite inequality is true because this function is *convex*. The fact that we naturally got a convex function shows that young people are indeed risk-prone.

How we can use this conclusion: ideas. Based on the fact that young people prefer risky situations, in which the results depend on a random selection, a good strategy is to introduce as much randomness into teaching as possible, e.g.:

- use surprise quizzes – in addition to normally scheduled tests and quizzes;
- for class activities, group students into randomly selected groups – instead of trying to group them into most effective groups;
- assign larger homeworks, with more problems than a Teaching Assistant and/or a professor can grade before the next class – with an understanding that only problems with randomly selected numbers will be graded.

Possibilities are unlimited, and, as our experience shows, excitement (and hence improvement) is guaranteed.

Possible future work. Our main focus was on the explanation. Of course, since we have a quantitative model, the next step would not just to give qualitative recommendations, but to use his model to provide *quantitative* recommendations: how much randomness should be introduce to make education most efficient.

References

1. K. Villaverde, O. Kosheleva, Why are students risk-prone, in *Proceedings of the World Conference on Soft Computing*, ed. by R.R. Yager, M.Z. Reformat, S.N. Shahbazova, S. Ovchinnikov (California, San Francisco, 2011), pp. 23–26
2. K. Villaverde, O. Kosheleva, Why are young people risk-prone. Int. J. Innov. Manag. Inf. Product. (IJIMIP) **2**(1), 118–125 (2012)
3. R. Cheu, V. Kreinovich, Y.-C. Chiu, R. Pan, G. Xiang, S.V. Bhupathiraju, S. Rao Manduva, *Strategies for Improving Travel Time Reliability*, Texas Department of Transportation, Research Report 0-5453-R2 (2007)
4. P.C. Fishburn, *Utility Theory for Decision Making* (Wiley, New York, 1969)
5. P.C. Fishburn, *Nonlinear Preference and Utility Theory* (The John Hopkins Press, Baltimore, 1988)
6. R.L. Keeney, H. Raiffa, *Decisions with Multiple Objectives* (Wiley, New York, 1976)
7. R.D. Luce, H. Raiffa, *Games and Decisions: Introduction and Critical Survey* (Dover, New York, 1989)
8. H. Raiffa, *Decision Analysis* (Addison-Wesley, Reading, Massachusetts, 1970)
9. D.J. Sheskin, *Handbook of Parametric and Nonparametric Statistical Procedures* (Chapman & Hall/CRC, Boca Raton, Florida, 2007)
10. L. Hurwicz, Optimality criteria for decision making under ignorance, cowles commission discussion paper, Statistics. **370**, (1951)

Chapter 31
How to Assess Individual Contributions to a Group Project

In modern education, a lot of students' efforts goes into group projects. In many real-life situation, the only information that we have for estimating the individual contributions E_j to a group project consists of individual estimates e_{ij} of contributions of other participants j. In this chapter, we describe a new faster algorithm for estimating individual contributions to a group project based on the estimates e_{ij}.

The results from this chapter first appeared in [1].

Formulation of the practical problem. How can we estimate individual contributions to a group project? This problem is important in education when several students work together on a project, it is important in the business environment when several people work together on a joint project.

In all such situations, we need to know the relative contributions $E_1, \ldots, E_n$ of all n participants, relative in the sense that they represent the fraction of the overall credit – and thus, the sum of these contributions should be equal to 1:

$$\sum_{i=1}^{n} E_i = 1. \tag{31.1}$$

Available information for solving the problem. In many practical situation, the only available information for estimating contributions consists of the estimates that different participants give to each other's contribution. In this case, we have n^2 values e_{ij} ($1 \le i, j \le n$) – estimates made by the i-th participate of the contribution of the j-th participant.

Ideal case when all estimates are unbiased. In the ideal case when all estimates are unbiased, for each participant j, we have n estimates $e_{1j}, \ldots, e_{nj}$ for the desired value E_j. In this case, we have n approximate equalities to find E_j:

$$E_j \approx e_{1j}, \quad \ldots, \quad E_j \approx e_{nj}. \tag{31.2}$$

© Springer-Verlag GmbH Germany 2018
O. Kosheleva and K. Villaverde, *How Interval and Fuzzy Techniques Can Improve Teaching*, Studies in Computational Intelligence 750,
https://doi.org/10.1007/978-3-662-55993-2_31

279

To find a reasonable estimate E_j from these approximate equalities, a natural idea is to use the Least Squares technique and find the value E_j for which the sum

$$\sum_{i=1}^{n}(E_j - e_{ij})^2 \tag{31.3}$$

is the smallest possible. This is a textbook use of the Least Squares method, to combine several estimates of the same quantity, and the solution to this optimization problem is well known – it is the arithmetic average of these estimates:

$$E_j = \frac{e_{1j} + \ldots + e_{nj}}{n}. \tag{31.4}$$

Problem: self-estimates are biased. In practice, estimates of others' contributions are often unbiased, but it is very difficult to get an unbiased estimate of one's own contribution in comparison with the contributions of others.

In other words, while the estimates e_{ij} for $i \neq j$ are really unbiased, the estimates e_{ii} are too subjective and biased to be useful.

Available information for solving the problem: revisited. Because of the bias, we can say that the only available information about each value E_j consists of estimates e_{ij} with $i \neq j$.

What was known before. The problem of estimating the values E_j based on the estimates e_{ij} with $i \neq j$ was considered earlier; see, e.g., [2, 3] and references therein. In particular, [3] describes an algorithm for estimating E_j.

Limitations of the previous approaches. The algorithm from [3] is based on solving a highly non-linear optimization problem and is, therefore, reasonably time-consuming.

What we do in this chapter. In this chapter, we use ideas from [2] to come up with a much faster algorithm for estimating E_j, an algorithm whose most time-consuming step is solving a system of linear equations.

Towards the formulation of the problem in precise terms. In principle, each participant i provides *relative* estimates $e_{i1}, \ldots, e_{in}$ which add up to 1. Once we have *absolute* estimates $a_{i1}, \ldots, a_{in}$, we get the relative contributions e_{ij} by dividing each value a_{ij} by the total contribution $a_{i1} + \ldots + a_{in}$:

$$e_{ij} = \frac{a_{ij}}{a_{i1} + \ldots + a_{in}}. \tag{31.5}$$

If all these estimates were unbiased, then each estimate e_{ij} would provide an unbiased estimate of E_j.

In particular, in the idealized case when all estimates are exact, we have $a_{ij} = c_i \cdot E_j$ for some value c_i. Thus, for the normalized values, we get

$$e_{ij} = \frac{c_i \cdot E_j}{c_i \cdot E_1 + \ldots + c_i \cdot E_n} = \frac{E_j}{E_1 + \ldots + E_n}, \tag{31.6}$$

hence $e_{ij} = E_j$ and

$$e_{i1} + \ldots + e_{in} = E_1 + \ldots + E_n = 1. \tag{31.7}$$

In practice, as we have mentioned, each self-estimate a_{ii} is biased: $a_{ii} \neq c_i \cdot E_i$. As a result, even when the i-th participant provides the exact *absolute* values of the contributions $c_i \cdot E_j$ for $i \neq j$, the corresponding *relative* (normalized) contribution

$$e_{ij} = \frac{a_{ij}}{a_{i1} + \ldots + a_{in}} =$$

$$\frac{c_i \cdot E_j}{c_i \cdot E_1 + \ldots + c_i \cdot E_{i-1} + a_{ii} + c_i \cdot E_{i+1} + \ldots + c_i \cdot E_n} = \frac{E_j}{s_i} \tag{31.8}$$

(where

$$s_i = E_1 + \ldots + E_{i-1} + \frac{a_{ii}}{c_i} + E_{i+1} + \ldots + E_n,) \tag{31.9}$$

is different from E_j: instead of the correct value

$$E_j = \frac{E_j}{E_1 + \ldots + E_{i-1} + E_i + E_{i+1} + \ldots + E_n} = \frac{E_j}{1}, \tag{31.10}$$

we get the new value

$$e_{ij} = \frac{E_j}{s_i}, \tag{31.11}$$

where, due to $a_{ii} \neq c_i \cdot E_i$, we have

$$s_i \overset{\text{def}}{=} E_1 + \ldots + E_{i-1} + \frac{a_{ii}}{c_i} + E_{i+1} + \ldots + E_n \neq 1. \tag{31.12}$$

Thus, the relative estimates e_{ij} of j-th contribution will be equal not to E_j, but to the normalized value (31.11). Since the estimates are only approximate, we have an approximate equality

$$e_{ij} \approx \frac{E_j}{s_i}, \tag{31.13}$$

or, equivalently,

$$E_j \approx e_{ij} \cdot s_i. \tag{31.14}$$

Here, the values e_{ij} are known, while the estimates E_j and the values s_i are unknown. The Least Squares approach now means that we minimize the sum of the squares of the discrepancies

$$\sum_{i \neq j} (E_j - e_{ij} \cdot s_i)^2 \tag{31.15}$$

under the constraint (31.1). Thus, we arrive at the following problem.

Precise formulation of the problem. Minimize the expression (31.15) under the constraint (31.1).

Towards solving the problem. First, let us use the Lagrange multiplier method to reduce the above constrained optimization problem to the following un-constrained one: minimize

$$J \overset{\text{def}}{=} \sum_{i \neq j} (E_j - e_{ij} \cdot s_i)^2 + \lambda \cdot \left(\sum_{i=1}^{n} E_i - 1 \right), \tag{31.16}$$

where the Lagrange multiplier λ must be chosen in such a way that the constraint (31.1) is satisfied.

Since the function J attains minimum, its partial derivatives must be equal to 0. Differentiating J with respect to E_j and equating the derivative to 0, we conclude that

$$\frac{\partial J}{\partial E_j} = 2 \sum_{i \neq j} (E_j - e_{ij} \cdot s_i) + \lambda = 0. \tag{31.17}$$

Dividing both sides of this equality by 2 and taking into account that E_j is repeated for each $i \neq j$, i.e., $n - 1$ times, we conclude that

$$(n-1) \cdot E_j - \sum_{i \neq j} e_{ij} \cdot s_i = -\frac{\lambda}{2}. \tag{31.18}$$

This equation can be somewhat simplified if we take $e_{ii} \overset{\text{def}}{=} 0$; then, the sum over all $i \neq j$ can be simply described as the sum over all i:

$$(n-1) \cdot E_j - \sum_{i=1}^{n} e_{ij} \cdot s_i = -\frac{\lambda}{2}. \tag{31.19}$$

It should be mentioned that the values $e_{ii} = 0$ are selected *only* for the purpose of simplifying computations; these values *do not* mean that we somehow think that each participate estimates his or new own contribution as 0.

Differentiating J with respect to s_i and equating the derivative to 0, we conclude that

$$\frac{\partial J}{\partial s_i} = 2 \sum_{j\neq i} (E_j - e_{ij} \cdot s_i) \cdot e_{ij} = 0. \tag{31.20}$$

Dividing both sides of this equality by 2, we get

$$s_i \cdot \sum_{j\neq i} e_{ij}^2 = \sum_{j\neq i} E_j \cdot e_{ij}, \tag{31.21}$$

or, equivalently, that

$$s_i = \frac{\sum\limits_{\ell\neq i} E_\ell \cdot e_{i\ell}}{\sum\limits_{m\neq i} e_{im}^2}. \tag{31.22}$$

(Here, for convenience of the following transformations, we renamed the indices in the two sums into two different ones.) By taking $e_{ii} = 0$, we can simplify this expression into the following one:

$$s_i = \frac{\sum\limits_{\ell=1}^{n} E_\ell \cdot e_{i\ell}}{\sum\limits_{m=1}^{n} e_{im}^2}. \tag{31.23}$$

Substituting the expression (31.23) into the formula (31.19), we conclude that

$$(n-1)\cdot E_j - \sum_{i=1}^{n} e_{ij} \cdot \frac{\sum\limits_{\ell=1}^{n} E_\ell \cdot e_{i\ell}}{\sum\limits_{m=1}^{n} e_{im}^2} = -\frac{\lambda}{2}, \tag{31.24}$$

i.e.,

$$(n-1)\cdot E_j - \sum_{\ell=1}^{n} c_{j,\ell} \cdot E_\ell = -\frac{\lambda}{2}, \tag{31.25}$$

where

$$c_{j,\ell} \stackrel{\text{def}}{=} \sum_{i=1}^{n} \frac{e_{ij} \cdot e_{i\ell}}{c_i}, \tag{31.26}$$

where

$$c_i \stackrel{\text{def}}{=} \sum_{m=1}^{n} e_{im}^2. \tag{31.27}$$

The values $c_{j,\ell}$ can be explicitly computed from the known values e_{ij}. To solve the resulting system of linear equations (31.25) with the unknown value λ, it is sufficient to solve it for $\lambda = -2$, when $-\dfrac{\lambda}{2} = 1$, and then multiply the resulting values e_j by a common factor in such a way that the new values add up to 1: i.e., take

$$E_j = \frac{e_j}{S}, \tag{31.28}$$

where

$$S \overset{\text{def}}{=} \sum_{i=1}^{n} e_i. \tag{31.29}$$

Thus, we arrive at the following algorithm.

Algorithm for solving the problem. Given the values e_{ij} for $i \neq j$, we first take $e_{ii} \overset{\text{def}}{=} 0$. Then, we compute the values

$$c_i \overset{\text{def}}{=} \sum_{m=1}^{n} e_{im}^2 \tag{31.30}$$

and

$$c_{j,\ell} \overset{\text{def}}{=} \sum_{i=1}^{n} \frac{e_{ij} \cdot e_{i\ell}}{c_i}, \tag{31.31}$$

and solve the following system of linear equations

$$(n-1) \cdot e_j - \sum_{\ell=1}^{n} c_{j,\ell} \cdot e_\ell = 1. \tag{31.32}$$

After this, we compute the sum

$$S = \sum_{i=1}^{n} e_i. \tag{31.33}$$

and then, the desired estimates as

$$E_j = \frac{e_j}{S}. \tag{31.34}$$

Discussion. In this algorithm, the most time-consuming step is solving a system of linear equations. Thus, this algorithm is indeed much faster than the algorithm from [3] that requires a solution of the non-linear optimization problem.

References

1. J. Sappakitkamjorn, W. Suriyakat, W. Suracherdkiati, O. Kosheleva, How to estimate individual contributions to a group project. J. Uncertain Syst. **4**(4), 301–305 (2010)
2. V. Kreinovich, How to Pay for Collective Creative Activity? (Technical Report, Informatika Research Center, St. Petersburg, Russia, 1989)
3. Y.-W. Leung, Least-squar-error estimate of individual contribution in group project. IEEE Trans. Educ. **41**(4), 282–284 (1998)

Chapter 32
How to Access Students's Readiness for the Next Class

How many points should we allocate to different assignments and tests? to different problems on a test? Usually, professors use subjective judgment to allocate points. In this chapter, for classes which are pre-requisites for others, we provide an objective procedure for allocating points.

The results from this chapter first appeared in [1].

Points need to be allocated. The overall grade for a class is formed by adding the grades for different assignments and tests. In the syllabus, it is usually described that, e.g., the first test is worth 10 points, the second test is worth 25 points, etc., with the total of 100 points.

Example For a better understanding, let us start with a simple example. Suppose that we allocate:

- 10 points to the first test,
- 40 points to the second test, and
- 50 points to the final exam.

Suppose that a student s got:

- $g_{s1} = 80$ points out of 100 on the first test,
- $g_{s2} = 75$ points out of 100 on the second test, and
- $g_{s3} = 90$ points out of 100 on the final exam.

In this case, the student's overall grade g_s for this class is

$$g_s = p_1 \cdot g_{s1} + p_2 \cdot g_{s2} + p_3 \cdot g_{s3} =$$

$$0.1 \cdot 80 + 0.4 \cdot 75 + 0.5 \cdot 90 = 8.0 + 30.0 + 45.0 = 83, \qquad (32.1)$$

© Springer-Verlag GmbH Germany 2018
O. Kosheleva and K. Villaverde, *How Interval and Fuzzy Techniques Can Improve Teaching*, Studies in Computational Intelligence 750,
https://doi.org/10.1007/978-3-662-55993-2_32

where

$$p_1 = \frac{10}{100} = 0.1, \quad p_2 = \frac{40}{100} = 0.4, \quad p_3 = \frac{50}{100} = 0.5. \qquad (32.2)$$

Points need to be allocated (cont-d). Similarly, the grade for a test is formed by adding the grades earned on each of the problems. In the text of the test, it is usually described how many points each problem is worth.

Allocating points is important. An appropriate allocation of points is very important. For example, if we allocate almost all the points to the final exam, some students will see no reason to study hard during the semester. They will try to cram the material during the finals, and as a result, even if they pass the finals, their knowledge of the material will not be as good as the knowledge of the students who studied diligently during the semester.

On the other hand, if we allocate too few points to the final exam, then some students will have no incentive to review the class material for the final exam. As a result, their knowledge of the material that was studied in the beginning of the semester will be not as good as the knowledge of those students who did review this material at the end of the class.

How points are allocated now. At present, the points for different assignments, tests, and problems are allocated based on the teacher's subjective experience. As a result, there is a high variety of point allocation.

Formulation of the problem. Since it is very important to properly allocate points for different assignments, tests, and problems, it is desirable to find a more objective ways for such allocation.

Such an objective way is presented in this chapter. This chapter capitalizes on the preliminary ideas described in [2].

Our main idea. Most classes are prerequisites for other classes (or for some qualification exams). For example, Pre-calculus is a prerequisite for Calculus I, Calculus I is a prerequisite for Calculus II, etc. For such classes, it is desirable to allocate the points in such a way that a success in this class will be a good indication of the success in the next class.

If we grade too easily, students may be happier with their good grades, but some of them will pass the class without acquiring the knowledge needed for the next class – and so they may fail this next class. On the other hand, if we grade too harsh, we unnecessarily fail many students who may have not learned all the details but whole knowledge is actually good enough to successfully pass the next class.

From this viewpoint, the best way to allocate points is to select the allocations for which the resulting grade is the best predictor for the grade in the next class.

Available data. To find the best allocation of points, we must use the grades that the students got for different assignments and tests, and the grades they got in the next class.

Let us denote the total number of assignments and tests by T, and let us denote the total number of students who took this class in the past by S. Let us denote the grade of student s ($1 \leq s \leq S$) on assignment or test t ($1 \leq t \leq T$) by g_{st}. The grade of student s at the next class (for which this class is a prerequisite) will be denoted by n_s.

From the idea to the algorithm. We want to predict the value n_s based on the grades $g_{s1}, \ldots, g_{sT}$. For such a prediction, it is natural to start with a linear regression

$$n_s \approx a_0 + a_1 \cdot g_{s1} + \ldots + a_T \cdot g_{sT}. \tag{32.3}$$

The coefficients a_t can be found from the Least Squares method (see, e.g., [3]), by minimizing the sum

$$\sum_{s=1}^{S} (a_0 + a_1 \cdot g_{s1} + \ldots + a_T \cdot g_{sT} - n_s)^2. \tag{32.4}$$

The standard Least Squares formulas lead to the following system of linear equations for determining the coefficients a_t:

$$a_0 + a_1 \cdot \overline{g_1} + \ldots + a_T \cdot \overline{g_T} = \overline{n}; \tag{32.5}$$

$$a_t + a_1 \cdot \overline{g_1 \cdot g_t} + \ldots + a_T \cdot \overline{g_T \cdot g_t} = \overline{n \cdot g_t} \quad 1 \leq t \leq T, \tag{32.6}$$

where

$$\overline{g_t} \stackrel{\text{def}}{=} \frac{1}{S} \sum_{s=1}^{S} g_{st}, \quad \overline{n} \stackrel{\text{def}}{=} \frac{1}{S} \sum_{s=1}^{S} n_s, \quad \overline{g_t \cdot g_{t'}} \stackrel{\text{def}}{=} \frac{1}{S} \sum_{s=1}^{S} g_{st} \cdot g_{st'}. \tag{32.7}$$

The resulting coefficients are not exactly points, since the values $a_1, \ldots, a_T$ do not necessarily add up to 1. To get the desired values of the points $p_1, \ldots, p_T$, we therefore need to normalize these values and take

$$p_t = \frac{a_t}{a}, \tag{32.8}$$

where

$$a \stackrel{\text{def}}{=} a_1 + \ldots + a_T. \tag{32.9}$$

For these values, we have $a_t = p_t \cdot a$ and thus, the above linear regression formula takes the form

$$n_s \approx a_0 + a_1 \cdot g_{s1} + \ldots + a_T \cdot g_{sT} = a_0 + a \cdot g_s, \tag{32.10}$$

where

$$g_s \overset{\text{def}}{=} p_1 \cdot g_{s1} + \ldots + p_T \cdot g_{sT}. \tag{32.11}$$

Thus, we arrive at the following algorithm.

Allocating points: main algorithm. Based on the grades $g_{s1}, \ldots, g_{sT}$, and n_s, we first compute the averages

$$\overline{g_t} \overset{\text{def}}{=} \frac{1}{S} \sum_{s=1}^{S} g_{st}, \quad \overline{n} \overset{\text{def}}{=} \frac{1}{S} \sum_{s=1}^{S} n_s, \quad \overline{g_t \cdot g_{t'}} \overset{\text{def}}{=} \frac{1}{S} \sum_{s=1}^{S} g_{st} \cdot g_{st'}. \tag{32.12}$$

Then, we solve the following system of $T+1$ linear equations to find $T+1$ unknowns $a_0, a_1 \ldots, a_T$:

$$a_0 + a_1 \cdot \overline{g_1} + \ldots + a_T \cdot \overline{g_T} = \overline{n}; \tag{32.13}$$

$$a_t + a_1 \cdot \overline{g_1 \cdot g_t} + \ldots + a_T \cdot \overline{g_T \cdot g_t} = \overline{n \cdot g_t} \ 1 \le t \le T, \tag{32.14}$$

After that, we compute the sum

$$a \overset{\text{def}}{=} a_1 + \ldots + a_T \tag{32.15}$$

and we compute the desired point values p_t as

$$p_t = \frac{a_t}{A}. \tag{32.16}$$

For each student s, the resulting class grade

$$g_s \overset{\text{def}}{=} p_1 \cdot g_{s1} + \ldots + p_T \cdot g_{sT} \tag{32.17}$$

can be used to predict the grade n_s in the next class as

$$n_s \approx a_0 + a \cdot g_s. \tag{32.18}$$

Discussion. One might expect that we would aim at finding the points for which $n_s \approx g_s$. However, it is reasonable to expect some decrease in knowledge between the classes, so $n_s \approx a_0 + a \cdot g_s$, with $a < 1$, is a more reasonable idea.

For example, in the University of Texas at El Paso's Computer Science (CS) undergraduate program, the passing grade for some classes is D (corresponding, crudely speaking, to 60 points out of 100). However, for the classes that serve as prerequisites to other classes, the passing grade is C (70 points out of 100), because it is understood that usually, student forget. So, to make sure that they have at least a D level knowledge of the material by the time they reach the next class, we must make sure that they have a C level by the time they finish the previous class.

Two iterations may be needed. In the previous text, we assumed that the grades for different assignments reflect the students' level of knowledge for different parts of the material. This is usually true when there are enough points allocated to this assignment. However, if too few points are allocated to an assignment, students may not spend as much time on it and thus, the corresponding grades may be low. If based on the above procedure, we allocate more points to this assignment, students may take it more seriously and their grades for this assignment may improve. This change in grades may require a re-allocation of points.

So, ideally, after we determine the initial grade allocations, we should then collect new grades data. After we have collected enough grade data, we should then repeat the same procedure – to see if re-allocation of points is necessary.

From the main algorithm to algorithms that cover more complex situations. In the derivation of the above algorithm, we assumed that we have a class with a well-defined next class, and that for this next class, the current class is the only prerequisite. We also assumed that linear regression is adequate.

Let us discuss what to do in more complex situations.

Case of multiple prerequisites. In some situations, there are several prerequisites for a class. For example, in the above-mentioned CS program, there are two prerequisites for the Data Structures class: Discrete Mathematics (DM) and Elementary Algorithms and Data Structures (CS2).

In such cases, we need to determine points for both prerequisite classes. Here, in addition to the grades $g_{s1}, \ldots, g_{sT}$ for the first class, we also have grades $g'_{s1}, \ldots, g'_{sT'}$ for the second class. We want to find the coefficients $a_0, a_1, \ldots, a_T, a'_1, \ldots, a'_{T'}$ for which

$$n_s \approx a_0 + a_1 \cdot g_{s1} + \ldots + a_T \cdot g_{sT} + a'_1 \cdot g'_{s1} + \ldots + a'_{T'} \cdot g_{sT'}. \tag{32.19}$$

The coefficients a_t and $a'_{t'}$ can be found from the Least Squares method, by minimizing the sum

$$\sum_{s=1}^{S} (a_0 + a_1 \cdot g_{s1} + \ldots + a_T \cdot g_{sT} + a'_1 \cdot g'_{s1} + \ldots + a'_{T'} \cdot g'_{sT'} - n_s)^2. \tag{32.20}$$

The standard Least Squares formulas lead to the following system of linear equations for determining the coefficients a_t and $a'_{t'}$:

$$a_0 + a_1 \cdot \overline{g_1} + \ldots + a_T \cdot \overline{g_T} + a'_1 \cdot \overline{g'_1} + \ldots + a_T \cdot \overline{g'_{T'}} = \overline{n}; \tag{32.21}$$

$$a_t + a_1 \cdot \overline{g_1 \cdot g_t} + \ldots + a_T \cdot \overline{g_T \cdot g_t} + a'_1 \cdot \overline{g'_1 \cdot g_t} + \ldots + a'_{T'} \cdot \overline{g'_{T'} \cdot g_t} = \overline{n \cdot g_t}; \tag{32.22}$$

$$a'_{t'} + a_1 \cdot \overline{g_1 \cdot g'_{t'}} + \ldots + a_T \cdot \overline{g_T \cdot g'_{t'}} + a'_1 \cdot \overline{g'_1 \cdot g'_{t'}} + \ldots + a'_{T'} \cdot \overline{g'_{T'} \cdot g'_{t'}} = n \cdot \overline{g'_{t'}},$$

$$(32.23)$$

where, in addition to (32.7), we have

$$\overline{g'_{t'}} \stackrel{\text{def}}{=} \frac{1}{S} \sum_{s=1}^{S} g'_{st'}, \quad \overline{g_t \cdot g'_{t'}} = \overline{g'_{t'} \cdot g_t} \stackrel{\text{def}}{=} \frac{1}{S} \sum_{s=1}^{S} g_{st} \cdot g'_{st'}, \qquad (32.24)$$

$$\overline{g'_t \cdot g'_{t'}} \stackrel{\text{def}}{=} \frac{1}{S} \sum_{s=1}^{S} g'_{st} \cdot g'_{st'}. \qquad (32.25)$$

To get the desired values of the points $p_1, \ldots, p_T, p'_1, \ldots, p'_{T'}$, we normalize these values and take

$$p_t = \frac{a_t}{a}, \quad p'_{t'} = \frac{a'_{t'}}{a'}, \qquad (32.26)$$

where

$$a \stackrel{\text{def}}{=} a_1 + \ldots + a_T; \quad a' \stackrel{\text{def}}{=} a'_1 + \ldots + a'_{T'}. \qquad (32.27)$$

For these values, we have $a_t = p_t \cdot a$ and $a'_{t'} = p'_{t'} \cdot a'$ and thus, the above linear regression formula takes the form

$$n_s \approx a_0 + a \cdot g_s + a' \cdot g'_s, \qquad (32.28)$$

where

$$g_s \stackrel{\text{def}}{=} p_1 \cdot g_{s1} + \ldots + p_T \cdot g_{sT}; \quad g'_s \stackrel{\text{def}}{=} p'_1 \cdot g'_{s1} + \ldots + p'_T \cdot g'_{sT'}. \qquad (32.29)$$

Case of several follow-up classes. Another realistic situation is when for a given class, there are several follow-up classes. For example, in the above-mentioned CS program, the Data Structures class is a pre-requisite for several junior- and senior-level classes. In this case, for each student s, we have several grades $n_{s1}, \ldots, n_{sC}$ for the follow-up classes.

To apply the above procedure, we can then take, as n_s,

- either the *smallest* of these follow-up grades

$$n_s = \min(n_{s1}, \ldots, n_{sC}); \qquad (32.30)$$

this will guarantee that the student is successful in *all* follow-up classes,
- or the *average* of these follow-up grades

$$n_s = \frac{n_{s1} + \ldots + n_{sC}}{C}; \qquad (32.31)$$

this will guarantee that the student is successful *on average* in the follow-up classes.

Case of non-linear regression. If we have reasons to believe that linear regression does not adequately capture the students' success in the next classes, we may want to use quadratic regression, i.e., find the coefficients a_t and $a_{tt'}$ for which

$$n_s \approx a_0 + \sum_{t=1}^{T} a_t \cdot g_{st} + \sum_{t=1}^{T} \sum_{t'=1}^{T} a_{tt'} \cdot g_{st} \cdot g_{st'}. \tag{32.32}$$

The optimal values of the coefficients a_t and $a_{tt'}$ can also be obtained from the Least Squares method. In this case, the grade for the class will be obtained not as a weighted average of the grades for different assignments and tests, but rather as a non-linear combination of these grades.

This non-linearity may sound unusual, but it is actually used in grading. For example, in some CS classes, to get a C, students need to get at least a C average for the lab assignments (assignments $t = 1, \dots, \ell$) and at least a C average for all the tests (assignments $t = \ell + 1, \dots, T$). Thus, in effect, we have a non-linear formula for the class grade:

$$g_s = \min \left(\frac{g_{s1} + \dots + g_{s\ell}}{\ell}, \frac{g_{s,\ell+1} + \dots + g_{s,T}}{T - \ell} \right). \tag{32.33}$$

This non-linearity prepares us for the next chapter, where non-linearity will be analyzed in its most general form.

References

1. S. Niwitpong, M. Chiangpradit, O. Kosheleva, Towards optimal allocation of points to different assignments and tests. J. Uncertain Syst. **4**(4), 291–295 (2010)
2. O. Kosheleva, V. Kreinovich, How to assigns weights to different tests?, in *Proceedings of the Sun Conference on Teaching and Learning*, El Paso, Texas, February 27, 2009
3. D.J. Sheskin, *Handbook of Parametric and Nonparametric Statistical Procedures* (Chapman & Hall/CRC, Boca Raton, 2007)

Chapter 33
How to Assess Students: Beyond Weighted Average

In many practical situations, it is desirable that the students learn all the parts of the material. It is therefore desirable to set up a grading scheme that encourages such learning. We show that the usual scheme of computing the overall grade for the class – as a weighted average of grades for different assignments and exams – does not always encourage such learning. Each such intermediate grade describes the student's knowledge of a certain part of the material. From the viewpoint of fuzzy logic, the degree to which the student knows the first part of the material *and* the second part of the material, etc., can be naturally described as a result of applying a t-norm ("and"-operation) to intermediate degrees (intermediate grades) – e.g., as the minimum of the intermediate grades. It turns out that this fuzzy-motivated min grading scheme indeed encourages students to learn all the material – and vice versa, the only grading scheme that provides such encouragement is the minimum of the intermediate grades.

The results from this chapter first appeared in [1, 2].

It is often important that the students learn all the material. The material taught in a typical semester-long class consists of several parts. In many cases, it is important that a student get reasonable knowledge of all the parts of the material. This is clear for such disciplines as medicine – we want a medical doctor to have basic knowledge of all types of diseases – but is also important in many other disciplines as well.

The desired level of knowledge may be different in different applications. For example, a medical doctor who just starts his internship under the mentorship of a skilled professional may have satisfactory knowledge of some parts, since the mentor is there to help. On the other hand, when the doctor is certified as capable to start his or her own medical practice, we would like the doctor to have good knowledge of *all* parts of the required material.

The grading scheme should reflect this requirement. It is desirable that the grading scheme not only gauge how well the students learn the material; the grading scheme should also encourage the students to learn *all* the parts of the material.

© Springer-Verlag GmbH Germany 2018
O. Kosheleva and K. Villaverde, *How Interval and Fuzzy Techniques Can Improve Teaching*, Studies in Computational Intelligence 750,
https://doi.org/10.1007/978-3-662-55993-2_33

Towards formalizing this idea: how a student plans his or her studies. A student has a limited time t that can allocated to learning the material. This time needs to be distributed between n different parts of the material, i.e., the student must select, for each part $i = 1, 2, \ldots, n$, the time $t_i \geq 0$ that is allocated for studying this part. The selected times t_i should add up to the given amount t: $t_1 + t_2 + \ldots + t_n = t$.

How to quantify knowledge. For each part of the material, it is reasonable to describe the student's knowledge in terms of a proportion of the material that the student learned, i.e., by a number from 0 to 1 such that 0 means no knowledge at all, and 1 means perfect knowledge. This can be estimated, e.g., as the proportion of correct answers on a comprehensive exam.

How student learn. Let us assume that for each value $t \geq 0$, we know the amount of knowledge $a(t)$ that a student will achieve if he or she studies the corresponding part for time t.

The more time the student learns, the more knowledge he or she acquires – unless the student already achieved the perfect knowledge $a(t) = 1$. In mathematical terms, this means that the function $a(t)$ is (non-strictly) increasing:

$$\text{if } t \leq t', \text{ then } a(t) \leq a(t').$$

It is also reasonable to assume that if a student slightly changes the time amount, the resulting knowledge will also change only slightly. In mathematical terms, this means that the function $a(t)$ is continuous.

The function $a(t)$ may differ from one group of students to others: some students have steeper learning curve, some learn slowly, etc.

Grading: general idea. Let us also assume that the tests, labs, and home assignments correctly gauge this knowledge. As a result, for each part of the material, we know the level of knowledge $a_i \stackrel{\text{def}}{=} a(t_i)$ acquired by the student i.

A grading scheme is a method F that transforms the n values $a_1, \ldots, a_n$ describing the student's knowledge of different parts of the material into a single grade

$$a = F(a_1, \ldots, a_n).$$

If all the grades a_i are the same, i.e., if $a_1 = a_2 = \ldots = a_n$, then it is reasonable to take this common grade as the grade for the class, i.e., to assume that

$$F(a, \ldots, a) = a.$$

In mathematical terms, this means that the function F should be *idempotent*.

The more students know about each part of the material, the better should be the overall grade. In mathematical terms, this means that the function F should be monotonic, i.e.:

$$\text{if } a_i \leq a_i' \text{ for all } i, \text{ then we should have } F(a_1, \ldots, a_n) \leq F(a_1', \ldots, a_n').$$

It is also reasonable to require that if a_i changes a little bit, the resulting grade should not change much. In precise terms, this means that the function F should be continuous.

How grading is done now. Usually, the overall grade is computed as the weighted average of grades corresponding to different parts of the material:

$$F(a_1, \ldots, a_n) = \sum_{i=1}^{n} w_i \cdot a_i$$

for some weights $w_i \geq 0$ for which $\sum_{i=1}^{n} w_i = 1$. The weighted average function is clearly monotonic and continuous.

How students decide how much time to allocate to different topics. When a student allocates time t_i to topic i, the student's grade for topic i becomes $a_i = a(t_i)$, and his or her overall grade is equal to $a = F(a(t_1), \ldots, a(t_n))$.

The student wants to maximize this grade, i.e., select the allocations $t_1, \ldots, t_n$ for which this value $F(a(t_1), \ldots, a(t_n))$ is the largest possible.

What we want. Let us denote by a_0 the desired degree of knowledge in every topic. We then want to make sure that the grading scheme (i.e., the function F) is such that if it is possible to find time allocation for which $a(t_i) \geq t_0$ for all i, then the allocation selected by the student will satisfy this property.

Ideally, this should be true for all types of students, with different functions $a(t)$.

The desired quantity is not always satisfied for the current grading system. Indeed, suppose that we have several parts of the material, and we want to get a level a_0 on all these parts – e.g., a satisfactory level $a_0 = 0.7$. Suppose also that the student's learning curve is $a(t) = t^2$ (describing a steep learning curve). In this case, to achieve the desired level of knowledge on each of n topics, the student needs to spend time t_0 for which $t_0^2 = a_0$, i.e., time $t_0 = \sqrt{a_0}$. After spending this amount of time on each of n topics, the student spends the total time $t = n \cdot t_0 = n \cdot \sqrt{a_0}$.

In this case, the student earns the grade a_0 on all the topics. Let us assume that all n grades are equally weighted, i.e., the overall grade is the arithmetic average of all n grades $a_i = a(t_i)$. In this case, for the desired student behavior, we get the overall grade

$$g = \frac{a_0 + \cdots + a_0}{n} = a_0.$$

Alternative, the student can spend this time by spending time $t_i = 1$ on $n \cdot \sqrt{a_0}$ out of n topics, and no time on remaining topics. For selected topics, this student get $a_i = a(t_i) = 1$, the perfect knowledge. For other topics, the student gets 0 knowledge $a_i = a(t_i) = a(0) = 0$. The overall grade is thus equal to $\frac{n \cdot \sqrt{a_0}}{n} = \sqrt{a_0}$.

For $a_0 < 1$, we have $\sqrt{a_0} > a_0$, so the student prefers the alternative learning strategy.

Heuristic idea motivated by fuzzy logic. We want the student to know:

- the first part of the material *and*
- the second part *and*
- ...*and*
- the nth part.

For each i, we know the degree a_i to which the student knows the ith part of the material. Thus, according to the fuzzy logic methodology (see, e.g., [3–5]), the degree a to which the desired requirement is satisfied can be found by applying a fuzzy "and"-operation (t-norm) to these degrees.

The requirement that $F(a, a) = a$ is satisfied only by one fuzzy "and"-operation – the minimum $\min(a, b)$ [3, 4]. If we use this "and"-operation, we get the grading scheme $a = \min(a_1, \ldots, a_n)$.

In the above example, the new grading scheme leads to the desired student behavior. Indeed, when the student spends the same amount of time $\sqrt{a_0}$ on each topic and get grades $a_1 = \cdots = a_n = a_0$, his overall grade – according to the new grading scheme – is $\min(a_0, \ldots, a_0) = a_0$.

Alternatively, if the students gets a perfect grade $a_i = 1$ on $n \cdot \sqrt{a_0}$ topics and $a_i = 0$ on all other topics, his or her overall grade is $\min(1, \ldots, 1, 0, \ldots, 0) = 0$. Since $0 < a_0$, the student will clearly prefer the desired learning strategy.

What we do in this paper. In this paper, we show the above-described behavior of the min grading scheme is not accidental. Specifically, we prove two results:

- that if we use the fuzzy-motivated min grading scheme, then the student would always prefers to equally distribute effort between different topics – exactly what we want to achieve;
- second, we prove that min grading is the only grading scheme with this property.

Need for definitions. To describe these results in precise terms, let us first define the problem in precise terms.

Definition 33.1 We say that a function $a(t_1, \ldots, t_n)$ is *(non-strictly) increasing* if whenever $t_1 \leq t_1'$, ..., and $t_n \leq t_n'$, we have $a(t_1, \ldots, t_n) \leq a(t_1', \ldots, t_n')$.

Definition 33.2 By a *learning curve*, we mean a continuous non-strictly increasing function $a(t)$ from non-negative real numbers to the interval $[0, 1]$.

Definition 33.3 We say that a function $F(a_1, \ldots, a_n)$ is *idempotent* if for every a, we have $F(a, \ldots, a) = a$.

Definition 33.4 For every integer $n \geq 2$, by a *n-grading scheme*, we mean a continuous non-strictly increasing idempotent function $F : [0, 1]^n \to [0, 1]$.

Comment As an example of the n-grading scheme, we have considered the min grading scheme $F(a_1, \ldots, a_n) = \min(a_1, \ldots, a_n)$.

Definition 33.5 Let $t > 0$ be a positive real number and let $n \geq 2$ be an integer. By a (t, n)-*learning strategy*, we mean a tuple of non-negative values $t_1, \ldots, t_n$ for which

$$t_1 + \cdots + t_n = t.$$

Comment. Not all (t, n)-learning strategies are always possible. For example, we may make sure that the students study in class. In this case, instead of the set of all possible (t, n)-learning strategies, we may be restricting ourselves to a *set* $\mathscr{S}$ of such strategies.

Definition 33.6 Let $a(t)$ be a learning curve, let $n \geq 2$ be an integer, let $F(a_1, \ldots, a_n)$ be an n-grading scheme, let $t > 0$ be a positive real number, let $(t_1, \ldots, t_n)$ be a (t, n)-learning strategy, and let $a_0 > 0$ be a positive real number.

- For every i from 1 to n, by the *grade for the ith assignment*, we mean the value $a(t_i)$.
- We say that the learning strategy is *uniformly a_0-successful* if $a(t_i) \geq a_0$ for all i.
- By an *overall grade*, we mean the value $F(a(t_1), \ldots, a(t_n))$.
- Let $\mathscr{S}$ be a set of (t, n)-leaning strategies. We say that this learning strategy from this set is $(\mathscr{S}, F)$-*optimal* if its overall grade is large than or equal to the overall grade of all other (t, n)-learning strategies from the set $\mathscr{S}$.

Definition 33.7 Let $n \geq 2$ be an integer, and let $F(a_1, \ldots, a_n)$ be an n-grading scheme. We say that this grading scheme *encourages students to learn all the material* if for every learning curve $a(t)$, for every two positive real numbers t and a_0, and for every set $\mathscr{S}$ of (t, n)-learning strategies,

- if, in the set $\mathscr{S}$, there exists a uniformly a_0-successful (t, n)-learning strategy,
- then every $(\mathscr{S}, F)$-optimal (t, n)-learning strategy is uniformly a_0-successful.

Theorem 33.1 *For every integer $n \geq 2$:*

- *the* min *grading scheme $F(a_1, \ldots, a_n) = \min(a_1, \ldots, a_n)$ encourages students to learn all the material;*
- *vice versa, if an n-grading scheme $F(a_1, \ldots, a_n)$ encourages students to learn all the material, then it coincides with the* min *grading scheme.*

Proof of the Theorem.

$1°$. Let us first prove that the min grading scheme encourages students to learn all the material, i.e., that if there exists a uniformly a_0-successful (t, n)-learning strategy, then every min-optimal learning strategy is uniformly a_0-successful.

Indeed, for a uniformly a_0-successful strategy, by definition, we have $a_i = a(t_i) \geq a_0$ for all i. Thus, the overall grade $a = F(a_1, \ldots, a_n) = \min(a_1, \ldots, a_n)$ corresponding to this strategy also has the property $a \geq a_0$.

For the optimal strategy, the grade, by definition, is $\geq a$ and thus, $\geq a_0$. So, for this strategy $(t_1, \ldots, t_n)$, we have $\min(a(t_1), \ldots, a(t_n)) \geq a_0$. Since each of n

numbers $a(t_i)$ is larger than or equal to the smallest of them $\min(a(t_1), \ldots, a(t_n))$, we thus conclude that $a(t_i) \geq a_0$ for all i – i.e., the optimal learning strategy is indeed uniformly a_0-successful.

$2°$. Let us now prove the second part of our theorem. Let us assume that a grading scheme $F(a_1, \ldots, a_n)$ encourages students to learn all the material. Let us prove that in this case, the grading scheme F coincides with the min grading scheme, i.e., that

$$F(a_1, \ldots, a_n) = \min(a_1, \ldots, a_n).$$

$2.1°$. It is sufficient to prove the above formula for the case when all the values a_i are positive.

Indeed, once we prove this formula for all positive a_i, we can use continuity to extend it to the case when some of the values a_i are equal to 0.

In view of this observation, in the remaining part of Part 2 of this proof, we will assume that $a_i > 0$ for all i.

$2.2°$. Let us prove that for every $m > 0$, for every $\varepsilon \in (0, m)$, and for every integer i from 1 to n, we have $F(1, \ldots, 1 \ (i - 1 \text{ times}), m - \varepsilon, 1 \ldots, 1) < m$.

Indeed, let us consider the function $a(t)$ for which:

- $a(0) = 0$,
- $a(1 - \varepsilon) = m - \varepsilon$,
- $a(1) = m$, and
- $a\left(1 + \dfrac{\varepsilon}{n-1}\right) = 1$.

Let us also assume that that the function $a(t)$ is piece-wise linear, i.e., that it is linear on the intervals $(0, 1 - \varepsilon)$, $(1 - \varepsilon, 1)$, and $\left(1, 1 + \dfrac{\varepsilon}{n-1}\right)$. For $t \geq 1 + \dfrac{\varepsilon}{n-1}$, we have $a(t) = 1$. One can easily check that this function $a(t)$ is continuous and non-strictly increasing.

Let us assume that the threshold is $a_0 = m$. If we take $t_1 = \ldots = t_n = 1$, then, due to our selection of the function $a(t)$, we get $a(t_1) = \ldots = a(t_n) = m$. Thus, by spending the time $t_1 + \ldots + t_n = n \cdot 1 = n$, we get a uniformly m-successful (n, n)-learning strategy. For this strategy, the overall grade is equal to $F(m, \ldots, m) = m$.

Let us now consider another (n, n)-learning strategy $(t'_1 \ldots, t'_n)$, in which $t'_i = 1 - \varepsilon$ and

$$t'_1 = \cdots = t'_{i-1} = t'_{i+1} = \cdots = t'_n = 1 + \frac{\varepsilon}{n-1}.$$

In this case,

$$t'_1 + \cdots + t'_n = (t'_1 + \cdots + t'_{i-1} + t'_{i+1} + \ldots + t'_n) + t'_i =$$

$$(n-1) \cdot \left(1 + \frac{\varepsilon}{n-1}\right) + (1 - \varepsilon) =$$

$$(n-1) + \varepsilon + 1 - \varepsilon = n,$$

i.e., the same time as before: $t' = n$.

Under this learning strategy, the grades for different assignments are equal to $a(t'_i) = a(1 - \varepsilon) = m - \varepsilon$ and

$$a(t'_1) = \cdots = a(t'_{i-1}) = a(t'_{i+1}) = \ldots = a(t'_n) =$$

$$a\left(1 + \frac{\varepsilon}{n-1}\right) = 1.$$

Thus, the overall grade is equal to

$$F(a(t'_1), \ldots, a(t'_{i-1}), a(t'_i), a(t'_{i+1}), \ldots, a(t'_n)) =$$

$$F(1, \ldots, 1 \ (i-1 \ \text{times}), m - \varepsilon, 1 \ldots, 1).$$

Since $a(t'_i) = m - \varepsilon < m$, this (n, n)-learning strategy is *not* uniformly m-successful.

Let us consider the 2-element set $\mathscr{S} \stackrel{\text{def}}{=} \{(t_1, \ldots, t_n), (t'_1, \ldots, t'_n)\}$. This set contains a uniformly m-successful (n, n)-learning strategy $(t_1, \ldots, t_n)$. By definition of a grading scheme that encourages students to learn all the material, this means that every $(\mathscr{S}, F)$-optimal (n, n)-learning strategy is uniformly m-successful. Since we have shown that the learning strategy $(t'_1, \ldots, t'_n)$ is *not* uniformly m-successful, we can conclude that this strategy cannot be $(\mathscr{S}, F)$-optimal. Thus, the overall grade corresponding to the learning strategy $(t'_1, \ldots, t'_n)$ must be smaller than the overall grade corresponding to the strategy $(t_1, \ldots, t_n)$, i.e.:

$$F(1, \ldots, 1 \ (i-1 \ \text{times}), m - \varepsilon, 1 \ldots, 1) < m.$$

The statement is proven.

2.3°. We have just proven that the inequality from Part 2.2 holds for every $\varepsilon > 0$. Since the grading scheme $F(a_1, \ldots, a_n)$ is a continuous function, in the limit $\varepsilon \to 0$, we get $F(1, \ldots, 1 \ (i-1 \ \text{times}), m, 1 \ldots, 1) \leq m$.

2.4°. Let us now conclude the proof of Part 2, by proving that

$$F(a_1, \ldots, a_n) = \min(a_1, \ldots, a_n).$$

Indeed, let m denote the minimum $\min(a_1, \ldots, a_m)$, and let i denote the smallest value a_i, for which $a_i = m$ and $m \leq a_j$ for all $j \neq i$.

2.4.1°. Since $a_j \leq 1$, by monotonicity, we conclude that

$$F(a_1, \ldots, a_{i-1}, a_i, a_{i+1}, \ldots, a_n) =$$

$$F(1, \ldots, 1 \ (i - 1 \ \text{times}), a_i, 1 \ldots, 1) =$$

$$F(1, \ldots, 1 \ (i - 1 \ \text{times}), m, 1 \ldots, 1) \leq m.$$

2.4.2°. Similarly, since $m = a_i \leq a_j$ for all j, by monotonicity, we get

$$F(m, \ldots, m) \leq F(a_1, \ldots, a_{i-1}, a_i, a_{i+1}, \ldots, a_n).$$

Since the function F is idempotent, we get $F(m, \ldots, m) = m$ and hence, $m \leq F(a_1, \ldots, a_{i-1}, a_i, a_{i+1}, \ldots, a_n)$.

2.4.3°. From

$$F(a_1, \ldots, a_{i-1}, a_i, a_{i+1}, \ldots, a_n) \leq m$$

and

$$m \leq F(a_1, \ldots, a_{i-1}, a_i, a_{i+1}, \ldots, a_n),$$

we can now conclude that

$$F(a_1, \ldots, a_{i-1}, a_i, a_{i+1}, \ldots, a_n) = m,$$

i.e., that

$$F(a_1, \ldots, a_{i-1}, a_i, a_{i+1}, \ldots, a_n) = \min(a_1, \ldots, a_n).$$

The theorem is proven.

Resulting recommendations are not that unusual. The resulting recommendation is to take, as an overall grade for the class, the smallest of the grades gained for each module. At first, this may sound like a very radical idea, it is in line with what is usually done in grading.

For example, in our university, for a student to pass Calculus I, the student has to get satisfactory grade on each module. This corresponds to minimum. In some computer science classes, in order to pass a class, the student has to get a satisfactory grade both on the tests and on the labs.

Similarly, to get a degree, it is not sufficient for a student to have a good GPA, the student must get satisfactory grades on all required classes.

References

1. O. Kosheleva, How to make sure that the grading scheme encourages students to learn all the material: fuzzy-motivated solution and its justification, in *Proceedings of the World Conference on Soft Computing*, ed. by R.R. Yager, M.Z. Reformat, S.N. Shahbazova, S. Ovchinnikov (CA, San Francisco, 2011), pp. 23–26

2. O. Kosheleva, How to make sure that the grading scheme encourages students to learn all the material: fuzzy-motivated solution and its justification. Int. J. Intell. Technol. Appl. Stat. (IJITAS) **10**(2), 7–19 (2017)
3. G.J. Klir, B. Yuan, *Fuzzy Sets and Fuzzy Logic: Theory and Applications* (Prentice-Hall, Upper Saddle River, New Jersey, 1995)
4. H.T. Nguyen, E.A. Walker, *A First Course in Fuzzy Logic* (CRC Press, Boca Raton, Florida, 2006)
5. L.A. Zadeh, Fuzzy sets. Information and Control **8**, 338–353 (1965)

Chapter 34
How to Assess a Class

Once we have assessed the individual performance of all the students in the class, it is desirable to combine these estimates into a class assessment. A natural way to assess the class is to provide the average grade and some measure of deviation from this average grade. There are many ways to measure this deviations; which one should we choose?

This problem was analyzed by W.J. Tastle and M.J. Wierman in a more general situation, when we have several different expert estimates, and we need to decide how close these estimates are to each other, i.e., to what extent these estimates show consensus and to what they show dissention. They proposed heuristic formulas for degrees of consensus and dissent, and showed that these formulas indeed capture the intuitive ideas of consensus and dissent. In this chapter, we show that the Tastle–Wierman formulas can be naturally derived from the basic formulas of fuzzy logic. So, it is reasonable to use theses formulas in education, to describe how different the grades of different students are.

The results from this chapter first appeared in [1].

Gauging consensus: a problem. In many practical situations, we have to use expert estimates to gauge the value of a quantity. Expert estimates rarely agree exactly.

- Sometimes, the expert estimates $x_1, \ldots, x_n$ mostly agree with each other, so we can say that they are in consensus.
- Sometimes, the expert estimates strongly disagree.

To compare different situations like this, it is desirable to come up with numerical measures of dissention and consensus.

Possible applications to education. In a typical class, we have students at different levels of knowledge, students with different ability to learn the material. In the ideal world, we should devote unlimited individual attention to all the students and make sure that everyone learns all the material. In real life, our resources are finite. Based

© Springer-Verlag GmbH Germany 2018
O. Kosheleva and K. Villaverde, *How Interval and Fuzzy Techniques Can Improve Teaching*, Studies in Computational Intelligence 750,
https://doi.org/10.1007/978-3-662-55993-2_34

on this finite amount of resources, what is the best way to distribute efforts between different students. In other words, how should we distribute efforts between different students so as to get the best results?

In order to answer this question, we must formalize what the term "the best results" means.

The success of each individual student i can be naturally gauged by this student's grade x_i. So, for two different teaching strategies T and T', we know the corresponding grades $x_1, \ldots, x_n$ and $x'_1, \ldots, x'_{n'}$. Which strategy is better?

In some cases, the answer to this question is straightforward. For example, when $n' = n$, $x_i \leq x'_i$ for all i and $x_i < x'_i$ for some i, then clearly the strategy T' is better.

In practice, however, the comparison is rarely that straightforward. Often, when we change a strategy, some grades decrease while some other grades increase. In this case, how do we usually decide whether a new method is better or not?

In pedagogical research, the decision is usually made based on the comparison of the average grades

$$\bar{x} \stackrel{\text{def}}{=} \frac{x_1 + \cdots + x_n}{n} \tag{34.1}$$

and

$$\bar{x'} \stackrel{\text{def}}{=} \frac{x'_1 + \cdots + x'_{n'}}{n'}. \tag{34.2}$$

Usually, it is assumed that if the average grade is larger, this means that the corresponding teaching method is better.

However, the average grade is not always the most adequate way to gauging the success of a pedagogical strategy. Whether the average grade is a good criterion or not depends on our objective.

Let us illustrate this dependence on a simplified example. Suppose that after using the original teaching method T, we get the grades $x_1 = 60$ and $x_2 = 90$. The average value of these grades is

$$\bar{x} = \frac{60 + 90}{2} = 75. \tag{34.3}$$

Suppose that the new teaching method T' leads to the grades $x'_1 = x'_2 = 70$. The average of the new grades is $\bar{x'} = 70$.

Since the average grade decreases, the traditional conclusion would be that the new teaching method T' is not as efficient as the original method T. However, one possible objective may be to decrease the failing rate. Usually, 70 is the lowest grade corresponding to C, and any grade below C is considered failing. In this case,

- in the original teaching method, one of the two students failed, while
- in the new teaching method, both students passed the class.

Thus, with respect to this objective, the new teaching method is better.

This inadequacy of the mean $\bar{x}$ is related to the fact the mean does not provide us any information about the "spread" of the grades, i.e., the information about how much the grades deviate from the mean. We therefore need to supplement the mean with a second numerical criterion – a criterion that describes how different the estimates x_i are from each other – i.e., equivalently, how different they are from the mean $\bar{x}$ of these estimates.

These are exactly the motivations for measures of consensus and dissention. So, such measures can be very useful in education as well – as a way to gauge the spread of the grades.

Tastle–Wierman (TW) dissention and consensus measures. In their papers [2–4], W.J. Tastle and M.J. Wierman describe a new consensus measure – and provide reasonable arguments that this measure indeed reflects our intuitive sense of dissention and consensus. Specifically, they define the measure of dissention as the mean value of the quantity

$$-\log_2\left(1 - \frac{|x_i - \bar{x}|}{d_x}\right),$$

(34.4)

where

$$\bar{x} \overset{\text{def}}{=} \frac{1}{n} \cdot \sum_{i=1}^{n} x_i,$$

(34.5)

and $d_x \overset{\text{def}}{=} x^+ - x^-$ is the width of the interval $[x^-, x^+]$ of possible values of the estimated quantity.

In other words, Tastle and Wierman define a dissention measure $D(x)$ related to the tuple $x = (x_1, \ldots, x_n)$ of expert estimates as

$$D(x) \overset{\text{def}}{=} -\frac{1}{n} \cdot \sum_{i=1}^{n} \log_2\left(1 - \frac{|x_i - \bar{x}|}{d_x}\right).$$

(34.6)

Comment. When several experts come up with the same estimate, i.e., if we have estimates $x_1, \ldots, x_m$ and the frequency $p_1, \ldots, p_m$ of experts who come up with these estimates, this formula can be reformulated as

$$D(x) = -\sum_{j=1}^{m} p_i \cdot \log_2\left(1 - \frac{|x_j - \bar{x}|}{d_x}\right),$$

(34.7)

where

$$\bar{x} = \sum_{j=1}^{m} p_j \cdot x_j.$$

(34.8)

Consensus measure. A consensus is, intuitively, an opposite to dissention. So, Wierman and Tastle define the degree of consensus $C(x)$ as simply 1 minus the degree of dissention:

$$C(x) = 1 - D(x).$$

(34.9)

Remaining problem. While Wierman and Tastle convincingly show that their dissention and consensus measure capture the intuitive meaning of consensus, it is not clear, from their analysis, whether these are the only possible measures that capture this intuition – and if not, what other possible measures capture this same intuition.

What we do in this chapter. In this chapter, we show that the TW measures can be naturally derived from a fuzzy logic formalization of the intuitive ideas behind dissention and consensus. Specifically, we show that the TW measures naturally appear if we use one of the simplest t-conorms – algebraic sum – and one of the simplest membership functions – a triangular one (see, e.g., [5, 6]).

We also explain what will happen if we use more complex t-conorms and/or membership functions.

How we can use fuzzy logic to justify the TW measures (or any other formula). Fuzzy logic has been originally invented as a tool for processing intuitive, semi-heuristic, "fuzzy" statements like "x is small", statements formulated in terms of a natural language. Fuzzy logic assigns, to these statements, real numbers – degrees to which these statements are true, and then uses appropriate operations to process these degrees.

Thus, in order to apply fuzzy logic to our situation, we must first reformulate the main ideas behind dissention and consensus in intuitive terms (i.e., in terms of a natural language). Once we have this reformulation, we will be able to use fuzzy logic techniques to assign and process the corresponding numerical characteristics.

Let us therefore start by reformulating the main ideas of dissention and consensus in terms of words from a natural language – i.e., by providing an intuitive common sense idea.

Intuitive idea behind dissention. In the ideal case of complete consensus, all expert estimates $x_1, \ldots, x_n$ coincide with each other

$$x_1 = \cdots = x_n \tag{34.10}$$

and therefore, they all coincide with the mean

$$\bar{x} = \frac{1}{n} \cdot \sum_{i=1}^{n} x_i \tag{34.11}$$

of these estimates:

$$(x_1 = \bar{x}) \,\&\, \cdots \,\&\, (x_n = \bar{x}). \tag{34.12}$$

Dissention means that at least some of these values are different, i.e., that:

- either x_1 is different from $\bar{x}$,
- or x_2 is different from $\bar{x}$,
- ...
- or x_n is different from $\bar{x}$.

Using fuzzy techniques to transform the above common-sense understanding of dissention into numerical degrees: a general idea. According to the above description, the degree to which the tuple $x = (x_1 \ldots, x_n)$ corresponds to dissention can be described as the degree to which the following statement holds:

$$(x_1 \text{ is different from } \bar{x}) \vee \cdots \vee (x_n \text{ is different from } \bar{x}).$$

According to the general fuzzy methodology, to assign a degree to this statement, we must do the following:

- first, we should assign reasonable degrees $d_{\neq}(a, b)$ to statements of the type "a is different from b";
- then, we should select an appropriate t-conorm ("or"-operation) $t_{\vee}(a, b)$;
- finally, we should use this t-conorm to combine the degrees $d_{\neq}(x_i, \bar{x})$ of the statements

$$\text{"}x_i \text{ is different from } \bar{x}\text{"}$$

corresponding to different $i = 1, 2, \ldots, n$ into a single degree – the desired degree

$$d(x) = t_{\vee}(d_{\neq}(x_1, \bar{x}), \ldots, d_{\neq}(x_n, \bar{x})) \qquad (34.13)$$

with which the above combined statement holds.

Let us use the simplest possible techniques. One of the general ideas of using fuzzy methodology is that, out of all possible techniques which are consistent with our intuition, we should use the computationally simplest techniques.

Indeed, fuzzy knowledge is, by definition, imprecise. In principle, we can have several slightly different representations of the same common sense knowledge. If a simple formula already captures the meaning of this knowledge, there is no sense in using more complex formulas, formulas which would require more computations.

For example, if our knowledge can be well described by a simple triangular membership function, why should we use a more complex membership function? If our understanding of an "and"-operation (t-norm) is well captured by a simple algebraic product operation $t_{\&}(a, b) = a \cdot b$, why use more complex t-norms?

In line with this general idea, let us analyze the above two tasks:

- the task of describing the membership function corresponding to the notion "a is different from b", and
- the task of selecting the appropriate t-conorm ("or"-operation).

Selecting the membership function corresponding to "different". We would like to describe, for each two real values a and b, the degree $d_{\neq}(a, b)$ to which a is different from b. In other words, we would like to describe a function of two variables a and b.

In order to select such a function, let us first observe that this problem can be simplified into selecting a function of only one variable. Indeed, a is different from b if and only if the absolute value $|a - b|$ of the difference $a - b$ is different from

0. Thus, the degree $d_{\neq}(a, b)$ to which a is different from b is equal to the degree $\mu_{\neq 0}(|a - b|)$ with which the difference $|a - b|$ is different from 0:

$$d_{\neq}(a, b) = \mu_{\neq 0}(|a - b|). \tag{34.14}$$

Thus, to describe the degree $d_{\neq}(a, b)$ to which a is different from b, it is sufficient to describe the degree $\mu_{\neq 0}(c)$ to which a non-negative number c is different from 0.

The value $c = 0$ is *not* different from 0. Thus, for $c = 0$, the statement "c is different from 0" is absolutely false, and the degree to which this statement is true should be therefore equal to 0. In other words, we should have

$$\mu_{\neq 0}(0) = 0. \tag{34.15}$$

The larger c, the more c is different from 0. Thus, $\mu_{\neq 0}(c)$ should be an increasing function of c. In accordance with the above general idea, we select the simplest possible monotonic function. The simplest possible functions – as we mentioned – are linear ones, so we take $\mu_{\neq 0}(c) = m + k \cdot c$. Since $\mu_{\neq 0}(0) = 0$, we have $m = 0$ and $\mu_{\neq 0}(c) = k \cdot c$.

The coefficient k can be found from a natural requirement that when the values a and b are as far away from each other as possible, the degree with which these values are different should be equal to 1. When both values a and b belong to the range $[x^-, x^+]$, the difference $|a - b|$ attains the largest possible value when a and b are two different endpoints of this range, i.e., when

- either $a = x^-$ and $b = x^+$,
- or $a = x^+$ and $b = x^-$.

In both cases, we have $|a - b| = x^+ - x^- = d_x$. Thus, our requirement is that for $c = d_x$, we should have

$$\mu_{\neq 0}(c) = k \cdot c = 1. \tag{34.16}$$

Substituting $c = d_x$ into this formula, we conclude that

$$k = \frac{1}{d_x}. \tag{34.17}$$

Thus, the degree that a non-negative number c is different from 0 is equal to

$$\mu_{\neq 0}(c) = k \cdot c = \frac{c}{d_x}. \tag{34.18}$$

As we have mentioned, the degree $d_{\neq}(a, b)$ (to which two numbers are different) is equal to the degree $\mu_{\neq 0}(|a - b|)$ that the value $|a - b|$ is different from 0, hence

$$d_{\neq}(a, b) = \mu_{\neq 0}(|a - b|) = \frac{|a - b|}{d_x}. \tag{34.19}$$

Therefore, for each $i = 1, \ldots, n$, the degree $d_{\neq}(x_i, \bar{x})$ to which this value x_i is different from the mean $\bar{x}$ is equal to

$$d_{\neq}(x_i, \bar{x}) = \frac{|x_i - \bar{x}|}{d_x}. \tag{34.20}$$

Selecting and applying the t-conorm. In accordance with the above idea of using the simplest possible formulas, let us use the simplest possible "algebraic sum" t-conorm

$$t_{\vee}(a, b) = a + b - a \cdot b. \tag{34.21}$$

Comment. To be more precise, the maximum t-conorm $t_{\vee}(a, b) = \max(a, b)$ is even computationally simpler, but, as we will show, it does not quite capture the intuitive meaning of dissention.

Indeed, with this t-conorm, we have identical degrees for the following two drastically different situations with the same range $[x^-, x^+] = [-1, 1]$ (and $d_x = x^+ - x^- = 2$):

- In the first situation, half of the experts selected 1 and half -1, i.e., the expert estimates form a sequence $x = (1, \ldots, 1, -1, \ldots, -1)$. For this sequence, the mean $\bar{x}$ is 0, so for each i, the degree with which each value is different from 0s is equal to 0.5. In this case, the total degree of dissention is

$$t_{\vee}(0.5, \ldots, 0.5) = \max(0.5, \ldots, 0.5) = 0.5. \tag{34.22}$$

- In the second situation, one expert selected 1, one -1, and all other experts selected 0. In this case, the expert estimates form a sequence $x = (1, -1, 0, \ldots, 0)$. For this sequence, the mean $\bar{x}$ is still 0. So, the degrees with which each value is different from 0s are equal, correspondingly, to 0.5, 0.5, 0, ..., 0. In this case, the total degree of dissention is

$$t_{\vee}(0.5, 0.5, 0, \ldots, 0) = \max(0.5, 0.5, 0, \ldots, 0) = 0.5. \tag{34.23}$$

In the first case, there is as much dissention as possible; in the second case, only two experts disagree – but the resulting degrees

$$t_{\vee}(d_{\neq}(x_1, \bar{x}), \ldots, d_{\neq}(x_n, \bar{x}))$$

are the same.

This example shows that $t_{\vee}(a, b) = \max(a, b)$ is not an adequate choice for our problem.

Resulting expression. With the triangular membership function

$$d_{\neq}(x_i, \bar{x}) = \frac{|x_i - \bar{x}|}{d_x}. \tag{34.24}$$

and the algebraic sum t-conorm

$$t_\vee(a, b) = a + b - a \cdot b, \tag{34.25}$$

we get the following expression for the desired degree

$$d(x) = t_\vee(d_{\neq}(x_1, \bar{x}), \ldots, d_{\neq}(x_n, \bar{x})): \tag{34.26}$$

$$d(x) = t_\vee\left(d_{\neq}\left(\frac{|x_1 - \bar{x}|}{d_x}, \ldots, \frac{|x_n - \bar{x}|}{d_x}\right)\right). \tag{34.27}$$

Relation between this expression and the TW measures. At first glance, the above expression looks drastically different from the TW dissention measure $D(x)$ (formula (34.6)). However, they are actually very closely related: namely, we have

$$D(x) = -\frac{1}{n} \cdot \log_2(1 - d(x)). \tag{34.28}$$

This formula comes from the fact that

$$\log_2(1 - t_\vee(a, b)) = \log_2(1 - a) + \log_2(1 - b), \tag{34.29}$$

which, in its turn, follows from the fact that

$$1 - t_\vee(a, b)) = 1 - (a + b - a \cdot b) = (1 - a) \cdot (1 - b), \tag{34.30}$$

and thus,

$$\log_2(1 - t_\vee(a, b)) = \log_2((1 - a) \cdot (1 - b)) =$$

$$\log_2(1 - a) + \log_2(1 - b). \tag{34.31}$$

The transformation

$$d(x) \Rightarrow -\frac{1}{n} \cdot \log_2(1 - d(x)) \tag{34.32}$$

is a monotonic (increasing) function. Thus,

- while the *numerical* values of $d(x)$ and $D(x)$ are different,
- the measures $d(x)$ and $D(x)$ lead to the same *relations* between different tuples x and x' (of the same size n):
  - $d(x) = d(x')$ if and only if $D(x) = D(x')$; and
  - $d(x) < d(x')$ if and only if $D(x) < D(x')$.

For many applications, we thus have the desired justification of the TW measures. In many applications – e.g., in the above education application – we are not directly interested in the numerical value of the dissention measure $D(x)$, only in comparing dissention values corresponding to different tuples. For these applications, we can thus conclude that we do have the desired fuzzy justification of the TW measures.

Reminder. In the previous section, we provided a justification for the TW measures of consensus and dissention based on fuzzy logic.

Remaining problem. The above justification is based on a rather ad hoc use of a special function $-\log_2(1-a)$. What remains unclear is how unique is this function – and, correspondingly, how unique are the TW formulas.

Towards the formulation of a precise mathematical problem. We are looking for a function $z(x)$ that has the following property:

$$z(t_\vee(a, b)) = z(a) + z(b), \tag{34.33}$$

for $t_\vee(a, b) = a + b - a \cdot b$.

In other words, we are looking for a "measure" $z(x)$ for which:

- the measure that "a or b" is true is equal to
- the sum of the measures that a is true and that b is true.

Example This makes sense, e.g., for describing the difference between the two vectors (x_1, x_2) and (x_1', x_2'). The two vectors are different if either x_1 is different from x_1', or x_2 is different from x_2'. Thus, the degree to which (x_1, x_2) is different from (x_1', x_2') is equal to the result of applying the "or" operation to the following two degrees:

- the degree to which x_1 is different from x_1', and
- the degree to which x_2 is different from x_2'.

It is also reasonable to be able to transform these degrees into a "measure of the difference" $z(d)$ which, due to the independence, should have the following property:

- the measure corresponding to two-coordinate vectors should be equal to
- the sum of the measures corresponding to both coordinates.

Now, we are ready for exact formulations.

Proposition 34.1 *Let* $t_\vee(a, b) = a + b - a \cdot b$. *A monotonic function* $z : [0, 1] \to \mathbb{R}$ *satisfies the property*

$$z(t_\vee(a, b)) = z(a) + z(b), \tag{34.34}$$

for every a and b if and only if $z(x) = -k \cdot \log_2(x)$ *for some constant k.*

Proof We already know, from the previous section, that the function $z(x) = -\log_2(x)$ satisfies the desired property (34.34). It is easy to see that for an arbitrary real number k, the corresponding function $z(x) = -k \cdot \log_2(x)$ also satisfies this property.

To complete the proof, it is therefore sufficient to prove that every monotonic function with the property (34.34) is of the desired form $z(x) = -k \cdot \log_2(x)$.

For this, let us consider an auxiliary function

$$t(a) \overset{\text{def}}{=} z(1-a). \qquad (34.35)$$

Substituting $a = 1 - x$ into this definition, we conclude that

$$t(1-x) = z(1-(1-x)) = z(x), \qquad (34.36)$$

i.e.,

$$z(x) = t(1-x). \qquad (34.37)$$

Substituting this expression for z in terms of t into the formula (34.34), we conclude that

$$t(1-(a+b-a \cdot b)) = t(1-a) + t(1-b). \qquad (34.38)$$

Here,

$$1 - (a+b-a \cdot b) = 1 - a - b + a \cdot b = (1-a) \cdot (1-b), \qquad (34.39)$$

so the formula (34.38) takes the form

$$t((1-a) \cdot (1-b)) = t(1-a) + t(1-b) \qquad (34.40)$$

for every a and b. In particular, for every $x, y \in [0, 1]$, we can take $a = 1 - x$ and $b = 1 - y$. For these values a and b, we have $1 - a = x$ and $1 - b = y$ and thus, the formula (34.40) takes the form

$$t(x \cdot y) = t(x) + t(y). \qquad (34.41)$$

For arbitrary $X \geq 0$ and $Y \geq 0$, we can take $x = \exp(-X)$ and $y = \exp(-Y)$. Then,

$$x \cdot y = \exp(-X) \cdot \exp(-Y) = \exp(-(X+Y)). \qquad (34.42)$$

Thus, the formula (34.40) takes the form

$$t(\exp(-(X+Y))) = t(\exp(-X)) + t(\exp(-Y)). \qquad (34.43)$$

In order words, for an auxiliary function

$$T(X) \overset{\text{def}}{=} t(\exp(-X)), \qquad (34.44)$$

we have

$$T(X + Y) = T(X) + T(Y). \tag{34.45}$$

It is well known that every monotonic solution of this functional equation is $T(X) = c \cdot X$ for some real number c. Thus,

$$T(X) = t(\exp(-X)) = k \cdot X. \tag{34.46}$$

To find $t(x)$, we must therefore find X for which $\exp(-X) = x$. By taking logarithms of both sides of this equality, we conclude that $-X = \ln(x)$, hence $X = -\ln(x)$. For this x, due to $\exp(-X) = x$, the formula (34.46) takes the form

$$t(x) = -c \cdot \ln(x). \tag{34.47}$$

Due to the formula $z(x) = t(1 - x)$ (formula (34.37)), we thus have

$$z(x) = t(1 - x) = -c \cdot \ln(1 - x). \tag{34.48}$$

If we take into account that

$$\ln(x) = \frac{\log_2(x)}{\log_2(e)}, \tag{34.49}$$

we conclude that

$$z(x) = -k \cdot \ln(1 - x), \tag{34.50}$$

with $k = \dfrac{c}{\log_2(e)}$.
  The proposition is proven.

What if we use a different t-conorm and/or a different membership function?
What if we use a different (more complex) t-conorm $t_\vee(a, b)$ and/or a different (more complex) membership function $\mu_{\neq 0}(c)$?

t-conorms: reminder. Let us start the corresponding analysis by recalling the basic facts about t-conorms. In fuzzy logic, several different types of t-conorms are used. One of the mostly widely used classes of t-norms are *Archimedean* t-conorms. It is known that these t-conorms $t_\vee(a, b)$ are isomorphic to the algebraic sum t-conorm $t_{\mathrm{alg}}(a, b) = a + b - a \cdot b$ – isomorphic in the following sense:
  There exists a monotonic 1-1 function $f : [0, 1] \to [0, 1]$ for which

$$t_\vee(a, b) = f^{-1}(t_{\mathrm{alg}}(f(a), f(b)) = f^{-1}(f(a) + f(b) - f(a) \cdot f(b)), \tag{34.51}$$

where $f^{-1}(x)$ denoted the inverse function (for which $f^{-1}(x) = y$ if and only if $f(y) = x$).

Not all t-conorms are Archimedean. For example, the above-mentioned maximum t-conorm $\max(a, b)$ is not Archimedean. It is known that a general t-conorm can be obtained:

- by setting Archimedean t-conorms on several (maybe infinitely many) subintervals of the interval $[0, 1]$, and
- by using $\max(a, b)$ as the value of $t_\vee(a, b)$ for the cases when a and b do not belong to the same Archimedean subinterval.

From this general classification theorem for t-conorms, we can conclude that for every t-norm and for every $\varepsilon > 0$, there exists an ε-close Archimedean t-conorm. (Crudely speaking, we can replace the corresponding non-Archimedean max-parts with a close Archimedean t-conorm, e.g., with $(a^p + b^p)^{1/p}$ for a sufficiently large p).

Thus, for an arbitrary accuracy $\varepsilon > 0$ and for an arbitrary t-conorm, we can have an Archimedean t-conorm which is, within this accuracy, indistinguishable from the original one. So, from the practical viewpoint, we can always safely assume that the t-conorm is Archimedean, i.e., that is has the form (34.51).

Towards resulting formulas. In general, the expression for the degree to which a tuple $x = (x_1, \ldots, x_n)$ represents consensus is equal to

$$d(x) = t_\vee(\mu_{\neq 0}(|x_1 - \bar{x}|), \ldots, \mu_{\neq 0}(|x_1 - \bar{x}|)). \tag{34.52}$$

Since $t_\vee(a, b) = f^{-1}(t_{\text{alg}}(f(a), f(b)))$, we conclude that

$$t_\vee(a_1, \ldots, a_n) = f^{-1}(t_{\text{alg}}(f(a_1), \ldots, f(a_n))), \tag{34.53}$$

hence

$$d(x) = f^{-1}(t_{\text{alg}}(f(\mu_{\neq 0}(|x_1 - \bar{x}|)), \ldots, f(\mu_{\neq 0}(|x_n - \bar{x}|)))), \tag{34.54}$$

or, equivalently,

$$f(d(x)) = t_{\text{alg}}(f(\mu_{\neq 0}(|x_1 - \bar{x}|)), \ldots, f(\mu_{\neq 0}(|x_n - \bar{x}|))) \tag{34.55}$$

and

$$f(d(x)) = t_{\text{alg}}(F(|x_1 - \bar{x}|), \ldots, F(|x_n - \bar{x}|)), \tag{34.56}$$

where we denoted

$$F(z) \stackrel{\text{def}}{=} f(\mu_{\neq 0}(z)). \tag{34.57}$$

We already know, from our analysis of the case of the algebraic sum t-conorm, that in this case, we have
$$D(x) = -\log_2(1 - f(d(x))) =$$

$$-\log_2(1 - F(|x_1 - \bar{x}|)) - \ldots - \log_2(1 - F(|x_b - \bar{x}|)). \tag{34.58}$$

Thus, we arrive at the following formulas:

Resulting formula. For a general t-conorm and a general membership function $\mu_{\neq 0}(c)$, it is reasonable to describe the degree of dissention as

$$D(x) = -\frac{1}{n} \cdot \sum_{i=1}^{n} \log_2(1 - F(|x_i - \bar{x}|)), \tag{34.59}$$

where $F(z) = f(\mu_{\neq 0}(z))$ and $f(z)$ is a function for which

$$t_\vee(a, b) = f^{-1}(f(a) + f(b) - f(a) \cdot f(b)).$$

Comment. This formula is a natural generalization of the TW formula (we added a factor $1/n$ to keep this relation).

Corresponding mathematical result. The above mathematical result can also be naturally extended from the algebraic sum t-conorm to an arbitrary Archimedean t-conorm $t_\vee(a, b)$:

Proposition 34.2 *Let* $t_\vee(a, b) = f^{-1}(f(a) + f(b) - f(a) \cdot f(b))$ *be an Archimedean t-conorm. A monotonic function* $z : [0, 1] \to \mathbb{R}$ *satisfies the property*

$$z(t_\vee(a, b)) = z(a) + z(b), \tag{34.60}$$

for every a and b if and only if $z(x) = -k \cdot \log_2(1 - f(x))$ *for some constant k.*

Proof Substituting the explicit expression for the t-conorm into the desired property (34.60), we conclude that

$$z(f^{-1}(f(a) + f(b) - f(a) \cdot f(b))) = z(a) + z(b). \tag{34.61}$$

Let us introduce an auxiliary function

$$Z(X) \overset{\text{def}}{=} z(f^{-1}(X)). \tag{34.62}$$

For an arbitrary x, for $X = f(x)$, we have $x = f^{-1}(X)$, thus we get

$$z(x) = Z(f(x)). \tag{34.63}$$

Substituting this expression for $z(x)$ in terms of Z into the formula (34.61), we conclude that

$$Z(f(a) + f(b) - f(a) \cdot f(b))) = Z(f(a)) + Z(f(b)). \tag{34.64}$$

For arbitrary two numbers A and B, we can take $a = f^{-1}(A)$ and $b = f^{-1}(B)$ for which $f(a) = A$ and $f(b) = B$. Thus, from the formula (34.64), we can conclude that the following is true for all A and B:

$$Z(A + B - A \cdot B) = Z(A) + Z(B). \tag{34.65}$$

We already know, from Proposition 34.1, that all solutions of this functional equation have the form $Z(A) = -k \cdot \log_2(1-A)$. Thus, from $z(x) = Z(f(x))$ formula (34.63), we conclude that $z(x) = -k \cdot \log_2(1 - f(x))$.
The proposition is proven.

Conclusions. To estimate how close the estimates of different experts are, W.J. Tastle and M.J. Wierman proposed numerical measures of dissention and consensus, and showed that these measures indeed capture the intuitive ideas of dissent and consensus.

In this chapter, we show that the Tastle–Wierman formulas can be naturally derived from fuzzy logic. We also show that the Tastle–Wierman measures of dissention and consensus can be used in education, to gauge how different the grades of different students are.

References

1. K. Villaverde, O. Kosheleva, Towards a new justification of the Tastle–Wierman (TW) dissention and consensus measures (and their potential role in education), in *Proceedings of the Annual Conference of the North American Fuzzy Information Processing Society NAFIPS'2010*, Toronto, Canada, July 12–14, 2010, pp. 110–116
2. W.J. Tastle, M.J. Wierman, Consensus and dissention: a new measure of agreement, in *Proceedings of the 24th International Conference of the North American Fuzzy Information Processing Society NAFIPS'2005*, Ann Arbor, Michigan, June 22–25, 2005, pp. 585–588
3. W.J. Tastle, M.J. Wierman, Consensus and dissention: a measure of ordinal dispersion. Int. J. Approx. Reason. **45**, 531–545 (2007)
4. M.J. Wierman, W.J. Tastle, Consensus and dissention: theories and properties, in *Proceedings of the 24th International Conference of the North American Fuzzy Information Processing Society NAFIPS'2005*, Ann Arbor, Michigan, June 22–25, 2005, pp. 75–79
5. G.J. Klir, B. Yuan, *Fuzzy Sets and Fuzzy Logic: Theory and Applications* (Prentice-Hall, Upper Saddle River, 1995)
6. H.T. Nguyen, E.A. Walker, *A First Course in Fuzzy Logic* (CRC Press, Boca Raton, 2006)

Chapter 35
How to Assess Teachers

Sometimes, the efficiency of a class is assessed by assessing the amount of knowledge that the students have after taking this class. However, this amount depends not only on the quality of the class, but also on how prepared were the students when they started taking this class. A more adequate assessment should therefore be *value-added*, estimating the added value that the class brought to the students.

In pedagogical practice, there are many value-added assessment models. However, most existing models have two limitations. First, they model the effect of the class as an additive factor independent on the initial knowledge. In reality, the amount of knowledge learned depends on the amount of the initial knowledge. Second, the existing models are statistical, they implicitly assume that the assessment values are objective – and are subject to random measurement errors and noises. In reality, many assessment values are subjective. Thus, fuzzy techniques provide, in our opinion, a more adequate way of processing these values.

In this chapter, we describe how the use of fuzzy techniques can help us overcome both limitations of the existing value-added assessments.

The results from this chapter first appeared in [1, 2].

Assessment is important. In order to improve the efficiency of education, it is important to assess this efficiency, i.e., to describe this efficiency in quantitative terms. This is important on all education levels: from elementary schools to middle and high schools to universities.

Quantitative description is needed because it allows natural comparison of different strategies of teaching and learning – and selection of the best strategy.

© Springer-Verlag GmbH Germany 2018
O. Kosheleva and K. Villaverde, *How Interval and Fuzzy Techniques Can Improve Teaching*, Studies in Computational Intelligence 750, https://doi.org/10.1007/978-3-662-55993-2_35

Traditional assessment. Sometimes, the efficiency of a class is assessed by assessing the amount of knowledge that the students have after taking this class. For example, we can take the average score of the students on some standardized test – this is actually how the quality of elementary and high school classes is now estimated in the US.

Limitation of the traditional assessment. The main problem with the traditional (outcome-only) assessment is that the class outcome depends

- not only on the quality of the class, but
- also on how prepared were the students when they started taking this class.

The idea of value-added assessment. Since the outcome depends on the initial level of students' knowledge and skills, a more adequate assessment should therefore be *value-added*, estimating the *added value* that the class brought to the students.

There exist several value-added assessment techniques. In pedagogical practice, there are many value-added assessment models; see, e.g., [3–5] and references therein.

Main idea behind the existing techniques. The main objective of these techniques is to estimate the added value. It therefore seems reasonable to evaluate this added value by subtracting the outcome from the input. For example, we can subtract the average grade after the class (on the post-test) on the average grade on similar questions asked before the class (on the pre-test).

This is, of course, the simplest possible approach. The existing techniques take into account additional parameters influencing learning. However, most existing models model the effect of the class as an additive factor independent on the initial knowledge.

Independence on the input is a limitation. In reality, the amount of knowledge learned depends on the amount of the initial knowledge. It is therefore desirable to take this dependence into account in value-added assessment.

Additional limitation. The existing models for value-added assessment are statistical. This means that these models implicitly assume that the assessment values are objective – and thus, are influencesd by well-defined random measurement errors and noises.

In reality, many assessment values come from grading, and are therefore somewhat subjective.

Natural idea: using fuzzy techniques. Since assessment are subjective, it is natural to use fuzzy techniques (see, e.g., [6, 7]) to process the corresponding values.

In this chapter, we describe how the use of fuzzy techniques can help us overcome both limitations of the existing value-added assessments.

Traditional approach to valued-added assessment: reminder. Value-added assessment described how the post-test result y depends on the pre-test result x.

As we have mentioned earlier, in the traditional approach we, in effect, assume that the post-test result y is obtained from the pre-test result x by adding a certain amount a of new knowledge (and new skills):

$$y \approx x + a.$$

Here, we say that y is only approximately equal to $x + a$, to take into account measurement errors, random fluctuations, and the effect of factors that we do not take into account in this simple model.

Traditional approach to valued-added assessment: graphical description. To make our text easier to understand, we will try to graphically illustrate all the dependencies. In these graphical explanation, we will assume – in line with the usual fuzzy techniques – that both the pre-test assessment x and the post-test assessment y take values from the interval $[0, 1]$, with 0 corresponding to the complete lack of knowledge and 1 to perfect knowledge.

Of course, if a student's pre-test knowledge is perfect or almost perfect, there is no sense for this student to take a class. Thus, we will assume that the pre-test value x only goes until a certain threshold t.

In these terms, the additive approximate dependence can be graphically represented as follows:

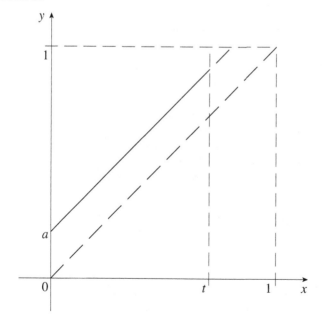

The actual dependence is non-additive. As we have mentioned, the actual dependence of the post-test value y on the pre-test value x is more complex, because the difference $y - x$ changes with x. To describe this dependence, we therefore need to use more general formulas than $y = x + a$.

First approximation: linear dependence. The natural next approximation is to use the general linear dependence of the post-test value y on the pre-test value x:

$$y \approx m \cdot x + a.$$

How to access the efficiency of the class under the new assessment model. With existing value-added assessment models, accessing the efficiency of a class appears straightforward: the higher the new knowledge amount a added, the better. For these models, the comparison of different teaching strategies is straightforward: we find the amount a corresponding to different strategies, and we select the strategy for which this amount a is the largest.

With the new teaching assessment models proposed in this chapter, accessing the efficiency of a class is a little bit more complex, since there are more parameters now. Specifically, the resulting efficiency of different teaching strategies depends not only on the strategy itself, but also on the prior knowledge of the class. For example, for two linear functions $f_1(x) = m_1 \cdot x + a_1$ and $f_2(x) = m_2 \cdot x + a_2$ corresponding to two different teaching strategies, we may have $f_1(x_1) < f_2(x_1)$ for some x_1 and $f_1(x_2) > f_2(x_2)$ for some $x_2 > x_1$. In this case,

- for weaker students, with prior knowledge x_1, the second strategy is better, while
- for stronger students, with prior knowledge $x_2 > x_1$, the first strategy is better.

Thus, the new model provides a more nuanced – and hence, more realistic – comparison between different teaching strategies.

In general, once we know the pre-test values $x_1, \ldots, x_n$ of different students of the class, we can use the known functions $f_1(x)$, $f_2(x)$, ..., describing different teaching strategies, and predict the post-test values $y_{1,j} = f_j(x_1)$, $y_{2,j} = f_j(x_2)$, ..., $y_{j,n} = f_j(x_n)$ for each strategy j. Then, for each teaching strategy j, we evaluate the value of our objective function – e.g., the mean post-test grade or a more sophisticated function (see examples below) – and select the strategy for which this value is the largest.

Linear dependence: examples. To better understand possible types of a linear dependence, let us describe and illustrate several examples of such a dependence.

Ideal case: perfect learning. In the ideal case, no matter what the original knowledge is, the resulting knowledge is perfect, $y \equiv 1$. The resulting constant function is a particular case of the general linear dependence, with $a = 1$ and $m = 0$:

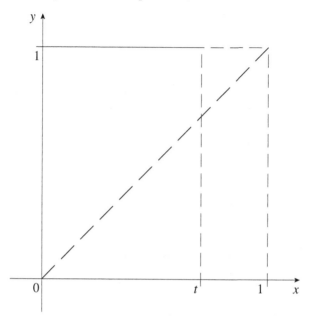

Example from middle schools and high schools. In many middle schools and high schools, one of the main objectives is to minimize the failure rate. Schools with high failure rate get penalized – and even disbanded.

Failure is most probable for students who start with the low starting knowledge, i.e., in our terms, with small values of x. Thus, to avoid failure, we must concentrate on the students with low x.

Since the amount of resources is limited, this means that only a few efforts are allocated to students with the originally higher level of x. As a result, the knowledge of students with a low level of x increases drastically, while the knowledge of students with high original knowledge level x does not increase that much (compared to the students with low original knowledge).

This behavior corresponds to a linear dependence in which $m < 1$. Here is a graphical illustration of such a dependence:

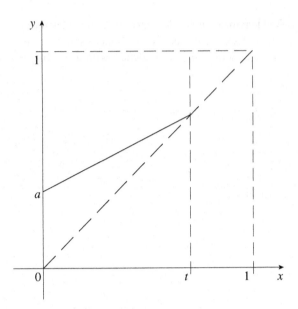

Typical top school strategy. In selective top schools, the emphasis is often on the top students. Due to the limited resources, this means that the knowledge of the top students, with $x \approx t$, increases drastically, practically to perfect knowledge, while the knowledge of the bottom students does not increase that much (compared to the students with high original knowledge).

In terms of a linear dependence, this means that we take $m > 1$. Here is a graphical illustration of such a dependence:

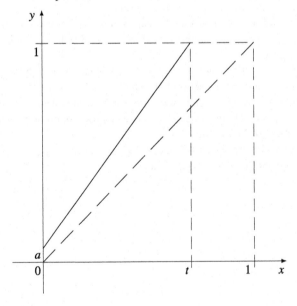

Problem: reminder. In the linear model, to quantitatively describe the success of the learning process, we must determine the parameters m and a of the corresponding linear dependence $y \approx m \cdot x + a$.

Crisp case. Let us start with a simple case when we know the exact values of the pre-test grades $x_1, \ldots, x_n$ and the exact values of the post-test grades $y_1, \ldots, y_n$. In this case, the problem is to find the values m and a for which

$$y_i \approx m \cdot x_i + a$$

for all $i = 1, 2, \ldots, n$.

Natural idea: use Least Squares method. We would like to make all n differences $e_i \stackrel{\text{def}}{=} y_i - (m \cdot x_i + a)$ close to 0. In other words, we want the vector $e = (e_1, \ldots, e_n)$ to be as close to the 0 vector $0 = (0, \ldots, 0)$ as possible.

The Euclidean distance between the vectors e and 0 is equal to

$$\sqrt{\sum_{i=1}^{n} e_i^2}.$$

Thus, the vector is the closest to 0 if this distance (or, equivalently, its square) is the smallest possible:

$$\sum_{i=1}^{n} e_i^2 \to \min,$$

or, equivalently,

$$\sum_{i=1}^{n} (y_i - (m \cdot x_i + a))^2 \to \min_{m,a}.$$

The resulting Least Squares approach is a standard approach in statistics; see, e.g., [8], because it is the optimal approach when the error are independent and normally distributed.

There exist good algorithms for solving the Least Squares problem. In our case, we can simply differentiate the minimized expression with respect to the unknowns m and a and equate the resulting derivatives to 0. As a result, we get explicit formulas for m and a:

$$m = \frac{C(x, y)}{V(x)},$$

where

$$C(x, y) \stackrel{\text{def}}{=} \frac{1}{n} \cdot \sum_{i=1}^{n} (x_i - E(x)) \cdot (y_i - E(y));$$

$$V(x) \stackrel{\text{def}}{=} \frac{1}{n} \cdot \sum_{i=1}^{n} (x_i - E(x))^2;$$

$$E(x) \stackrel{\text{def}}{=} \frac{1}{n} \cdot \sum_{i=1}^{n} x_i; \quad E(y) \stackrel{\text{def}}{=} \frac{1}{n} \cdot \sum_{i=1}^{n} y_i;$$

and

$$a = E(y) - m \cdot E(x).$$

As an example, let us show the values obtained based on the results of teaching CS2, the second class in the introductory Computer Science sequence, in Fall 2009:

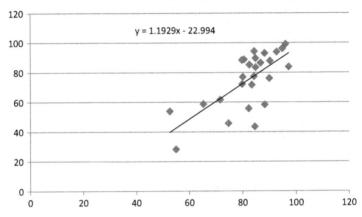

Case of interval uncertainty: description. In the previous section, we assumed that we know the numerical grade on the exam represents an exact measure of the student knowledge. In practice, however, the number grades are reasonably subjective.

Usually, instructors allocate certain number of points to different questions and problems on the test and to different aspects of the same question or problem. As a result, when the answer to each of the problems or questions is either absolutely correct or absolutely wrong (or missing), the resulting grade is uniquely determined. The subjectivity comes when the answer is partly correct, and we need to decide how much partial credit this answer deserves. Some such situations can be described from the very beginning, but often, it is not practically possible to foresee all possible mistakes and thus, to decide how much partial credit the student deserves.

Often, when two instructors co-teach a class or teach two different sections of the same class, their grades for similar mistakes can slightly differ – because of the slightly different allocation of partial credit. Even the same instructor, when grading two different student papers with similar mistakes, can sometimes assign two slightly different numerical grades.

As a result of this subjectivity, the numerical grade given to the test is not an exact measure of the student knowledge – because other instructors may assign a slightly different number grade to the same test results.

This subjectivity is well understood by instructors. This is one of the reasons why student transcripts usually list not the exact overall number grades, but rather the letter grades.

For example, usually, a letter grade A is assigned to all the numerical grades from 90 to 100, and a letter grade B is assigned to all numerical grades between 80 and 89. This assignment is in good accordance with the fact that

- while the difference between, say, 85 and 95 is meaningful and most probably not subjective, and a student with a grade of 95 has a higher knowledge level than a student with a grade of 85,
- the difference between, say 92 and 93 can be caused by the subjective reasons – and thus, a student with a grade of 93 does not necessarily know the material better than a student with a grade of 92.

The traditional letter grades may provide too crude a picture. In many cases, the distinction between, say, low 90s and high 90s also makes sense. To emphasize such a difference, some schools, in addition to usual letter grades, also use signed letter-type grades like A− or B+. Letter grades from the resulting set correspond to intervals which are narrower than the width-10 intervals describing the usual letter grades.

Because the distinction within each interval may be caused by the subjectivity of an individual instructor grading, it makes sense, when describing how well the students learned, to use not the original numerical grades x_i, but rather the corresponding letter grades – i.e., in other words, the intervals $\mathbf{x}_i = [\underline{x}_i, \overline{x}_i]$ that describe possible values of the student knowledge.

Comment. Education is, of course, not the only area where intervals appear. Intervals appear in many measurement situations where we only know the upper bound Δ_i on the measurement inaccuracy $\Delta x_i \stackrel{\text{def}}{=} \widetilde{x}_i - x_i$, i.e., on the difference between the measurement result $\widetilde{x}_i$ and the actual (unknown) value x_i of the measured quantity: $|\Delta x_i| \le \Delta_i$. In such cases, the only information that we have about the desired value x_i is that this value belongs to the interval $\mathbf{x}_i = [\widetilde{x}_i - \Delta_i, \widetilde{x}_i + \Delta_i]$.

There exist many algorithms for processing such interval uncertainty; see, e.g., [9–12].

How to describe dependence between x and y under interval uncertainty. We want to describe a dependence between the pre-test grade x and the post-test grade y.

In the crisp case, we have an exact grade x and we want to predict the exact grade y. In the ideal case, to every value x, we would like to assign the corresponding value y. In mathematical terms, this means that we would like to have a *function* $y = f(x)$ that maps numbers (= pre-test grades) into numbers (= post-test grades).

In the interval case, we start with an interval pre-test grade $\mathbf{x}$, and we would like to predict the interval post-test grade $\mathbf{y}$. Thus, to every interval value $\mathbf{x}$, we would like to assign the corresponding interval value $\mathbf{y}$. In mathematical terms, this means that we would like to have a *function* $\mathbf{y} = f(\mathbf{x})$ that maps intervals (= pre-test grades) into intervals (= post-test grades).

Number-to-number case. To analyze which interval-to-interval functions f can represent the map from pre-test to post-test intervals, let us first consider a simplified situation, in which,

- for each student, we know the pre-test and post-test grades which exactly describe the student's knowledge, and
- the student's post-test grade is uniquely determined by his or her pre-test grade.

In this simplified situation, due to uniqueness, the dependence between the student's pre-test grade x and his or her post-test grade y can be described by a number-to-number function $f(x)$.

In principle, an arbitrary mathematical mapping from real numbers to real numbers can occur in real learning. One might argue that we probably should require that $f(x) \geq x$, since the knowledge at the end cannot be smaller than the starting knowledge.

In reality, however, even this requirement is not necessary: people forget, so it is quite possible that without repetitions, some students will score much worse on a post-test than on the pre-test.

For intervals, there are additional restrictions on interval-to-interval functions: example. In our more realistic description, we do not know the exact value of the characteristic describing the student's pre-class and post-class knowledge; instead, for each student, we only know:

- the pre-test interval grade $\mathbf{x}$ that describes the possible values of the student's pre-class knowledge, and
- the post-test interval grade $\mathbf{y}$ that describes the possible values of the student's post-class knowledge.

Within this description, an interval-to-interval function $f(\mathbf{x})$ describes the set of all possible post-test grades for all the students who pre-test grades are within the interval $\mathbf{x}$.

Let us show that, in contrast to a number-to-number case where every mathematical number-to-number function could be potentially interpreted as a pre-test-to-post-test function f, in the more realistic interval-to-interval case, not all mathematically possible interval-to-interval functions can be thus interpreted: only interval-to-interval functions that satisfy a certain restriction can be interpreted as pre-test-to-post-test functions.

To explain this restriction, let us start with a simple example. Suppose that we know that

- when the pre-test grades are from the interval $\mathbf{x}_1 = [80, 90]$, then the post-test grade is from the interval $\mathbf{y}_1 = f(\mathbf{x}_1) = [85, 95]$; and
- when the pre-test grades are from the interval $\mathbf{x}_2 = [90, 100]$, then the post-test grade is from the interval $\mathbf{y}_2 = f(\mathbf{x}_2) = [92, 100]$.

What if now we have a student whose pre-test grade is between 80 and 100, i.e., for whom $\mathbf{x} = [80, 100]$.

In general, a mathematically defined interval-to-interval function can have any value of $f(\mathbf{x}) = f([80, 100])$, a value which is not necessarily related to the values $f(\mathbf{x}_1)$ and $f(\mathbf{x}_2)$. For example, from the purely mathematical viewpoint, we can have $f([80, 100]) = [50, 100]$.

However, in our case, the value $f(\mathbf{x})$ has a meaning – it is the set of all possible post-test grades of all the students whose pre-test grades are in the interval $\mathbf{x}$. Let us show that this meaning imposes a restriction on the possible interval-to-interval functions. Indeed, the interval $\mathbf{x} = [80, 100]$ is a union of the previous two intervals $\mathbf{x} = \mathbf{x}_1 \cup \mathbf{x}_2$, meaning that the student who has the actual pre-test grade in the interval $\mathbf{x}$ either has the actual grade between 80 and 90, or between 90 and 100.

- In the first case, when $x \in \mathbf{x}_1$, we expect that the final grade y is in the corresponding interval $\mathbf{y}_1$.
- In the second case, when $x \in \mathbf{x}_2$, we expect that the final grade y is in the corresponding interval $\mathbf{y}_2$.

Thus, we can conclude that y belongs either to the interval $\mathbf{y}_1$ or to the interval $\mathbf{y}_2$, i.e., that it belongs to the union

$$\mathbf{y}_1 \cup \mathbf{y}_2 = f(\mathbf{x}_1) \cup f(\mathbf{x}_2) = [85, 95] \cup [92, 100] = [85, 100]$$

of these two intervals. Thus, for the pre-test interval $\mathbf{x} = \mathbf{x}_1 \cup \mathbf{x}_2$, the set $f(\mathbf{x})$ of all possible values of post-test grades should be equal to

$$f(\mathbf{x}) = f(\mathbf{x}_1) \cup f(\mathbf{x}_2) = [85, 100].$$

For intervals, there are additional restrictions on interval-to-interval functions: general formulas. In general, the pre-test-to-post-test mapping f from intervals to intervals must satisfy the following property:

$$f(\mathbf{x}_1 \cup \mathbf{x}_2) = f(\mathbf{x}_1) \cup f(\mathbf{x}_2).$$

Similar argument leads us to the conclusion that

$$f(\mathbf{x}) = \bigcup_{x \in \mathbf{x}} f([x, x]).$$

Towards a description of all interval-to-interval functions that satisfy the above property: analysis. According to the above formula, to describe a pre-test-to-post-test interval-to-interval function $f(\mathbf{x})$, it is sufficient to describe a numbers-to-intervals function $f([x, x])$ corresponding to degenerate intervals of the type $[x, x]$. For each such degenerate value, let us denote the lower endpoint of the interval $f([x, x])$ by $\underline{f}(x)$, and its upper endpoint by $\overline{f}(x)$.

In these terms, the interval $f([x, x])$ corresponding to a degenerate interval $[x, x]$ has the form

$$f([x, x]) = \left[\underline{f}(x), \overline{f}(x)\right].$$

Thus, we have

$$f([\underline{x}, \overline{x}]) = \bigcup_{x \in [\underline{x}, \overline{x}]} \left[\underline{f}(x), \overline{f}(x)\right].$$

When the take the union of intervals, we thus take the minimum of their lower endpoints, and the maximum of their upper endpoints. Thus, the union $f([\underline{x}, \overline{x}])$ has the form

$$f([\underline{x}, \overline{x}]) = [\underline{y}, \overline{y}],$$

where

$$\underline{y} = \min_{x \in [\underline{x}, \overline{x}]} \underline{f}(x);$$

$$\overline{y} = \max_{x \in [\underline{x}, \overline{x}]} \overline{f}(x).$$

Usually, the better the original knowledge, the better results. Thus, both function $\underline{f}(x)$ and $\overline{f}(x)$ should be increasing with x. So, the minimum of $\underline{f}(x)$ is attained at the smallest possible value $x = \underline{x}$ and the maximum of $\overline{f}(x)$ is attained at the largest possible value $x = \overline{x}$.

Thus, we arrive at the following formula:

Resulting formula:

$$f([\underline{x}, \overline{x}]) = \left[\underline{f}(\underline{x}), \overline{f}(\overline{x})\right].$$

What we want to compute: a reminder. In the interval case, we want to find two functions $\underline{f}(x)$ and $\overline{f}(x)$ for real numbers to real numbers.

We consider the case of a linear dependence, so we assume that both functions are linear: $\underline{f} = \underline{m} \cdot x + \underline{a}$ and $\overline{f} = \overline{m} \cdot x + \overline{a}$.

These two linear functions must satisfy the following condition for all x: $\underline{f}(x) \le \overline{f}(x)$, i.e.,

$$\underline{m} \cdot x + \underline{a} \le \overline{m} \cdot x + \overline{a}.$$

Since the functions are linear, it is sufficient to require that these conditions be satisfied for $x = 0$ and $x = t$, i.e., that

$$\underline{a} \le \overline{a}$$

and

$$\underline{m} \cdot t + \underline{a} \le \overline{m} \cdot t + \overline{a}.$$

Example. Here is a graphical illustration of the corresponding two functions:

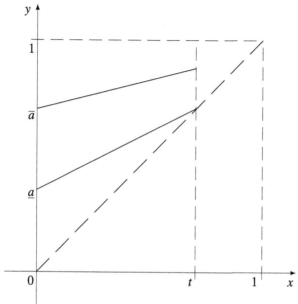

Comment. On this example, we see that it is possible to have $\underline{m} > \overline{m}$. Thus,

- while the corresponding estimates $\underline{a}$ and $\overline{a}$ for a do satisfy the inequality $\underline{a} \le \overline{a}$ and thus, form an interval $[\underline{a}, \overline{a}]$,
- the estimates $\underline{m}$ and $\overline{m}$ for m do not necessarily form an interval.

Estimating $\underline{m}$, $\underline{a}$, $\overline{m}$, and $\overline{a}$: a problem. We are given intervals $\mathbf{x}_i = [\underline{x}_i, \overline{x}_i]$ and $\mathbf{y}_i = [\underline{y}_i, \overline{y}_i]$. We would like to find the values of the parameters $\underline{m}$, $\underline{a}$, $\overline{m}$, and $\overline{a}$ for which

$$\mathbf{y}_i \approx f(\mathbf{x}_i)$$

for the corresponding interval-to-interval function f.

Due to the above representation of a general linear interval-to-interval function, this means that for every i, we must have

$$\underline{y}_i \approx \underline{m} \cdot \underline{x}_i + \underline{a}$$

and

$$\overline{y}_i \approx \overline{m} \cdot \overline{x}_i + \overline{a}.$$

Estimating $\underline{m}$, $\underline{a}$, $\overline{m}$, and $\overline{a}$: analysis. We see that we have, in effect, two independent sets of approximate equalities:

- to find $\underline{m}$ and $\underline{a}$, we use approximate equalities

$$\underline{y}_i \approx \underline{m} \cdot \underline{x}_i + \underline{a};$$

- to find $\overline{m}$ and $\overline{a}$, we use approximate equalities

$$\overline{y}_i \approx \overline{m} \cdot \overline{x}_i + \overline{a}.$$

For each set of approximate equalities, we can apply the same Least Squares approach as we described for the case of crisp estimates. As a result, we arrive at the following formulas.

Resulting algorithm. First, we compute

$$\underline{m} = \frac{C(\underline{x}, \underline{y})}{V(\underline{x})},$$

where

$$C(\underline{x}, \underline{y}) \stackrel{\text{def}}{=} \frac{1}{n} \cdot \sum_{i=1}^{n} (\underline{x}_i - E(\underline{x})) \cdot (\underline{y}_i - E(\underline{y}));$$

$$V(\underline{x}) \stackrel{\text{def}}{=} \frac{1}{n} \cdot \sum_{i=1}^{n} (\underline{x}_i - E(\underline{x}))^2;$$

$$E(\underline{x}) \stackrel{\text{def}}{=} \frac{1}{n} \cdot \sum_{i=1}^{n} \underline{x}_i; \quad E(\underline{y}) \stackrel{\text{def}}{=} \frac{1}{n} \cdot \sum_{i=1}^{n} \underline{y}_i;$$

and

$$\underline{a} = E(\underline{y}) - \underline{m} \cdot E(\underline{x}).$$

Then, we compute

$$\overline{m} = \frac{C(\overline{x}, \overline{y})}{V(\overline{x})},$$

where

$$C(\overline{x}, \overline{y}) \stackrel{\text{def}}{=} \frac{1}{n} \cdot \sum_{i=1}^{n} (\overline{x}_i - E(\overline{x})) \cdot (\overline{y}_i - E(\overline{y}));$$

$$V(\overline{x}) \stackrel{\text{def}}{=} \frac{1}{n} \cdot \sum_{i=1}^{n} (\overline{x}_i - E(\overline{x}))^2;$$

$$E(\overline{x}) \stackrel{\text{def}}{=} \frac{1}{n} \cdot \sum_{i=1}^{n} \overline{x}_i; \quad E(\overline{y}) \stackrel{\text{def}}{=} \frac{1}{n} \cdot \sum_{i=1}^{n} \overline{y}_i;$$

and

$$\overline{a} = E(\overline{y}) - \overline{m} \cdot E(\overline{x}).$$

Actual grades are fuzzy. In the previous section, we assumed that we while we do not know the exact value x characterizing the student's knowledge, we know the interval $[\underline{x}, \overline{x}]$ that is guaranteed to contain the actual (unknown) value x.

This assumption is also an idealization. In reality, the bounds that we know are "fuzzy", i.e., they contain x only with some degree of confidence α.

From the computational viewpoint, fuzzy simply means a family of intervals. Usually, we have different intervals $[\underline{x}(\alpha), \overline{x}(\alpha)]$ corresponding to different degrees α. The narrower the interval, the less confident we are that x belongs to this interval. Thus, we have a *nested family* of intervals corresponding to different values α.

Alternatively, for each possible value of x, we describe the degree $\mu_i(x)$ to which this value is possible. For each degree of certainty α, we can determine the set of values of x_i that are possible with at least this degree of certainty – the α-*cut* $\mathbf{x}(\alpha) = \{x \mid \mu(x) \geq \alpha\}$ of the original fuzzy set.

Vice versa, if we know α-cuts for every α, then, for each object x, we can determine the degree of possibility that x belongs to the original fuzzy set [6, 7, 13–15].

A fuzzy set can be thus viewed as a nested family of its (interval) α-cuts.

Formulation of the problem. We have:

• fuzzy numbers X_i describing the pre-test grades and
• fuzzy numbers Y_i describing the post-test grades.

We would like to describe the corresponding dependence of the post-test grade Y on a pre-test grade X.

Resulting algorithm. For each α, we can take the α-cuts $\mathbf{x}_i(\alpha)$ and $\mathbf{y}_i(\alpha)$ of the corresponding fuzzy numbers X_i and Y_i.

For each α, based on these intervals, we can now use the above Least Squares method to find the interval-values linear function

$$[\underline{m}(\alpha) \cdot x + \underline{a}(\alpha), \overline{m}(\alpha) \cdot x + \overline{a}(\alpha)]$$

corresponding to this α.

The list of all these interval-values linear functions corresponding to different values α forms the desired description of the dependence of the post-test grade y on the pre-test grade x – i.e., the desired value-added teacher assessment.

How to use the resulting fuzzy estimates to compare different teaching strategies.

• Once we know the fuzzy numbers $X_1, \ldots, X_n$ describing the pre-class knowledge of students from the class, we thus have α-cut intervals corresponding to different α.
• Now, based on the functions corresponding to different α, we can, for every possible teaching strategy j, form α-cuts and thus, fuzzy numbers $Y_{1,j}, \ldots, Y_{n,j}$ describing expected post-test results.

- For each j, we can then apply the objective function and get the fuzzy values V_j corresponding to different teaching strategies.
- We then use one of the known fuzzy optimization techniques to select the teaching strategy which is the most appropriate for a given situation.

References

1. K. Villaverde, O. Kosheleva, Towards more detailed value–added teacher assessments, in *Proceedings of the IEEE World Congress on Computational Intelligence WCCI'2010, Barcelona, Spain, 18–23 July 2010* (2010), pp. 3126–3133
2. K. Villaverde, O. Kosheleva, Towards more detailed value-added teacher assessments: how intervals can help. J. Uncertain Syst. **6**(2), 128–137 (2012)
3. C. Fincher, What is value-added education? Res. High. Educ. **22**(4), 395–398 (1985)
4. D.F. McCaffrey, J.R. Lockwood, D.M. Koretz, L.S. Hamilton, *Evaluating Value-Added Models for Teacher Accountability* (RAND Corporation, Santa Monica, 2003)
5. W.L. Sanders, S.R. Horn, The Tennessee value-added assessment system (TVAAS): mixed-model methodology in educational assessment. J. Pers. Eval. Educ. **8**(9), 299–311 (1994)
6. G.J. Klir, B. Yuan, *Fuzzy Sets and Fuzzy Logic: Theory and Applications* (Prentice-Hall, Upper Saddle River, 1995)
7. H.T. Nguyen, E.A. Walker, *A First Course in Fuzzy Logic* (CRC Press, Boca Raton, 2006)
8. D.J. Sheskin, *Handbook of Parametric and Nonparametric Statistical Procedures* (Chapman & Hall/CRC, Boca Raton, 2007)
9. Interval computations website, http://www.cs.utep.edu/interval-comp
10. L. Jaulin, M. Kieffer, O. Didrit, E. Walter, *Applied Interval Analysis, with Examples in Parameter and State Estimation, Robust Control and Robotics* (Springer, London, 2001)
11. V. Kreinovich, E. Johnson-Holubec, L.K. Reznik, M. Koshelev, Cooperative learning is better: explanation using dynamical systems, fuzzy logic, and geometric symmetries, in *Proceedings of the Vietnam–Japan Bilateral Symposium on Fuzzy Systems and Applications VJFUZZY'98, HaLong Bay, Vietnam, 30th Sept–2nd Oct 1998*, ed. by H.P. Nguyen, A. Ohsato (1988), pp. 154–160
12. R.E. Moore, R.B. Kearfott, M.J. Cloud, *Introduction to Interval Analysis* (SIAM Press, Philadelphia, 2009)
13. D. Dubois, H. Prade, Operations on fuzzy numbers. Int. J. Syst. Sci. **9**, 613–626 (1978)
14. R.E. Moore, W. Lodwick, Interval analysis and fuzzy set theory. Fuzzy Sets Syst. **135**(1), 5–9 (2003)
15. H.T. Nguyen, O. Kosheleva, V. Kreinovich, Is the success of fuzzy logic really paradoxical? or: towards the actual logic behind expert systems. Int. J. Intell. Syst. **11**, 295–326 (1996)

Chapter 36
How to Assess Teaching Teachniques

There are many papers that experimentally compare effectiveness of different teaching techniques. Most of these papers use traditional statistical approach to process the experimental results. The traditional statistical approach is well suited to numerical data but often, what we are processing is either intervals (e.g., A means anything from 90 to 100) or fuzzy-type perceptions, words from the natural language like "understood well" or "understood reasonably well". We show that the use of intervals and fuzzy techniques leads to more adequate processing of educational data.

The results from this chapter first appeared in [1, 2].

Practical problem: comparing teaching techniques. Teaching is very important, and teaching is not always very effective. There exist many different pedagogical techniques that help teach better, and new teaching techniques are being invented all the time. To select the techniques which are the most efficient for a given educational environment, we must experimentally compare effectiveness of different teaching techniques in this and similar environments.

Traditional approach to solving this problem. There exist numerous papers that perform this experimental comparison. The vast majority of these papers use traditional statistical techniques (see, e.g., [3, 4]) to process the experimental results.

Namely, usually, the results (grades, degree of satisfaction, etc.) are translated into numbers, and then these numbers are processed by using the standard statistical techniques.

Problems with the traditional approach: general description. The traditional statistical approach is well suited for processing numerical data. However, in processing educational data, often what we are processing is:

- either *intervals*: e.g., the A grade usually means anything from 90 to 100, the B grade means anything between 80 and 90, and the C grade mean anything between 70 and 80;

© Springer-Verlag GmbH Germany 2018
O. Kosheleva and K. Villaverde, *How Interval and Fuzzy Techniques Can Improve Teaching*, Studies in Computational Intelligence 750,
https://doi.org/10.1007/978-3-662-55993-2_36

- or *fuzzy*-type perceptions, words from the natural language like "understood well" or "understood reasonably well".

Problems with the traditional approach: example. In selecting a teaching method, it is important not only to make sure that the *average* results m are good – e.g., that the average grade on a standard test is good – but also to ensure that the results are *consistently* good – i.e., in statistical terms, that the standard deviation σ of the grade is low.

If the standard deviation σ is high, that would mean while some student learn really well under this technique, there are many others who are left behind, and we cannot afford that.

So, to compare several teaching techniques based on the grades the student got, we must compare not only their averages, but also the standard deviations.

The following simple example will show that when we replace an interval with a single value, we lose important information that could influence the computation of the standard deviation, and we could get erroneous results.

Suppose that in one method, all the students got Bs, while in the other method, half of the students got Bs and half of the students got As. Which of the two methods shows more stable results, with a smaller standard deviation?

In the traditional statistical approach, we interpret A as 4 and B as 3.

- In the first method, the resulting grades are $x_1 = \cdots = x_n = 3$, so the average grade is equal to $m = (x_1 + \cdots + x_n)/n = 3$, and the population variance is equal to $V = \sigma^2 = \dfrac{1}{n} \cdot \sum_{i=1}^{n} (x_i - m)^2 = 0$.

- In the second method, the average is equal to $m = (3+4)/2 = 3.5$, so for each i, we have $(x_i - m)^2 = 0.25$, hence the standard deviation is equal to

$$V = \sigma^2 = \frac{1}{n} \cdot \sum_{i=1}^{n} (x_i - m)^2 = 0.25.$$

So, if we use the traditional statistical approach, we conclude that while the second method has a higher average, it is less stable than the first one.

In reality, if we go back from the "interval" grades like A, B, and C to the original grades, it may turn out the second method is not only better on average, but also much more stable. Indeed, suppose that:

- in the first method, half of the students got a grade 80, and half got a grade 88; and
- in the second method, half of the students got a grade 89, and half of the student got a grade 91.

In terms of As and Bs, this is exactly the situation as described above. However, when we compute the standard deviation for these actual grades, we get a different result than when we process the letter grades:

- In the first method, the average is equal to $m = (80 + 88)/2 = 84$, so for each i, $(x_i - m)^2 = 16$, hence the standard deviation is equal to

$$V = \sigma^2 = \frac{1}{n} \cdot \sum_{i=1}^{n}(x_i - m)^2 = 16.$$

- In the second method, the average is equal to $m = (89 + 91)/2 = 90$, so for each i, $(x_i - m)^2 = 1$, hence the standard deviation is equal to

$$V = \sigma^2 = \frac{1}{n} \cdot \sum_{i=1}^{n}(x_i - m)^2 = 1 \ll 16.$$

What needs to be done. It is desirable to develop techniques for processing educational data that would take into account that the grades are not exactly equal to the corresponding numerical values but may differ from these values.

In other words, we need techniques that would provide guaranteed answers to questions like: Is the first method better than the second one? It is OK to have an answer "we do not know", but if the answer is "yes", we want to be sure that no matter what additional information we learn about these experiments the answer will remain the same.

Such techniques are outlined in this chapter.

Processing interval data: analysis of the situation. The main reason why we had the above problem is that letter grade ℓ represents not a *single* value of the number grade x, but rather an *interval* $\mathbf{x} = [\underline{x}, \overline{x}]$ of possible values of the number grade. For example:

- the letter grade A represents the interval $[90, 100]$;
- the letter grade B represents the interval $[80, 90]$;
- the letter grade C represents the interval $[70, 80]$.

Processing interval data: formulation of the problem. Our objective is, given a set of letter grades $\ell_1, \ldots, \ell_n$, to compute a certain statistical characteristic C such as average, standard deviation, correlation with other characteristics (such as the family income or the amount of time that a student spends on homeworks), etc.

The desired statistical characteristic is defined in terms of numerical values, as $C = C(x_1, \ldots, x_n)$. For example, the average is defined as $m = \dfrac{x_1 + \cdots + x_n}{n}$, the variance is defined as $V = \dfrac{1}{n} \cdot \sum_{i=1}^{n}(x_i - m)^2$, etc.

For the educational data, instead of the *exact* values x_i, we often only know the *intervals* $\mathbf{x}_i$ corresponding to the letter grade ℓ_i. For different possible values $x_1 \in \mathbf{x}_1$, ..., $x_n \in \mathbf{x}_n$, we get different values of the corresponding characteristic C.

Our objective is to to find the range of possible values of the desired characteristic when $x_i \in \mathbf{x}_i$, i.e., the interval

$$\mathbf{C} = \{C(x_1, \ldots, x_n) \mid x_1 \in \mathbf{x}_1, \ldots, x_n \in \mathbf{x}_n\}.$$

This problem is a particular case of the general problem of interval computations. The need to perform computations under interval uncertainty occurs in many areas of science and engineering. In many such areas, we therefore face the following problem:

- we know:
  - n intervals $\mathbf{x}_1, \ldots, \mathbf{x}_n$ and
  - an algorithm $y = f(x_1, \ldots, x_n)$ that transforms n real numbers (inputs) into a single number y (result of data processing);
- we must estimate the range of possible values of y, i.e., the interval

$$\mathbf{y} = \{f(x_1, \ldots, x_n) \mid x_1 \in \mathbf{x}_1, \ldots, x_n \in \mathbf{x}_n\}.$$

This problem is called the main problem of *interval computations*; see, e.g., [5–9].

We can therefore conclude that the problem of processing educational data under interval uncertainty is a particular case of the more general problem of interval computations.

How we can process interval data: general description. Many efficient techniques have been developed to solve generic interval computations problems; see, e.g., [5–9].

How we can process interval data: case of statistical characteristics. In particular, several algorithms have been developed for the case when the the function $f(x_1, \ldots, x_n)$ is one of the standard statistical characteristics such as average m or standard deviation V; see, e.g. [10, 11] and references therein.

Computing average under interval uncertainty. In particular, since the average is a monotonic function of each of its variables, its value is the largest when each x_i attains the largest possible value $x_i = \overline{x}_i$, and its value is the smallest when the variance attains its smallest possible value $\underline{x}_i$. Thus, for the average m, the interval takes the form $[\underline{m}, \overline{m}]$, where

$$\underline{m} = \frac{\underline{x}_1 + \cdots + \underline{x}_n}{n}; \quad \overline{m} = \frac{\overline{x}_1 + \cdots + \overline{x}_n}{n}.$$

If all the letter grades are A, B, C, or D, then the width $\overline{x}_i - \underline{x}_i$ of each corresponding interval is 10, so $\overline{m} = \underline{m} + 10$. In this situation, it is sufficient to compute *one* of the bounds $\overline{m}$ or $\underline{m}$, the other bound can be easily reconstructed from this one.

If one of the grades is a F grade, for which the interval of possible values is [0, 60] with a width $60 > 10$, then we must compute *both* bounds.

Computing variance under interval uncertainty. For the variance V, there exist efficient algorithms for computing the lower bound $\underline{V}$, but the problem of computing the upper bound $\overline{V}$ is, in general, NP-hard. However, for educational data, the intervals only intersect at a single point. For such data, there exist efficient algorithms for computing $\overline{V}$.

Specifically, to compute $\overline{V}$ in such a situation, we sort the grades into an increasing sequence for which $\underline{x}_1 \leq \underline{x}_2 \leq \cdots \leq \underline{x}_n$ and $\overline{x}_1 \leq \overline{x}_2 \leq \cdots \leq \overline{x}_n$. For every k from 1 to n, we pick $x_i = \underline{x}_i$ for $i \leq k$ and $x_i = \overline{x}_i$ for $i > k$; then, we compute the average $m = \dfrac{x_1 + \cdots + x_n}{n}$ of the selected x_i, and check whether this average satisfies the inequality $\underline{x}_k \leq m \leq \overline{x}_{k+1}$. If it does, then the population variance of the corresponding sequence $x_1, \ldots, x_n$ is exactly the desired upper bound $\overline{V}$.

To compute the lower bound $\underline{V}$, similarly, for every k, we select:

- $x_i = \overline{x}_i$ when $\overline{x}_i \leq \underline{x}_k$, and
- $x_i = \underline{x}_i$ when $\underline{x}_i \geq \overline{x}_k$.

We then compute the average m of the selected x_i and check whether this average satisfies the inequality $\underline{x}_k \leq m \leq \overline{x}_k$. If it does, then we assign $x_i = m$ for all the un-assigned value i, and the population variance of the corresponding sequence $x_1, \ldots, x_n$ is exactly the desired lower bound $\underline{V}$.

Numerical example. For three sorted grades C, B, and A, we get $\underline{x}_1 = 70, \overline{x}_1 = 80$, $\underline{x}_2 = 80, \overline{x}_2 = 90, \underline{x}_3 = 90, \overline{x}_3 = 100$. For this data, let us first compute $\overline{V}$.

For $k = 1$, we pick $x_1 = \underline{x}_1 = 70$, $x_2 = \overline{x}_2 = 90$, and $x_3 = \overline{x}_3 = 100$. Here,

$$m = \frac{x_1 + x_2 + x_3}{3} = 86\frac{2}{3}.$$

Since $\underline{x}_1 = 60 \leq m \leq \overline{x}_2 = 90$, the upper bound $\overline{V}$ is equal to the population variance

$$\frac{1}{n} \cdot \sum_i (x_i - m)^2$$

of the values $x_1 = 70, x_2 = 90$, and $x_3 = 100$, hence

$$\overline{V} = 102\frac{2}{9}.$$

For $\underline{V}$, we also start with $k = 1$. For this k, in accordance with the above algorithm, we assign the values $x_2 = \underline{x}_2 = 80$ and $x_3 = \underline{x}_3 = 90$. Their average $m = 85$ is outside the interval $[\underline{x}_1, \overline{x}_1] = [70, 80]$, so we have to consider the next k.

For $k = 2$, we assign $x_1 = \overline{x}_1 = 80$ and $x_3 = \underline{x}_3 = 90$. The average $m = 85$ of these two values satisfies the inequality $\underline{x}_2 = 80 \leq m \leq \overline{x}_2 = 90$; hence we assign $x_2 = 85$, and compute $\underline{V}$ as the population variance of the values $x_1 = 80, x_2 = 85$, and $x_3 = 90$. So, $\underline{V} = 16\frac{2}{3}$.

Computing other statistical characteristics under interval uncertainty. Similar algorithms are known for other statistical characteristic such as median, higher moments, covariance, etc. [10, 11].

Formulation of the problem. The main idea behind fuzzy uncertainty (see, e.g., [12, 13]) is that, instead of just describing which objects (in our case, grades) are possible, we also describe, for each object x, the degree $\mu(x)$ to which this object is possible. For each degree of possibility α, we can determine the set of objects that are possible with at least this degree of possibility – the α-cut $\{x \mid \mu(x) \geq \alpha\}$ of the original fuzzy set. Vice versa, if we know α-cuts for every α, then, for each object x, we can determine the degree of possibility that x belongs to the original fuzzy set [12–16].

A fuzzy set can be thus viewed as a nested family of its α-cuts.

How we can process fuzzy data: general idea. If instead of a (crisp) interval $\mathbf{x}_i$ of possible grades, we have a fuzzy set $\mu_i(x)$ of possible grades, then we can view this information as a family of nested intervals $\mathbf{x}_i(\alpha)$ – α-cuts of the given fuzzy sets.

Our objective is then to compute the fuzzy number corresponding to this the desired characteristic $C(x_1, \ldots, x_n)$.

In this case, for each level α, to compute the α-cut of this fuzzy number, we can apply the interval algorithm to the α-cuts $\mathbf{x}_i(\alpha)$ of the corresponding fuzzy sets. The resulting nested intervals form the desired fuzzy set for C.

How we can process fuzzy data: case of statistical characteristics. For statistical characteristics such as variance, more efficient algorithms are described in [17].

Towards combining probabilistic, interval, and fuzzy uncertainty: need for such a combination. In the case of interval uncertainty, we consider all possible values of the grades, and do not make any assumptions about the probability of different values within the corresponding intervals. However, in many cases, we can make commonsense conclusions about the frequency of different grades.

For example, if a student has almost all As but only one B, this means that this is a strong student, and most probably this B is at the high end of the B interval. On the other hand, if a student has almost all Cs but only one B, this means that this is a weak student, and most probably this B is at the lower end of the B interval. It is desirable to take such arguments into account when processing educational data.

Let us describe how we can do this.

Simplest case: normally distributed grades. Let us first consider the reasonable case when the actual number grades are normally distributed, with an (unknown) mean m and an unknown standard deviation σ. In other words, we assume that the cumulative probability distribution (cdf) $F(x) \stackrel{\text{def}}{=} \text{Prob}(\xi < x)$ has the form $F_0\left(\dfrac{x - m}{\sigma}\right)$, where $F_0(x)$ is the cdf of the standard Gaussian distribution with 0 mean and unit standard deviation. Our objective is to determine the values a and σ.

If we knew the values of the number grades x_i, then we could apply the above statistics and estimate a and $\sigma = \sqrt{V}$. In many situations, we do not know the values of the *number* grades, we only know the values of the *letter* grades. How can we then estimate a and σ based on these letter grades?

Case of normally distributed grades: towards an algorithm. Based on the letter grades, we can find, for the threshold values 60, 70, etc., the frequency with which we have the grade smaller that this threshold. If we denote by f the proportion of F grades, by d the proportion of D grades, etc., then the frequency of $x < 60$ is f, the frequency of $x < 70$ is $f + d$, the frequency of $x < 60$ is $f + d + c$.

It is well known that the probability can be defined as a limit of the corresponding frequency when the sample size n increases. Thus, when the sample size is large enough, we can safely assume that the corresponding frequencies are close to the corresponding probabilities, i.e., to the values $F(x)$. In other words, we conclude that:

$$F_0\left(\frac{60 - m}{\sigma}\right) \approx f; \quad F_0\left(\frac{70 - m}{\sigma}\right) \approx f + d;$$

$$F_0\left(\frac{80 - m}{\sigma}\right) \approx f + d + c; \quad F_0\left(\frac{90 - m}{\sigma}\right) \approx f + d + c + b.$$

If we denote by $\psi_0(t)$ the function that is inverse to $F_0(t)$, then, e.g., the first equality takes the form $60 - m/\sigma \approx \psi_0(f)$, i.e., $\sigma \cdot \psi_0(f) + m \approx 60$. Thus, to find the unknowns m and σ, we get a system of linear equations:

$$\sigma \cdot \psi_0(f) + m \approx 60; \quad \sigma \cdot \psi_0(f + d) + m \approx 70;$$

$$\sigma \cdot \psi_0(f + d + c) + m \approx 80; \quad \sigma \cdot \psi_0(f + d + c + b) + m \approx 90,$$

which can be solved, e.g., by using the Least Squares Method.

Comment. In some cases, the distribution is non-Gaussian, and we know its shape, i.e., we know that $F(x) = F_0((x - m)/\sigma)$, where $F_0(t)$ is a known function, and m and σ are unknown parameters. In this case, we can use the same formulas as above.

Simplified case when all the grades are C or above. In many cases, only C and above is an acceptable grade. In such situations, $f = d = 0$ and $c + b + a = 1$, so we get a simplified system of two linear equations with two unknowns:

$$\sigma \cdot \psi_0(c) + m = 80; \quad \sigma \cdot \psi_0(c + b) + m = 90.$$

Subtracting the first equation from the second one, we conclude that

$$\sigma = \frac{10}{\psi_0(b + c) - \psi_0(c)}.$$

This formula can be further simplified if the distribution $F_0(x)$ is symmetric (e.g., Gaussian distribution is symmetric), i.e., for every x, the probability $F_0(-x)$ that $\xi \le -x$ is equal to the probability $1 - F_0(x)$ that $\xi \ge x$. Thus, we can conclude that $\psi_0(1 - x) = -\psi_0(x)$ for every x. In particular, since $c + b + a = 1$, we conclude that $-\psi_0(c + b) = \psi_0(1 - (c + b)) = \psi_0(a)$. Thus, the formula for σ takes the form:

$$\sigma = -\frac{10}{\psi_0(a) + \psi_0(c)}. \tag{36.1}$$

Similarly, if we divide the equation $(90 - m)/\sigma = \psi_0(b + c)$ by $(80 - m)/\sigma = \psi_0(c)$, we conclude that

$$\frac{90 - m}{80 - m} = \frac{\psi_0(b + c)}{\psi_0(c)} = -\frac{\psi_0(a)}{\psi_0(c)},$$

hence

$$m = 80 + \frac{1}{10 + \dfrac{\psi_0(a)}{\psi_0(c)}}. \tag{36.2}$$

Relation to fuzzy logic. As we can see from the formulas (36.1) and (36.2), the standard deviation is an increasing function of the sum $\psi_0(a) + \psi_0(c)$, while the mean m is monotonically increasing with the ratio $\psi_0(a)/\psi_0(c)$. This makes sense of we take into account that $\psi_0(a)$ monotonically depends on the proportion a of grades in the A range: the more grades are in the A range and the fewer grades are in the C range, the larger the average grade m, so m should be kind of monotonically depending on the degree to which is is true that we have A grades and not C grades.

It is worth mentioning that the operations of sum as "or" and ratio as "a and not c" appear when we try to interpret neural networks in terms of fuzzy logic [18]; see also the sections at the end of the chapter.

Selecting an "or" operation. The degree of belief a in a statement A can be estimated as proportional to the number of arguments in favor of A. In principle, there exist infinitely many potential arguments, so in general, it is hardly probable that when we pick a arguments out of infinitely many and then b out of infinitely many, the corresponding sets will have a common element. Thus, it is reasonable to assume that every argument in favor of A is different from every argument in favor of B. Under this assumption, the total number of arguments in favor of A and arguments in favor of B is equal to $a + b$. Hence, the natural degree of belief in $A \vee B$ is proportional to $a + b$.

Selecting an "and" operation. Different experts are reliable to different degrees. Our degree of belief in a statement A made by an expert is equal to $w \mathbin{\&} a$, where w is our degree of belief in this expert, and a is the expert's degree of belief in the statement A. What are the natural properties of the "and"-operation?

First, since $A \mathbin{\&} B$ means the same as $B \mathbin{\&} A$, it is reasonable to require that the corresponding degrees $a \mathbin{\&} b$ and $b \mathbin{\&} a$ should coincide, i.e., that the "and"-operation be commutative.

Second, when an expert makes two statements B and C, then our resulting degree of belief in $B \vee C$ can be computed in two different ways:

- We can first compute *his* degree of belief $b \vee c$ in $B \vee C$, and then us the "and"-operation to generate our degree of belief $w \& (b \vee c)$.
- We can also first generate our degrees $w \& b$ and $w \& c$, and then use an "or"-operation to combine these degrees, arriving at $(w \& b) \vee (w \& c)$.

It is natural to require that both ways lead to the same degree of belief, i.e., that the "and"-operation be distributive with respect to $\vee$.

It is also reasonable to assume that the value $w \& a$ is a monotonically (non-strictly) increasing function of each its variables.

It can be shown [18] that every commutative, distributive, and monotonic operation $\& : R \times R \to R$ has the form $a \& b = C \cdot a \cdot b$ for some $C > 0$. This expression can be further simplified if we introduce a new scale of degrees of belief $a' \overset{\text{def}}{=} C \cdot a$; in the new scale, $a \& b = a \cdot b$.

Selecting a crisp truth value. We know that "true" and "true" is "true", and that "false" and "false" is "false". Thus, it is reasonable to call a positive degree of belief e_0 is a crisp value if $e_0 \& e_0 = e_0$.

This implies that $e_0 = 1$.

Selecting implication and negation. From the commonsense viewpoint, an implication $A \to B$ is a statement C such that if we add C to B, we get A. Thus, it is natural to define an *implication operation* as a function $\to : R \times R \to R$ for which, for all a and b, we have $(a \to b) \& a = b$. One can easily check that $a \to b = b/a$.

Negation $\neg A$ can be viewed as a particular case of implication, $A \to F$, for a crisp (specifically, false) value F. Thus, we can define negation operation as $a \to e_0$, i.e., as $1/a$.

References

1. O.M. Kosheleva, M. Ceberio, Processing educational data: from traditional statistical techniques to an appropriate combination of probabilistic, interval, and fuzzy approaches, in *Proceedings of the International Conference on Fuzzy Systems, Neural Networks, and Genetic Algorithms FNG'05*, 13–14 Oct 2005 (Tijuana, Mexico, 2005), pp. 39–48
2. O. Kosheleva, V. Kreinovich, L. Longpre, M. Tchoshanov, G. Xiang, Towards interval techniques for processing educational data, in *IEEE Proceedings of the 12th GAMM–IMACS International Symposium on Scientific Computing, Computer Arithmetic and Validated Numerics*, Duisburg, Germany, 26–29 Sept 2006
3. D.J. Sheskin, *Handbook of Parametric and Nonparametric Statistical Procedures* (Chapman & Hall/CRC, Boca Raton, Florida, 2007)
4. H.M. Wadsworth Jr. (ed.), *Handbook of Statistical Methods for Engineers and Scientists* (McGraw-Hill Publishing Co., New York, 1990)
5. L. Jaulin, M. Kieffer, O. Didrit, E. Walter, *Applied Interval Analysis, with Examples in Parameter and State Estimation, Robust Control and Robotics* (Springer, London, 2001)
6. R.B. Kearfott, *Rigorous Global Search: Continuous Problems* (Kluwer, Dordrecht, 1996)
7. R.B. Kearfott, V. Kreinovich (eds.), *Applications of Interval Computations* (Kluwer, Dordrecht, 1996)
8. V. Kreinovich, D. Berleant, M. Koshelev, website on interval computations, http://www.cs.utep.edu/interval-comp

9. R.E. Moore, *Methods and Applications of Interval Analysis* (SIAM, Philadelphia, 1979)

10. S. Ferson, L. Ginzburg, V. Kreinovich, L. Longpré, M. Aviles, Exact bounds on finite populations of interval data. Reliab. Comput. **11**(3), 207–233 (2005)

11. V. Kreinovich, G. Xiang, S.A. Starks, L. Longpré, M. Ceberio, R. Araiza, J. Beck, R. Kandathi, A. Nayak, R. Torres, J. Hajagos, Towards combining probabilistic and interval uncertainty in engineering calculations: algorithms for computing statistics under interval uncertainty, and their computational complexity. Reliab. Comput. **12**(6), 471–501 (2006)

12. G.J. Klir, B. Yuan, *Fuzzy Sets and Fuzzy Logic: Theory and Applications* (Prentice-Hall, Upper Saddle River, New Jersey, 1995)

13. H.T. Nguyen, E.A. Walker, *A First Course in Fuzzy Logic* (CRC Press, Boca Raton, Florida, 2006)

14. G. Bojadziev, M. Bojadziev, *Fuzzy Sets, Fuzzy Logic, Applications* (World Scientific, Singapore, 1995)

15. R.E. Moore, W. Lodwick, Interval analysis and fuzzy set theory. Fuzzy Sets Syst. **135**(1), 5–9 (2003)

16. H.T. Nguyen, O. Kosheleva, V. Kreinovich, Is the success of fuzzy logic really paradoxical? or: towards the actual logic behind expert systems. Int. J. Intell. Syst. **11**, 295–326 (1996)

17. D. Dubois, H. Fargier, J. Fortin, The empirical variance of a set of fuzzy intervals, in *Proceedings of the 2005 IEEE International Conference on Fuzzy Systems FUZZ–IEEE'2005*, Reno, Nevada, 22–25 May 2005, pp. 885–890

18. S. Dhompongsa, V. Kreinovich, H.T. Nguyen, How to interpret neural networks in terms of fuzzy logic?, in *Proceedings of the Second Vietnam–Japan Bilateral Symposium on Fuzzy Systems and Applications VJFUZZY'2001*, Hanoi, Vietnam, 7–8 Dec 2001, pp. 184–190

Chapter 37
How to Assess Universities: Defining Average Class Size in a Way Which Is Most Adequate for Teaching Effectiveness

When students select a university, one of the important parameters is the average class size. This average is usually estimated as an arithmetic average of all the class sizes. However, it has been recently shown that to more adequately describe students' perception of a class size, it makes more sense to average not over classes, but over all students – which leads to a different characteristics of the average class size. In this chapter, we analyze which characteristic is most adequate from the viewpoint of efficient learning. Somewhat surprisingly, it turns out that the arithmetic average *is* the most adequate way to describe the average student's gain due to a smaller class size. However, if we want to describe the effect of *deviations* from the average class size on the teaching effectiveness, then, instead of the standard deviation of the class size, a more complex characteristic is most appropriate.

The results from this chapter first appeared in [1].

Average class size is an important characteristic of a university. The fewer students in a class, the more individual attention can a student get, and thus, the better the student learns. Thus, for students, expected class size is an important criterion for selecting a university.

How average class size is usually estimated. Usually, the average class size is estimated by taking the arithmetic average of all the class sizes $s_1, \ldots, s_c$, i.e., as

$$E \stackrel{\text{def}}{=} \frac{s_1 + \cdots + s_c}{c} = \frac{1}{c} \cdot \sum_{i=1}^{c} s_i. \tag{37.1}$$

Problem with the usual definition of an average class size, and a more adequate definition. Recent papers [2–4] show that the student's perception of an average class size does not always coincide with the above quantity E.

© Springer-Verlag GmbH Germany 2018
O. Kosheleva and K. Villaverde, *How Interval and Fuzzy Techniques Can Improve Teaching*, Studies in Computational Intelligence 750,
https://doi.org/10.1007/978-3-662-55993-2_37

Indeed, students gauge an average class size by averaging the sizes of the classes in which they are enrolled. At any given moment of time, we have s_1 students enrolled in a class of size s_1, s_2 students enrolled in a class of size s_2, etc. So, to find the average class size from the student prospective, we must add all these numbers and divide by the total number of students. The resulting estimate is:

$$E_s = \frac{s_1 + \cdots + s_1 \,(s_1 \text{ times}) + \cdots + s_c + \cdots + s_c \,(s_c \text{ times})}{s_1 + \cdots + s_c}.$$

By combining the terms equal to each s_i, we get an equivalent expression

$$E_s = \frac{s_1^2 + \cdots + s_c^2}{s_1 + \cdots + s_c},$$

which can be described as the ratio $\dfrac{M_2}{E}$, where the second moment M_2 is defined as

$$M_2 \overset{\text{def}}{=} \frac{1}{c} \cdot \sum_{i=1}^{c} s_i^2.$$

As usual in statistics, we can represent M_2 as $M_2 = E^2 + V$, where the variance V is defined as

$$V \overset{\text{def}}{=} \frac{1}{c} \cdot \sum_{i=1}^{c} (s_i - E)^2. \tag{37.2}$$

Thus, the student-based estimate of the size of a class can be described as

$$E_s = \frac{M_2}{E} = E + \frac{V}{E}. \tag{37.3}$$

From this formula, we can see that since $V \geq 0$, this student-based average is always not smaller – and often larger – than the usual estimate (37.1).

This explains why, in the student's opinion, the official estimate (37.1) of the class size is usually an underestimation.

What we do in this chapter. The main objective of a student is to get a better education. From this viewpoint, what is the best estimate of the average class size that the student should use when selecting a university? This is a problem that we will solve in this chapter.

How class size affects education. A student's gain from the class consists of two parts:

- There is knowledge that this student gains during the lectures and classes. The amount a of this knowledge does not depend on the class size.

- There is also knowledge that comes from an individual contact with an instructor; since the instructor's time t is limited, this knowledge is proportional to the time that the instructor can spare for this particular student. This time is equal to $\dfrac{t}{s_i}$ and thus, the resulting knowledge is also inverse proportional to s_i, i.e., equal to $\dfrac{b}{s_i}$, for some constant b.

Thus, the overall student gain (= utility) from studying in a class of size s_i is equal to $a + \dfrac{b}{s_i}$.

Resulting average gain. Now that we have s_1 students with gain $a + \dfrac{b}{s_1}$, s_2 students with gain $a + \dfrac{b}{s_2}$, …, s_c students with gain $a + \dfrac{b}{s_c}$, we can compute the average gain u as

$$u = \frac{s_1 \cdot \left(a + \dfrac{b}{s_1}\right) + \cdots + s_c \cdot \left(a + \dfrac{b}{s_c}\right)}{s_1 + \cdots + s_c}.$$

Multiplying each term in the numerator, we conclude that

$$u = \frac{a \cdot s_1 + b + \cdots + a \cdot s_c + b}{s_1 + \cdots + s_k} = \frac{a \cdot \sum\limits_{i=1}^{c} s_i + b \cdot c}{\sum\limits_{i=1}^{c} s_i} = a + b \cdot \frac{c}{\sum\limits_{i=1}^{c} s_i},$$

or, in terms of the average (37.1):

$$u = a + \frac{b}{E}. \tag{37.4}$$

Conclusion. The average effect of class size on students is inverse proportional to the average class size E – as it is computed usually, by formula (37.1). This is a somewhat unexpected result since, as we have mentioned, the average class class as perceived by students is different from E.

Gauging deviations is important. Different classes have different sizes. So, for a student, it it important to know not only the *average* class size (or, alternatively, the average gain), but also the *deviations* from the average class size.

Variance and standard deviation as natural measures of deviation. In statistics, deviation from the average is usually gauged by the variance V or by its square root – standard deviation $\sigma = \sqrt{V}$; see, e.g., [5]

Variance of the class size: traditional approach. In the traditional approach, deviation is described by the variance (37.2).

Variance: student-based approach. In the student-based approach, the average is equal to $E + \dfrac{V}{E}$. So, for all s_i students from the ith class, the square of the difference is equal to $\left(s_i - \left(E + \dfrac{V}{E}\right)\right)^2$. Thus, the mean value of this square is equal to

$$V_s = \frac{1}{\displaystyle\sum_{i=1}^{c} s_i} \cdot \sum_{i=1}^{c} s_i \cdot \left(s_i - \left(E + \frac{V}{E}\right)\right)^2.$$

By using the fact that $s_i = (s_i - E) + E$, we can represent the expression V_s as the sum of two terms $V_s = V_1 + V_2$, where

$$V_1 = \frac{1}{\displaystyle\sum_{i=1}^{c} s_i} \cdot \sum_{i=1}^{c} (s_i - E) \cdot \left(s_i - \left(E + \frac{V}{E}\right)\right)^2$$

and

$$V_2 = \frac{1}{\displaystyle\sum_{i=1}^{c} s_i} \cdot \sum_{i=1}^{c} E \cdot \left(s_i - \left(E + \frac{V}{E}\right)\right)^2.$$

In the expression for V_1, we can explicitly separate $s_i - E$ in the squared term, thus getting

$$V_1 = \frac{1}{\displaystyle\sum_{i=1}^{c} s_i} \cdot \sum_{i=1}^{c} (s_i - E) \cdot \left((s_i - E) - \frac{V}{E}\right)^2.$$

By explicitly describing the square of the difference, we get

$$\frac{1}{\displaystyle\sum_{i=1}^{c} s_i} \cdot \sum_{i=1}^{c} (s_i - E) \cdot \left((s_i - E)^2 - 2 \cdot (s_i - E) \cdot \frac{V}{E} + \frac{V^2}{E^2}\right) =$$

$$\frac{1}{\displaystyle\sum_{i=1}^{c} s_i} \cdot \sum_{i=1}^{c} \left((s_i - E)^3 - 2 \cdot (s_i - E)^2 \cdot \frac{V}{E} + (s_i - E) \cdot \frac{V^2}{E^2}\right).$$

Here, the average of $s_i - E$ is 0, the average of $(s_i - E)^2$ is V; so, by defining the skewness

$$S \stackrel{\text{def}}{=} \frac{1}{c} \cdot \sum_{i=1}^{c} (s_i - E)^3, \tag{37.5}$$

we conclude that

$$V_1 = \frac{S - 2V \cdot \dfrac{V}{E}}{E} = \frac{S}{E} - 2\frac{V^2}{E^2}.$$

In the expression for V_2, we move the common factor E outside the sum, getting

$$V_2 = E \cdot \frac{1}{\displaystyle\sum_{i=1}^{c} s_i} \cdot \sum_{i=1}^{c} \left(s_i - \left(E + \frac{V}{E} \right) \right)^2.$$

By explicitly performing the squaring, we conclude that

$$V_2 = E \cdot \frac{1}{\displaystyle\sum_{i=1}^{c} s_i} \cdot \sum_{i=1}^{c} \left(s_i^2 - 2 \cdot s_i \cdot \left(E + \frac{V}{E} \right) + \left(E + \frac{V}{E} \right)^2 \right).$$

The average of s_i^2 is equal to $M_2 = V + E^2$, the average of s_i is equal to E, so we get

$$V_2 = E \cdot \frac{1}{E} \cdot \left(V + E^2 - 2E \cdot \left(E + \frac{V}{E} \right) + \left(E + \frac{V}{E} \right)^2 \right) =$$

$$\frac{1}{E} \cdot \left(V + E^2 - 2E^2 - 2V + E^2 + 2V + \frac{V^2}{E^2} \right).$$

The terms E and $\dfrac{1}{E}$ cancel each other, and so do the terms $2V$ and $-2V$, and the terms E^2, $-2E^2$; so, we get

$$V_2 = V + \frac{V^2}{E^2}.$$

By combining the formulas for V_1 and V_2, we get the following formula for the student-based variance $V_s = V_1 + V_2$:

$$V_s = V + \frac{S}{E} - \frac{V^2}{E^2}. \tag{37.6}$$

Variance: utility approach. As we have shown, the utility of each of s_i students enrolled in the $i$the class is equal to $a + \dfrac{b}{s_i}$, and the average utility is equal to $a + \dfrac{b}{E}$. Thus, for each of these students, the square of the difference between the actual and average utility is equal to

$$b^2 \cdot \left(\frac{1}{s_i} - \frac{1}{E} \right)^2 = b^2 \cdot \left(\frac{1}{s_i^2} - 2 \cdot \frac{1}{s_i} \cdot \frac{1}{E} + \frac{1}{E^2} \right).$$

So, the corresponding variance is equal to

$$V_u = b^2 \cdot \frac{1}{\sum\limits_{i=1}^{c} n_i} \cdot \sum_{i=1}^{c} s_i \cdot \left(\frac{1}{s_i^2} - 2 \cdot \frac{1}{s_i} \cdot \frac{1}{E} + \frac{1}{E^2} \right),$$

i.e., to

$$V_u = b^2 \cdot \frac{1}{\sum\limits_{i=1}^{c} s_i} \cdot \sum_{i=1}^{c} \left(\frac{1}{s_i} - 2 \cdot \frac{1}{E} + \frac{s_i}{E^2} \right).$$

We know that the average of the values s_i is E. Let us denote

$$M_{-1} \overset{\text{def}}{=} \frac{1}{c} \cdot \sum_{i=1}^{c} \frac{1}{s_i}. \tag{37.7}$$

Then, the above formula takes the form

$$V_u = b^2 \cdot \frac{1}{E} \cdot \left(M_{-1} - 2 \cdot \frac{1}{E} + \frac{E}{E^2} \right),$$

i.e.,

$$V_u = b^2 \cdot \left(\frac{M_{-1}}{E} - \frac{1}{E^2} \right). \tag{37.8}$$

Comment. In the above formula, we used the expression M_{-1}, moment of order -1. This moment is closely related to an alternative way of describing the mean of several numbers $s_1, \ldots, s_c$, namely, to the *harmonic mean*

$$\frac{c}{\dfrac{1}{s_1} + \cdots + \dfrac{1}{s_c}}.$$

How to meaningfully compare the utility-based variance with other variances. The first two variances are in terms of number of students, while the utility variance is in terms of its inverse. How can we compare the utility-based variance with other variances?

A variance V means that instead of the exact value E, we have $E + k \cdot \sigma$, where $\sigma = \sqrt{V}$ and k is small number – for normal distributions, with certainty 95%, its absolute value is smaller than 2. The utility variance means that instead of the original value of the utility $u = a + \dfrac{b}{E}$, we have $u + k \cdot \sigma_u$, where $\sigma_u = \sqrt{V_u}$. To meaningfully compare this change with other variances, it is desirable to come up with a value $\sigma_e = \sqrt{V_e}$ for which the corresponding change from E to $E + k \cdot \sigma_e$ will lead to exactly this change from u to $u + k \cdot \sigma_u$.

The change in E changes the original value of $u = a + \dfrac{b}{E}$ to the new value

$$u' = a + \frac{b}{E + k \cdot \sigma_e} = a + \frac{b}{E \cdot \left(1 + k \cdot \dfrac{\sigma_e}{E}\right)}.$$

When the deviation σ_e is small, we can ignore terms which are quadratic and higher order in σ_e and conclude that

$$\frac{1}{1 + k \cdot \dfrac{\sigma_e}{E}} \approx 1 - k \cdot \frac{\sigma_e}{E}.$$

Thus, we get the following expression:

$$u' \approx \frac{b}{E} \cdot \left(1 - k \cdot \frac{\sigma_e}{E}\right) = a + \frac{b}{E} - k \cdot \sigma_e \cdot \frac{b}{E^2}.$$

So, the difference $u' - u$ between this expression and the average utility $u = a + \dfrac{b}{E}$ has the form $k \cdot \sigma_u$, where we denoted $\sigma_u = \sigma_e \cdot \dfrac{b}{E^2}$. Since we know σ_u, we can therefore compute the equivalent standard deviation as $\sigma_e = \sigma_u \cdot \dfrac{E^2}{b}$, and the equivalent variance as $V_e = \sigma_e^2 = V_u \cdot \dfrac{E^4}{b^2}$. Substituting the above expression for V_u, we get

$$V_e = E^3 \cdot M_{-1} - E^2. \tag{37.9}$$

Conclusions. Our objective is to describe the effect of class size on *teaching effectiveness*. It turns out that to gauge average effectiveness, it is sufficient to know the arithmetic average of class sizes. This fact is somewhat unexpected: while this arithmetic average is the mostly used characteristic of the average class size, it is not the most adequate in describing *student perception* of class sizes.

Once we know how the *average* class size affects teaching effectiveness, a natural next question is how *deviations* from the average class size affect teaching effectiveness. At first glance, the above conclusion seems to imply that standard deviation of the class size would be the most adequate characteristic of the effect of deviations on the teaching effectiveness. Again, somewhat unexpectedly, it turns out not be the case: the most adequate characteristic is a more complex expression (37.9) – that uses both the arithmetic mean $E = \dfrac{s_1 + \cdots + s_c}{c}$ of class sizes *and* their harmonic mean $\dfrac{c}{\dfrac{1}{s_1} + \cdots + \dfrac{1}{s_c}}$.

References

1. O. Kosheleva, V. Kreinovich, How to define average class size (and deviations from the average class size) in a way which is most adequate for teaching effectiveness, in *Proceedings of the Workshop on Informatics and Information Technologies in Education: Theory, Applications, Didactics*, vol. 1, Novosibirsk, Russia, 26–29 Sept 2012, pp. 113–120
2. L. Lesser, Sizing up class size: a deeper classroom investigation of central tendency. Math. Teach. **103**(5), 376–380 (2009)
3. L. Lesser, Sizing up class size: additional insights. Math. Teach. **104**(2), 86–87 (2010)
4. L.M. Lesser, K. Kephart, Setting the tone: a discursive case study of problem–based inquiry learning to start a graduate statistics course for in–service teachers. J. Stat. Educ. **19**(3) (2011)
5. D.J. Sheskin, *Handbook of Parametric and Nonparametric Statistical Procedures* (Chapman & Hall/CRC, Boca Raton, Florida, 2007)

Chapter 38
Conclusions

Education is a very important process. We are all aiming to teach better. And in the 21st century, this means using computers to automate (and improve) as many aspects of teaching as possible.

There are many great teachers who are willing to explain to us how to teach better. The problem is that these explanations are usually formulated by using imprecise("fuzzy") words from natural language, such as "weak" "strong", "difficult". To translate these recommendations into precise teaching strategies, it is therefore important to use fuzzy logic techniques, techniques that were designed specifically to translate such fuzzy knowledge into precise terms.

Another reason why fuzzy techniques are useful is that when we evaluate the student's success, we also often use natural-language words such as "very good", "needs some work", etc. To be able to automatically process this information, we also need to translate these words into precise terms.

In this book. we show that indeed, fuzzy techniques can help in all the stages of the teaching process, from teaching itself to curriculum design to assessment. What we did is just scratching the surface: there are many important education-related problems in which, in our opinion, fuzzy techniques can be used. Yes, one of the objectives of this book is to present the results but our main objective is, by showing that fuzzy techniques can help, to inspire further research in this direction. Let us all do our best and use fuzzy and other techniques to improve teaching even further!

© Springer-Verlag GmbH Germany 2018
O. Kosheleva and K. Villaverde, *How Interval and Fuzzy Techniques Can Improve Teaching*, Studies in Computational Intelligence 750,
https://doi.org/10.1007/978-3-662-55993-2_38

Bibliography

1. M.G. Averill, K.C. Miller, G.R. Keller, V. Kreinovich, R. Araiza, S.A. Starks, Using expert knowledge in solving the seismic inverse problem. Int. J. Approx. Reason. **45**(3), 564–578 (2007)
2. M.G. Averill, A Lithospheric Investigation of the Southern Rio Grande Rift, University of Texas at El Paso, Department of Geological Sciences, Ph.D. Dissertation, 2007
3. S.M. Bai, S.M. Chen, A new approach for constructing concept maps based on fuzzy rules, in *Proceedings of the 20th International Conference on Industrial, Engineering and Other Applications of Applied Intelligent Systems, Kyoto, Japan, 2007* (2007), pp. 155–165
4. G. Barolat, Current status of epidural spinal cord stimulation. Neurosurg. Q. **5**(2), 98–124 (1995)
5. G.M. Barolat, J. He, S. Zeme, B. Ketcik, Mapping of sensory responses to epidural stimulation of the intraspinal neural structures in man. J. Neurosurg. **78**, 233–239 (1993)
6. I.I. Blekhman, *Vibrational Mechanics: Nonlinear Dynamic Effects, General Approach, Applications* (World Scientific, Singapore, 1999)
7. J. Bruner, *Going Beyond the Information Given* (Norton, New York, 1973)
8. J. Bruner, *Acts of Meaning* (Harvard University Press, Cambridge, 1990)
9. J.E. Gamez, F. Modave, O. Kosheleva, Selecting the most representative sample is NP–hard: need for expert (fuzzy) knowledge, in *Proceedings of the IEEE World Congress on Computational Intelligence WCCI'2008, Hong Kong, China, June 1–6, 2008* (2008), pp. 1069–1074
10. M. Gardner, Puzzles and number–theoretic problems arising from the curious fractions of Ancient Egypt, in *Scientific American* (1978)
11. L. Hong, S.E. Page, Problem solving by heterogeneous agents. J. Econ. Theory **97**(1), 123–163 (2001)
12. L. Hong, S.E. Page, Groups of diverse problem solvers can outperform groups of high-ability problem solvers. Proc. Natl. Acad. Sci. **101**(46), 16385–16389 (2004)
13. M. King, *Process Control: A Practical Approach* (Wiley, Chichester, 2010)
14. O.M. Kosheleva, On the optimal choice of quality metric in image compression: a soft computing approach. Soft Comput. **8**, 268–273 (2004)
15. O. Kosheleva, Degree–based (fuzzy) techniques in math and science education, in *Proceedings of the World Conference on Soft Computing, San Francisco, CA, May 23–26, 2011*, ed. by R.R. Yager, M.Z. Reformat, S.N. Shahbazova, S. Ovchinnikov (2011)
16. O. Kosheleva, How to explain usefulness of different results when teaching calculus: example of the mean value theorem. J. Uncertain Syst. **7**(3), 164–175 (2013)

© Springer-Verlag GmbH Germany 2018
O. Kosheleva and K. Villaverde, *How Interval and Fuzzy Techniques Can Improve Teaching*, Studies in Computational Intelligence 750,
https://doi.org/10.1007/978-3-662-55993-2

17. O. Kosheleva, Towards combining Freirean ideas and Russian experience in mathematics education, in *Teaching for Global Community*, ed. by C.A. Rossatto, A. Luykx, H.S. Garcia (Information Age Publishing, Charlotte, 2011), pp. 207–218

18. V. Kreinovich, E. Johnson-Holubec, L.K. Reznik, M. Koshelev, Cooperative learning is better: explanation using dynamical systems, fuzzy logic, and geometric symmetries, in *Proceedings of the Vietnam–Japan Bilateral Symposium on Fuzzy Systems and Applications VJFUZZY'98, HaLong Bay, Vietnam, 30th September–2nd October, 1998*, ed. by H.P. Nguyen, A. Ohsato (1998), pp. 154–160

19. V. Kreinovich, L. Longpré, S.A. Starks, G. Xiang, J. Beck, R. Kandathi, A. Nayak, S. Ferson, J. Hajagos, Interval versions of statistical techniques, with applications to environmental analysis, bioinformatics, and privacy in statistical databases. J. Comput. Appl. Math. **199**(2), 418–423 (2007)

20. C.C. Kuo, F. Glover, K.S. Dhir, Analyzing and modeling the maximum diversity problem by zero-one programming. Decision Sci. **24**(6), 1171–1185 (1993)

21. J.H. Miller, S.E. Page, *Complex Adaptive Social Systems: The Interest in Between* (Princeton University Press, Princeton, 2006)

22. H.T. Nguyen, O. Kosheleva, V. Kreinovich, Is the success of fuzzy logic really paradoxical? Or: towards the actual logic behind expert systems. Int. J. Intell. Syst. **11**, 295–326 (1996)

23. S.E. Page, *The Difference: How the Power of Diversity Creates Better Groups, Firms, Schools, and Societies* (Princeton University Press, Princeton, 2007)

24. J. Van Patten, C. Chao, C. Reigeluth, A review of strategies for sequencing and synthesizing instruction. Rev. Educ. Res. **56**(4), 437–471 (1986)

25. M.Q. Patton, *Qualitative Research and Evaluation Methods* (Sage Publishing, Thousand Oaks, 2002)

26. U. Rossi, J. Vernes, Epidural spinal electrical stimulation for pain control: a ten–year experience, in *Proceedings of the XII Annual Meeting of Australasian College of Rehabilitation Medicine* (1992), pp. 17–21

27. C. Suddath, Morning the death of handwriting, *Time Magazine* (2009). Accessed 03 Aug 2009

28. M. Tchoshanov, S. Blake, A. Duval, Preparing teachers for a new challenge: teaching calculus concepts in middle grades, in: *Proceedings of the Second International Conference on the Teaching of Mathematics (at the undergraduate level), Hersonissos, Crete, Greece, 2002* (2002)

29. K. Villaverde, O. Kosheleva, Towards more adequate value–added teacher assessments: how intervals can help, in *Abstracts of the 14th GAMM–IMACS International Symposium on Scientific Computing, Computer Arithmetic and Validated Numerics SCAN'2010, Lyon, France, September 27–30, 2010* (2010), pp. 140–141

30. L. Vygotsky, *Thought and Language* (M.I.T. Press, Cambridge, 1962)

31. A.M. Zimichev, V. Kreinovich et al., Automatic control system for psychological experiments. Final Report on Grant No. 0471–2213–20 from the Soviet Science Foundation, 1982 (in Russian)

32. A.M. Zimichev, V. Kreinovich et al., Automatic system for quick education of computer operators. Final Report on Grant No. 0471–0212–60 from the Soviet Science Foundation, 1980–82 (in Russian)

Index

© Springer-Verlag GmbH Germany 2018
O. Kosheleva and K. Villaverde, *How Interval and Fuzzy Techniques Can Improve Teaching*, Studies in Computational Intelligence 750,
https://doi.org/10.1007/978-3-662-55993-2

Printed in the United States
By Bookmasters